CYBERINFRASTRUCTURE TECHNOLOGIES AND APPLICATIONS

Cyberinfrastructure Technologies and Applications

Junwei Cao
Editor

Nova Science Publishers, Inc.
New York

For permission to use material from this book please contact us:
Telephone 631-231-7269; Fax 631-231-8175
Web Site: http://www.novapublishers.com

LIBRARY OF CONGRESS CATALOGING-IN-PUBLICATION DATA

Cyberinfrastructure technologies and applications / Junwei Cao (editor).
p. cm.
ISBN 978-1-60692-063-3 (hardcover)
1. Cyberinfrastructure. I. Cao, Junwei, 1970-
QA76.9.C92C93 2009
004--dc22 2008037243

Published by Nova Science Publishers, Inc. ✢ New York

CONTENTS

PREFACE

Cyberinfrastructure was proposed in a report of the NSF Blue-Ribbon advisory panel in 2003. Cyberinfrastructure will provide a unified environment to access and manage cyber resources, e.g. supercomputers, data archives, software services, scientific instruments and virtual organizations. In this book, the authors review latest research and development and discuss new technologies and applications involved in building Cyberinfrastructure. The purpose of this book is to provide a detailed summary of early experiences, practices and lessons leaned in building Cyberinfrastructure from multiple perspectives: software development and maintenance, resource integration and sharing, cyber environment construction, operation and management, testing and troubleshooting, application enabling, security and QoS ensuring. Consequently, this book can serve as a valuable source of reference and indispensable reading for researchers, educators, engineers, graduate students, and practitioners in the field of design and implementation of Cyberinfrastructure systems.

Chapter 1 - The recent emergence of multicore and hybrid microprocessor designs marks the beginning of a forced march toward an era of computing in which research applications must be able to exploit parallelism at an unprecedented scale. This chapter presents a new generation of dense linear algebra libraries that achieve the fastest possible time to an accurate solution on multicore systems by efficiently using all the processors that such systems will make available. To work within this design space and leverage the power of million way parallelism, it is necessary to combine new, highly parallelizable algorithms, a programming and execution model that can exploit massive task parallelism, and a flexible memory management facility that can help optimize data locality across a range of different platforms. The design space is also conditioned by the fact that, in order to support the broadest possible range of Computational Science, the resulting library frameworks must be able to scale both up and down, running at all levels of the platform development chain.

Chapter 2 - Cyberinfrastructure (CI) for instrument sharing is an infrastructure that aims to facilitate effective resource sharing of expensive scientific instruments, e.g., telescopes and observatories, through a system of grid services. This cyberinfrastructure consists of three components: an instrument pool alliance, instrument pools, and physical instruments. When a user submits an experiment to the CI environment,the instrument pool alliance is responsible to allocate instruments in related instrument pools to conduct the experiment. After the experiment is finished and results are returned, the user will appraise performance of corresponding services.

In this chapter, fuzzy random scheduling algorithms are proposed in instrument pools when a job is submitted to one of instruments within a pool. The randomness lies in the probability of which instrument be chosen for an experiment and the fuzziness origins from vagueness of users' feedback opinions on experimental results. Users' feedback information is utilized to improve overall quality of service (QoS) of an instrument CI. Several algorithms are provided to increase utilization of instruments providing higher QoS and decrease utilization of those with poor QoS. This is demonstrated in details using quantitative simulation results included in this chapter.

Chapter 3 - Information services are a critical piece of Grid infrastructure, they collect, aggregate, and organize sources of Grid resource information, and provide the information to applications to schedule computational tasks, and enable Virtual Organizations (VO) to share distributed computing resources. To be successful an information provider needs to be able to aggregate diverse information from numerous sources, publish information into different Grid monitoring systems, function under different Grid deployments, and be adaptable to different information schemas. Addressing these challenges would provide a platform for an interoperable information service.

In this chapter, the authors present the Modular Information Provider (MIP) that address the aforementioned challenges and eliminates the shortcomings of the existing Grid information providers. MIP adopts a XML-based data model to enable robust schema validation as well as to support the latest generation of service-oriented Grid monitoring software.

MIP provides a critical capability for production Grid environments focusing on achieving Grid interoperability and manageability. To validate this we conduct experiments on the Open Science Grid and Grid Australia. These experiments demonstrate MIP's Grid-independent approach is successful and can easily adapt to different software stacks and the heterogeneous configuration of several sites on a Grid. We use GISolve, a TeraGrid science gateway as an example to demonstrate how MIP is useful from the perspective of Grid application. GISolve provides a geospatial problem solving platform that hides the complexity of Grid information services based on MIP, which enables GISolve applications to access multiple Grids seamlessly. The authors illustrate how MIP is integrated within GISolve so that GISolve applications can benefit from the interoperable information service MIP provides.

Chapter 4 - This chapter addresses the dynamic scheduling of parallel jobs with QoS demands (soft-deadlines) in multiclusters and grids. Three performance metrics (over-deadline, makespan and idle-time) are combined with variable weights to evaluate the scheduling performance. These three metrics are used for measuring the extent of jobs' QoS demands compliance, resource throughput and resource utilization, respectively. Therefore, clusters situated in different administrative organizations can utilize different weight combinations to represent their different performance requirements. Two levels of performance optimisations are applied in the multicluster. At the multicluster level, a scheduler, (which the authors call MUSCLE), allocates parallel jobs with high packing potential to the same cluster; MUSCLE also takes the jobs' QoS requirements into account and employs a heuristic to allocate suitable workloads to each cluster to balance the performance. At the local cluster level, a workload manager, called TITAN, utilizes a genetic algorithm to further improve the scheduling performance of the jobs sent by MUSCLE. The extensive experimental studies are conducted to verify the effectiveness of the scheduling mechanism in MUSCLE; the results show that comparing with the traditional workload allocation policies in distributed systems (Dynamic Least Load and Weighted Random), the comprehensive scheduling performance (in terms of over-deadline, makespan and idle-time) of parallel jobs is significantly improved and well balanced across the multicluster.

Chapter 5 - Service-oriented Grid computing promises to change the way scientists will tackle future research challenges by offering advanced data and application services, providing transparent access to distributed heterogeneous data sources and to high-end computing facilities for performing computationally demanding and data-intensive modeling, simulation and analysis tasks. In this article the authors describe the Vienna Grid Environment (VGE), a service-oriented Grid infrastructure based on standard Web Services technologies for virtualizing scientific applications and data sources as Grid services that hide the details of the underlying software and hardware infrastructure. The VGE service provision framework adopts a component-based approach which supports the configuration of application and data services from a set of basic service components providing capabilities like job or query execution, data transfers, QoS negotiation, data staging, and error recovery. VGE relies on a business-oriented model to Grid computing based on a flexible QoS infrastructure, dynamic negotiation of service-level agreements, and on-demand access to Grid services. VGE has been developed and utilized in the context of several European projects for the realization of Grid infrastructures within medical and bio-medical application domains.

Chapter 6 - This chapter presents the design and implementation of seamless integration of two complex systems component-based distributed application framework ProActive and Gridbus Resource Broker. The integration solution provides: (i) the potential ability for component-based distributed applications developed using ProActive framework, to leverage the economy-based and data-intensive scheduling algorithms provided by the Gridbus Resource Broker; (ii) the execution runtime environment from ProActive for the Gridbus Resource Broker over component-based distributed applications. It also presents the evaluation of the integration solution based on examples provided by the ProActive distribution and some future directions of the current system.

Chapter 7 - In the past few years, the Grid computing paradigm has emerged as an instance of cyber infrastructure, promising to enable resource sharing and collaborating across multiple domains. In the research community there has been an intense interest in designing and studying of such system.

CROWN (China R and D Environment Over Wide-area Network) project is an e-Science project funded by China Natural Science Foundation Committee, and China 863 High-tech Program. The main goal of CROWN project is to empower in-depth integration of resources and cooperation of researchers nationwide and worldwide. CROWN was started in late 2003. The main goal of CROWN project is to build the middleware infrastructure and wide area testbed to support computation intensive, data centric e-Science applications.

Recently, with the evolution of Web services, the service-oriented architecture has become a significant trend for grid computing, with OGSA/ WSRF as the de facto standards. CROWN has adopted the service-oriented architecture, connecting large amount of services deployed in universities and institutes. Up till mid 2007, lots of applications in different domains have been deployed into CROWN grid, such as gene comparison in bioinformatics, climates pattern prediction in environment monitoring, etc. The long-range goal for CROWN is to integrate home user resources in a fully decentralized way with a robust, scalable grid middleware infrastructure.

In this chapter, based on a proposed Web service-based grid architecture, a service grid middleware system called CROWN is introduced. As the two kernel points of the middleware, the overlay-based distributed grid resource management mechanism is proposed, and the policy-based distributed access control mechanism with the capability of automatic negotiation of the access control policy and trust management and negotiation is also discussed in this chapter. Experience of CROWN testbed deployment and application development shows that the service-oriented middleware can support the typical scenarios such as computing-intensive applications, data-intensive applications and mass information processing applications.

Chapter 8 – The authors present the design and implementation of a semantics-enabled service discovery framework in a pan-European pharmaceutical Grid: SIMDAT, an industry-oriented Grid environment for integrating thousands of Grid-enabled biological data services and analysis services. The framework consists of three major components: the OWL-DL-based biological domain ontology, OWL-S-based service annotation, and semantic matchmaker based on the ontology reasoning. Built upon the framework, workflow technologies are extensively exploited in the SIMDAT to assist biologists in (semi)automatically performing *in silico* experiments. They present a typical usage scenario through the case study of a biological workflow: *IXodus*.

Chapter 9 - Grid computing is rapidly evolving into a service-oriented computing infrastructure that facilitates resource sharing and large-scale problem solving on the Internet. It is envisioned that many resources on the grid would be exposed as services for a wider use by the community. Service discovery and composition has thus become a vitally important component in utilizing grid facilities. This chapter focuses on service composition. One challenge in service composition is how to automate the composition process in terms of a large number of services (atomic services or component services) and a variety of user requests. A novel Hierarchical Two Directions Partial Order Planning (H2POP) algorithm is presented for discovery of composite services which are dynamically composed from component services. A use case is given to illustrate the application of the H2POP algorithm for travel planning service composition automation.

Chapter 10 - With the emergence of Service Oriented Computing, workflow has become an important method to compose services and reuse existing resources in the Cyberinfrastructure (CI) and the Grid. In this chapter, the authors first summarize research activities in the field of workflow in service oriented computing. They discuss five major research topics, i.e., languages and tools for service orchestration, automatic service composition, mediation-aided service composition, verification of service workflow, and decentralized execution of workflow. Although some of this work was originally targeted at the area of business process management, they can be adopted by the CI community with some modification or enhancement. In the second part of this chapter, the authors introduce a service-oriented workflow execution system, WS-Workflow, and explain the key modules in this system, including the data-driven composition module, the mediation-aided composition module, and the model fragmentation module. This system has been used in many projects to facilitate the flexible workflow composition and decentralized execution.

Chapter 11 - Dynamical and flexible resource aggregation tools are required in 21st century research. Scientists need to aggregate various digital equipments and cooperate with each other in different organizations through Virtual Organizations (VO) on the Internet in a flexible and dynamical way. In this cooperation and resource sharing process, trust evaluation is of great importance for flexible VO management. Traditional tools such as VOMS for grids are short in dynamism and trust evaluation. In this chapter, the authors propose a new scheme providing federal VO membership management based on trust evaluation, with which researchers can achieve appropriate trust relationships with each other and establish a particular VO dynamically to aggregate resources for their own purposes.

Chapter 12 - This chapter presents some of the key aspects of Cyberinfrastructure (CI) research and development targeting community-scale exploratory science. The motivation comes from the fact that successful software for CI is increasing scientific productivity of a single investigator, small groups of scientists as well as dispersed teams spanning multiple institutions. Community scale scientific activities and their informatics requirements are driving the development of new CI solutions. It becomes critical to follow CI design principles based on past, present and future efforts. In addition, data- and hypothesis-driven explorations are fundamental scientific activities leading to discoveries. In this work, our focus is on informatics requirements and CI design principles behind existing software. We have included our experiences and described several prototype CI solutions to support exploratory science.

Chapter 13 - Biomedical, translational and clinical research through increasingly complex computational modeling and simulation generate enormous potential for personalized medicine and therapy, and an insatiable demand for advanced cyberinfrastructure. Metascheduling that provides integrated interfaces to computation, data, and workflow management in a scalable fashion is essential to advanced pervasive computing environment that enables mass participation and collaboration through virtual organizations (VOs). Avian Flu Grid (AFG) is a VO dedicated to members from the international community to collaborate on antiviral drug discovery for potential pandemic influenza viruses. The complex and dynamic drug discovery workflow requirement in the AFG-VO is met through innovative service oriented architecture with metascheduling playing a key role. The community scheduler framework (CSF4) is a web service resource framework (WSRF)-compliant metascheduler with an extensible architecture for customized plugins that provide cross site scheduling, workflow management, and data-aware scheduling on the Grid Datafarm (Gfarm) global filesystem. The Opal web service toolkit enables existing scientific applications to be deployed as web services, accessible by various types of clients including advanced workflow management tools such as Vision and Kepler. Molecular dynamics and virtual screening applications exposed as Opal based services are metascheduled using CSF4 to access distributed resources such as the Pacific Rim Applications and Middleware Assembly (PRAMGA) grid and the TeraGrid. Emerging trends in multicore processors, virtualization and Web 2.0 continue to shape the pervasive computing environment in the years to come and pose interesting opportunities for metascheduling research and development.

Chapter 14 - This chapter presents a study on the feasibility of applying high performance computing (HPC) to the Bridging Domain Multiscale (BDM) method, so that featured scalable multiscale computations can be achieved. Wave propagation in a molecule chain through an entire computational domain is employed as an example to demonstrate its applicability and computing performance when multiscale-based simulations are conducted in a large-scale parallel computing environment. In addition, the conceptual idea and computing framework using Grid computing technologies is proposed to enhance future multiscale computations in nanotechnology.

Chapter 15 - During the last decade, billions of national investment has been spent in China on building agricultural information systems for data collection, integration, analysis and processing. Each province has built its own technology platform for information sharing and access. However, since data sources are separate and corresponding applications are developed by different organizations, cross-domain data and knowledge sharing becomes very difficult. A Cyberinfrastructure (CI) for agricultural data and knowledge sharing in China is proposed in this work and funded by the Ministry of Science and Technology of China under the national 863 high-tech R and D program. In this work, related work is summarized and our system structure is described in details. Heterogeneity of different operating systems and databases has to be addressed. System performance can be improved by avoiding large-scale data transferring by introducing an application server within each domain for local data processing.

In: Cyberinfrastructure Technologies and Applications ISBN: 978-1-60692-063-3
Editor: Junwei Cao

Chapter 1

PARALLEL DENSE LINEAR ALGEBRA SOFTWARE IN THE MULTICORE ERA

Alfredo Buttari[1], Jack Dongarra[2], Jakub Kurzak[3] and Julien Langou[4]

[1] Department of Electrical Engineering and Computer Science,
University of Tennessee, Knoxville, Tennessee
[2] Department of Electrical Engineering and Computer Science,
University of Tennessee, Knoxville, Tennessee
Oak Ridge National Laboratory, Oak Ridge, Tennessee
University of Manchester, Manchester UK
[3] Department of Electrical Engineering and Computer Science,
University of Tennessee, Knoxville, Tennessee
[4] Department of Mathematical and Statistical Sciences,
University of Colorado Denver, Denver, Colorado

ABSTRACT

The recent emergence of multicore and hybrid microprocessor designs marks the beginning of a forced march toward an era of computing in which research applications must be able to exploit parallelism at an unprecedented scale. This chapter presents a new generation of dense linear algebra libraries that achieve the fastest possible time to an accurate solution on multicore systems by efficiently using all the processors that such systems will make available. To work within this design space and leverage the power of million way parallelism, it is necessary to combine new, highly parallelizable algorithms, a programming and execution model that can exploit massive task parallelism, and a flexible memory management facility that can help optimize data locality across a range of different platforms. The design space is also conditioned by the fact that, in order to

[1] E-mail address: buttari@eecs.utk.edu
[2] E-mail address: dongarra@eecs.utk.edu
[3] E-mail address: kurzak@eecs.utk.edu
[4] E-mail address: julien.langou@ucdenver.edu

support the broadest possible range of Computational Science, the resulting library frameworks must be able to scale both up and down, running at all levels of the platform development chain.

INTRODUCTION

The recent emergence of multicore and hybrid microprocessor designs marks the beginning of a forced march toward an era of computing in which research applications must be able to exploit parallelism at an unprecedented scale. This confronts the scientific software community with both a daunting challenge and a unique opportunity. The challenge arises from the disturbing mismatch between the design of systems based on this new chip architecture — hundreds of thousands of nodes, a million or more cores, reduced bandwidth and memory available to cores — and the components of the traditional software stack, such as numerical libraries, on which scientific applications have relied for their accuracy and performance. So long as library developers could depend on ever increasing clock speeds and instruction level parallelism, they could also settle for incremental improvements in the scalability of their algorithms. But to deliver on the promise of tomorrow's petascale systems, library designers must find methods and algorithms that can effectively exploit levels of parallelism that are orders of magnitude greater than most of today's systems offer. This is an unsolved problem. Yet this problem also presents a rare opportunity because, in the wake of the multicore revolution, we are now compelled to rethink essential components of our software infrastructure that the normal constraints of business as usual otherwise render more or less untouchable.

The goal is to create a new generation of dense linear algebra libraries that achieve the fastest possible time to an accurate solution on multicore systems by efficiently using all the processors that such systems will make available. To work within this design space and leverage the power of million way parallelism, it is necessary to combine new, highly parallelizable algorithms, a programming and execution model that can exploit massive task parallelism, and a flexible memory management facility that can help optimize data locality across a range of different platforms. The design space is also conditioned by the fact that, in order to support the broadest possible range of Computational Science, the resulting library frameworks must be able to scale both up and down, running at all levels of the platform development chain. To achieve this goal, special focus must be put on the following four objectives:

- *Explore new, highly parallelizable algorithms:* Novel work [1–3] shows that a wide range of dense linear algebra algorithms can be expressed as algorithms by tiles. The PLASMA project (Parallel Linear Algebra Software for Multicore Architectures) shows how this new approach can be applied to Cholesky, LU and QR factorizations [1–3], greatly improving parallel performance of these operations on multicore processors. It is reasonable to expect that this concept can be extended to a broad range of linear algebra algorithms reduction to bi-diagonal, tri-diagonal and Hessenberg forms, eigensolver algorithms like QR iterations etc.
- *Develop an algorithm description abstraction for expressing parallelism:* Building on the well established concepts from dataflow architectures, it is possible to develop

high-level abstractions for algorithm description that make task dependencies explicit and therefore facilitate scheduling on multicore systems. Since all the information about parallelism in the algorithm is contained in its dependency graph, which is a Direct Acyclic Graph (DAG), graph-based algorithm definition can replace programming language definition. This language independent approach will greatly improve the process of development and maintenance of the numerical library.

- *Implement scalable methods of dynamic task scheduling and synchronization for multicore systems:* With heterogeneity becoming pervasive both at the chip level, as well as in supercomputer installations, it is necessary to develop scalable mechanisms for dynamic scheduling of tasks with lightweight synchronization, assuring load balance and exploiting communication overlapping. The goal here is not to reinvent such a system, but rather to leverage best existing practices to construct an event driven task scheduling system to carry out execution of the task graph at runtime.
- *Design a unified approach for different memory architectures:* It is also necessary to develop a unified approach for different memory architectures including symmetric multiprocessors (SMP), non-uniform memory access architectures (NUMA), distributed memory systems and processors with scratchpad memories (e.g., Cell processor). Initial results indicate that a similar execution model is applicable to SMP-like multicore processors, NUMA-based systems and the Cell processor. The purpose of this effort is to cover these types of memory architectures with a unified approach and also extend the model to distributed memory architectures.

The rationale for these objectives derives in large measure from an analysis of the major factors that blocked the road to further performance improvements for serial programs using traditional microprocessor designs.

BACKGROUND

The familiar story of exponential growth in uniprocessor performance during the last decades of the twentieth century has now been overtaken, as we begin the twenty-first, by a more sobering story of how the seemingly relentless increases in the serial performance finally came to an end. It is not that the phenomenon known as Moore's law [4, 5] has reached its limit, but rather that the ability of engineers to exploit the increasing number of components on a chip for the purpose of accelerating *serial performance* has been effectively blocked. The essence of the situation boils down to a simple, informal equation: *Power Wall + ILP Wall + Memory Wall = Brick Wall for serial performance* [6]. Since these "walls" constrain the microprocessor design space that the High Performance Computing community must soon confront, it is important to briefly survey the factors they bring into play.

The "power wall" represents an intractable physical barrier — too much heat, too much power consumption, and too much leaking voltage — to further increases in clock speeds. The explosive improvements in the performance of microprocessors we have witnessed over the past decade depended directly on a 4000-fold increase in clock speeds over that same period of time [7]. But power dissipation is proportional to processor clock frequency, and, as a result, when clock rates rise above 3 GHz, the heat generated and the power leakage at the gates becomes unmanageable, putting an end to performance improvements along this path.

At the same time, opportunities for improving the speed of a single thread of control, which was the primary source of performance acceleration in the twentieth century, are also blocked by the exhaustion of techniques based on instruction level parallelism (ILP). This is the "ILP wall." ILP based techniques increase overall execution speed by employing superscalar execution with multiple instruction issue and mechanisms such as branch prediction and speculation. The major advantage of this approach is that serial programs, taking no explicit account of parallelism, can receive the benefits of ILP-based speed up by simply being recompiled. But these techniques, which lead to the dominance of large monolithic processors, have now reached the point of diminishing returns in both issue width and pipeline depth; they have saturated their design space, leaving little instruction level parallelism left to be exploited. The "free ride" is over in the sense that software developers who want to receive the performance benefits of greater parallelism must, from here on, explicitly program for it in the form of task or data parallelism — very much a non-trivial task.

Finally, physical limits on the number and bandwidth of pins on a single chip means that the gap between processor speed and memory access time, which was already wide, will continue to widen. Moreover, further increases in the number and sophistication of levels of caches has also reached the point of diminishing returns. These phenomena, taken together, constitute a "Memory Wall" with which the scientific software community must find some way to cope.

In order to deal with the emergence of these limitations, around 2004 the microprocessor industry took a historic turn towards designs with multiple processing units in a single chip [8]. Chips following this design are better able to deal with power and heat dissipation issues that arise at high clock speeds. But from the point of view of increased performance, the most important consequence of this new design paradigm is that it shifts the focus from instruction level parallelism to thread level parallelism (TLP) and data level parallelism (DLP). The advantage of the multicore approach is a consequence of "Pollack's Rule," which holds that performance increases are roughly proportional to the square root of increases in complexity [9]. In other words, if the chip area is doubled to build a large monolithic processor, a performance increase of only 40% can be expected. If it is used, however, to build two smaller cores, then in principle, i.e. if TLP and/or DLP can be efficiently exploited, the performance can be doubled. This fact has generated a new corollary of Moore's law, according to which the number of cores is expected to double with each new generation of chips, roughly every two years. Dual-core commodity chips are already common, quad-core machines are available and 8-core machines are expected in 2008.

It also is worth noting that future chips are likely to go beyond simple multicore designs, bringing in heterogeneous cores, such as superscalar cores to handle control tasks and very long instruction word (VLIW), or single instruction multiple data (SIMD) cores to handle compute and data intensive tasks. The Sony/Toshiba/IBM Cell processor is an example of a commercially successful processor of this type. Future designs may also offer a reconfigurable memory system, where the memory attached to a core can act as either a coherent cache or private scratchpad memory. It is interesting to note that, in this respect, the STI Cell processor represents the extreme approach, basically being a distributed memory system with global address space. Solutions based on graphics processing units (GPUs) and field-programmable gate arrays (FPGAs) are also likely to come into the picture. Mixed hardware configurations have already appeared (e.g., Los Alamos RoadRunner, Cray BlackWidow).

As has been widely observed, accessing the power of these new designs will require a paradigm change in the programming community. Though a great deal of research has been directed to the development of auto-parallelizing compilers, these tools, which would relieve programmers of the need to recast their serial programs into a parallel form, have so far proved themselves to be efficient only in a highly restricted set of applications. Thus, high performance can be extracted from multicore architectures only if TLP can be exploited at the programming level. Software for multicore architectures must be designed with parallelism in mind and algorithms have to be reformulated or rewritten in order to exploit TLP or DLP.

Given the constraints just discussed, we believe that DLP is not a viable option going forward for dense linear algebra. In fact, in this context, dense linear algebra suffers from the legacy of the data-parallel model, which emphasizes asymptotic performance and isoscaling. This model relied on the fact that the impact of inefficient operations could always be overcome if the size of the systems could be increased sufficiently. Basically, because of the surface to volume effect, good performance could always be achieved if a large enough problem was used. But the explosion in computing power promised by the multicore revolution is unlikely to be followed by a concomitant explosion in memory capacity, memory bandwidth or the speed of interconnections between nodes in distributed memory architectures. These phenomena will render the data-parallel model incapable of delivering high performance on these new architectures and, consequently, it will be replaced a TLP-based approach.

Given these facts about the multicore design space, the traditional approach to dense linear algebra libraries must be changed. Traditionally, the class of problems under investigation is approached from the lower end of parallelism, by parallelizing the fastest sequential algorithm. However, the fastest serial algorithms may not be the best to parallelize. Frequently, alternatives exist, which, although slower sequentially, deliver faster time to solution when parallelized. This calls for the rethinking of not only programming models, but also algorithms. With parallel hardware becoming ubiquitous, it is time to approach parallelism from the high end and aim for maximum parallelism. It is time to replace the question "On how many processors can the algorithm run efficiently?" with the question "How efficiently can the algorithm run on all available processors?" The key to achieving this goal is the emphasis on strong scaling and the replacement of the data-parallel model with the task-parallel model and the employment of the following techniques:

- *Asychronicity and Dynamic Scaling:* The state-of-the-art software for dense linear algebra, such as LAPACK (Linear Algebra Package) and ScaLAPACK [10, 11] employs the fork-join parallel execution, a heavily synchronous approach, stemming from the data-parallel model. This solution allowed for writing of hundreds of thousands of lines of LAPACK and ScaLAPACK code in tractable and maintainable manner — a true marvel of software engineering. Unfortunately, today it is also the cause of rapidly declining performance of these libraries on new generations of processors. Not only is this approach incapable of exploiting heterogeneous systems, but also proves inadequate for delivering performance on homogeneous systems. Basically, parallel code written in such a manner enforces nonexistent dependencies, which prevents flexible scheduling of operations and causes unnecessary stalls. Alternatively, the producer/consumer philosophy can be applied, replacing collective operations with point-to-point communication and synchronization. Dense linear algebra is inherently well suited for dynamic execution. It is rich in diverse types of operations, with different performance characteristics, in terms of computation and memory intensity, and different levels of parallelism. Historically, the performance of dense linear algebra algorithms was improved by applying the concept of a look-ahead [12–16], which is nothing else than a manual adjustment of the order of operations, and can be seen as a very modest attempt at dynamic scheduling. Dynamic scheduling also addresses the problem of efficient execution in heterogeneous environments. One problem associated with dynamic execution is the overhead of task scheduling and synchronization. It is not a fundamental problem, however, but rather an issue of implementing efficient mechanisms to carry out this job. So far, dynamic scheduling has proven its benefits for multicore processors in a shared memory arrangement [1, 2, 17, 18]. It has also recently been applied to dense linear algebra algorithms on distributed memory systems [19], showing similar potential.

- *Fine granularity:* Overgrown caches are likely to become the thing of the past, as they have proven to bring diminishing returns in performance gains [20]. It has been argued that, in a power-efficient design, increase in the resource size, such as the cache, is only justifiable if it results in a proportional increase in performance. This is known as the "kill if less than linear" (KILL) rule of chip design [21]. Designs like the STI Cell processor, the Intel Polaris prototype and the Tilera chip already follow this path. On the other hand, fine granularity of operations is known to inhibit performance of dense linear algebra operations. It has been recognized that one of the reasons for this is of a technical nature — existing BLAS implementations, such as GotoBLAS [22], are not tuned for efficiency on small data sets [17]. There is however a more fundamental problem. The surface to volume ratio may not take effect for the underlying BLAS if the size of the elementary operation is too small. Nevertheless, finer granularity creates more parallelism, and parallelism will be the key to unlocking the performance of massively parallel chip multiprocessors. Obviously, caution has to be given when adjusting the level of granularity to the specifics of a particular architecture, which is a great area for exploration of auto-tuning techniques.
- *Locality of reference:* Locality of reference is a broad problem that can be tackled at many levels. One important aspect is the use of specialized data structures oriented towards maximizing locality of reference for linear algebra operations, such as the block data layout (BDL) [23, 24]. Another issue is affinity of data to processing units, typically strictly enforced for distributed memory systems, rather loosely enforced for shared memory systems, and basically meaningless for the Cell processor. It can be observed that affinity can be traded for the scheduling flexibility, in principle on any kind of system. Overall, it seems that distributed memory systems were granted the benefits of block data organization, but not shared memory systems. On the other hand the shared memory systems enjoyed the profits of dynamic scheduling, but not the distributed memory systems. It seems a natural thing to treat both classes with the same generosity.

METHODOLOGY

New Algorithms

Commonly used linear algebra algorithms rely on the concept of *block operations* [11] to achieve high performance on modern architectures. Based on the observation that linear algebra operations can be described as the application of a sequence of operators, block operations leverage the idea that operators can be accumulated in sets and applied at once. A block algorithm can, thus, roughly be described as a sequence of two main steps:

Panel reduction: at this step a set of *nb* (commonly referred to as the *block size*) operators are computed for a small portion of the matrix. This step typically involves Level-2 BLAS operations, which are memory bound operations and cannot be efficiently parallelized on shared memory machines due to the disproportion between CPU-memory bus bandwidth and CPU performance that characterizes virtually all modern architectures.

Trailing submatrix update: at this step all the *nb* operators computed in the panel reduction are applied at once to the rest of the matrix (i.e., the *trailing submatrix*). This step is typically performed by means of Level-3 BLAS operations.

If *nb* is much lower than the problem size *n*, then most of the floating point operations are done in Level-3 BLAS routines, which delivers high performance on memory-hierarchy systems thanks to the so called *surface-to-volume* effect.

Since the gap between the processor speed and the memory access time is likely to be increased by multicore technologies, block operations are still of key importance to achieve high performance in linear algebra computations. However, blocking of operations poses some limitations to fine-granularity parallelization of most of the commonly used linear algebra algorithms. From this observation stems the necessity to either reformulate traditional algorithms or develop new ones in order to achieve those important features discussed in Section 2.

Some novel work from the PLASMA project in this context focuses on the definition of "tile" algorithms [1–3] where a linear algebra operation can be described as a sequence of tasks that operate on "tiles", i.e., square blocks of data that can be arbitrarily small.

The Cholesky factorization represents one of the cases where the well known LAPACK algorithm can be easily reformulated in a tiled fashion [1, 3]. Each operation that defines an atomic step of the LAPACK algorithm can be broken into a sequence of tasks where the same algebraic operation is performed on smaller portions of data, i.e., the tiles.

In most of the cases, however, the same approach cannot be applied and novel algorithms must be introduced. Recent work [1, 2] proposes, for the QR and LU factorizations, algorithms based on the well known methods for updating factorizations that were introduced [25] and later on reformulated in terms of block-operations [26, 27], to implement out-of-core solvers. These updating factorization methods can be used to derive tile algorithms for LU and QR factorizations that provide very fine granularity of parallelism and the necessary flexibility that is required to exploit dynamic scheduling of the tasks according to the techniques described in Section 3.2. It is worth nothing, however, that, as in any case where such fundamental changes are made, trade-offs have to be taken into account. For instance, in the case of the LU factorization the tile algorithm replaces partial with block pairwise pivoting which results in slightly worse stability [26]. As we suggest below (Section 3.4.2), techniques such as iterative refinement can often be applied to remedy such problems.

High performance can be achieved when tile algorithms are used in combination with dynamic scheduling techniques (see Section 3.2) and high data locality storage formats such as BDL (see Section 2). Figure 1 shows the performance of the tile algorithms, with BDL and dynamic scheduling, for the LU, Cholesky and QR factorizations compared to the performance of the LAPACK block algorithm linked to a multithreaded BLAS library (ACML) and to that of two vendor implementations.

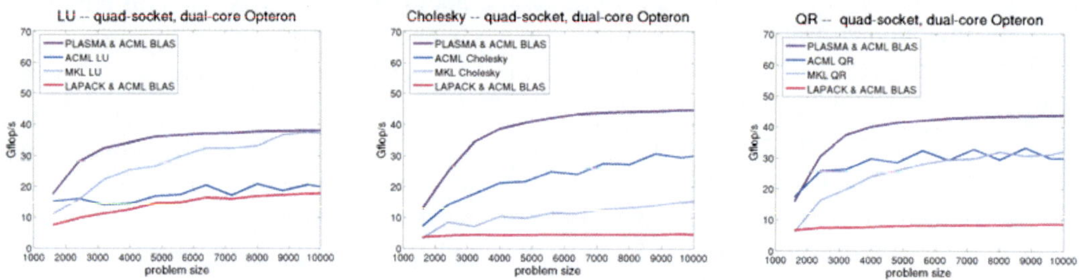

Figure 1. Performance comparison between different implementations of LU, Cholesky and QR algorithms on a quad-socket, dual-core Opteron system (ACML 4.0.0 and MKL 9.1 were used).

The applicability of this approach to a broader set of linear algebra algorithms including two-sided, invariant transformations like reductions to Hessenberg, tri-diagonal and bi-diagonal forms is still under investigation. Two-sided transformations are of key importance for the solution of eigenvalue or singular value decomposition problems but are very expensive, and the LAPACK block algorithm is rather inefficient. The reason for this inefficiency lies in the extremely high cost of the panel reduction which, moreover, makes these operations poorly scalable when parallelized with a fork-join paradigm (for example, using a multithreaded BLAS library).

The design and development of novel algorithms for these operations is a rather challenging task and the achievement of the properties discussed in Section 2 is hindered by a number of factors:

Fine granularity Techniques have been studied in the past [28, 29] to reduce a matrix to a block-condensed form. Earlier work [29] shows how a tiled algorithm can be used to reduce a matrix to a block-Hessenberg form. The same approach can be applied to the reduction to tridiagonal form. Even if such algorithms provide considerable benefits for parallelization and for achieving fine granularity, the block-Hessenberg and block-tridiagonal forms are of no interest to any application. For this reason, studies have been conducted [28] to reduce a block-Hessenberg or block-tridiagonal to Hessenberg and tridiagonal forms respectively through invariant transformations.

Asynchronicity Even if, in principle, a graph-driven execution model can be applied to any algorithm, its efficiency is strongly dependent on the complexity of the algorithm itself. The presence of many dependencies among the elementary tasks, i.e., many edges in the DAG that represents the operation, limits the flexibility of the scheduling and may result in a poor parallelization of the tasks.

Memory locality Memory locality can be improved by using specific data storage formats like BDL or by scheduling the tasks in the DAG according to policies that maximize the affinity of data to core memories (caches or scratchpad memories). It is very difficult to apply these techniques to algorithms for two-sided transformations. Two-sided transformations, in fact, are characterized by a complex pattern of access to memory that make very difficult the definition of tiled algorithms as well as the application of storage formats as BDL. Moreover, the algorithms for two-sided transformations are rather complicated, which translates in the presences of many dependencies and, thus, edges in the DAG. As a result, the flexibility of the task scheduling is seriously limited when memory affinity is exploited.

Algorithm Description and Execution Environment

It can be observed that parallel performance problems frequently stem from implicit expression of parallelism. Although today massive parallelism is mostly achieved through message passing, applications often hide parallelism behind software middle layers. This is, for instance, the case with ScaLAPACK, PBLAS [30], and BLACS [31], and also the case of LAPACK running on top of multi-threaded BLAS. The alternative is to provide a means for direct expression of parallelism at the algorithmic level based on the dataflow model. This can be achieved by an algorithm description abstraction relying on explicit definition of tasks and their dependencies. At the same time, such abstraction shall also facilitate unified treatment of different memory architectures.

Similar concepts were popular two decades ago. The early 90s introduced the *Parallel Virtual Machine* (PVM) model for programming distributed systems [32, 33], which was accompanied by a number of tools for graphical development of parallel programs. Examples include Phred [34, 35], Paralex [36], HeNCE [37–40] and CODE [41, 42] and their roots can be traced to a yet older project called SCHEDULE [43]. Although these tools had a lot in common, the straightforward approach taken by the HeNCE system appears particularly appealing. It seems that many solutions introduced by the HeNCE project can almost readily be carried over to a multicore programming framework.

The fundamental idea is an acyclic representation of the algorithm as a directed graph with procedures attached to the nodes. These procedures can be written in any programming language, but emphasis on commonly used imperative (procedural) programming languages, such as FORTRAN and C, is desired to facilitate software reuse and encourage the adoption of the technology by the community. The nodes are annotated with the list of input and output parameters of the associated routines. The system takes care of matching parameters between the nodes by searching the graph and handles the necessary data communication, also attempting to maximize the overlapping of communication with computation. In addition to simple dependency arcs, the system provides constructs to denote different types of control flow, such as conditionals, loops, fans, and pipes. These constructs can be thought of as graph rewriting primitives.

The system is built around a well defined textual representation of the graph, which can be manipulated without the use of a *graphical user interface* (GUI). Making the graphical interface an independent component allows for different implementations of the front-end and ensures portability.

Solutions such as OpenMP [44] and Cell/SMP SuperScalar [45, 46] attempt to exploit parallelism by instrumenting procedural programs with parallel constructs. In contrast, the principle proposed here is the augmentation of parallel constructs with pieces of procedural code.

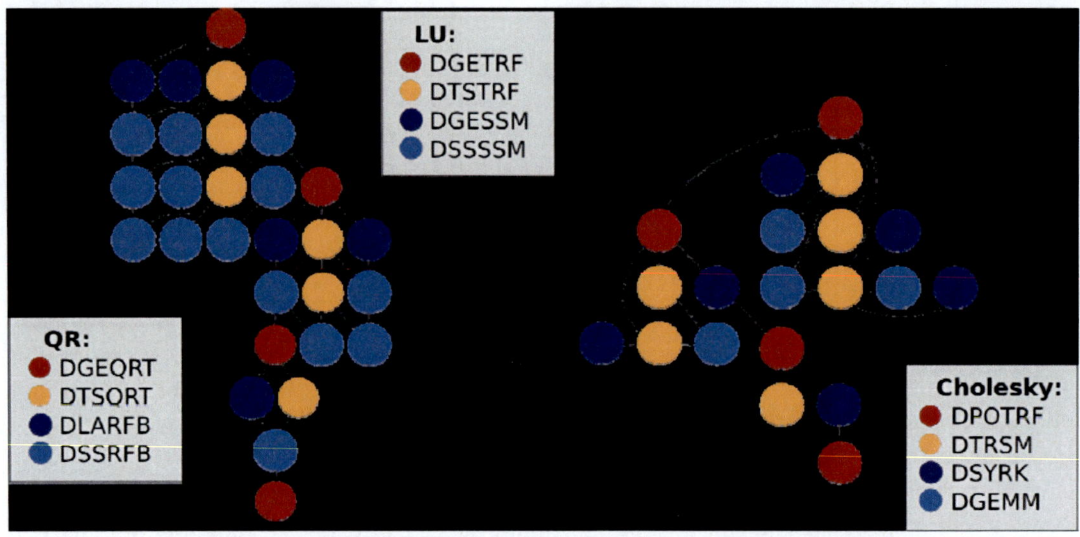

Figure 2. Task dependency graph for tile LU and QR factorizations (left) and Cholesky factorization (right). For clarity some dependencies are not shown.

The DAG scheduling problem poses many challenges, specifically in the context of distributed memory environments. Dynamic scheduling is a difficult problem in general, trading off maximum data reuse with the aggressive pursuit of the critical path. In many cases manual instrumentation, such as hinting the critical path or enforcing certain data affinity schemes (e.g., block-cyclic distribution), can provide a remedy. Another issue is DAG construction and exploration. Construction and storing of the entire DAG is not a feasible solution. It can be done for small problems, but in general it is not scalable. It is necessary to explore efficient methods for online and offline DAG construction and exploration, specifically exploration of the DAG in a "sliding window" fashion, with bounds on the memory footprint. A plausible solution in the initial stages of the project is static partitioning of the task graph and dynamic scheduling within each subdomain. A scalable solution for tracking the progress of the algorithm has to rely on distributed data structures with an emphasis on point to point producer/consumer notifications/synchronizations. Fundamental ideas to be exploited here are: compact DAG representation and its dynamic scheduling [47], bundle scheduling [48], work-stealing [49, 50], and alike.

Unified Approach for Different Memory Architectures

Currently, multicore processor architectures can be considered as "more of the same" due to their resemblance to traditional shared memory systems, specifically *symmetric multiprocessor* (SMP) designs. This approach is predicted to reach its scalability limits at the size of a few tens of cores [6]. It is likely that the model will be evolving towards the distributed memory concept. One path leads through reconfigurable memories, which may act both as shared and private memories as required. In some application areas, it may be beneficial to use local memories, sometimes referred to as scratchpad memories, completely under control of the programmer. This solution is also sometimes referred to as software-controlled cache. (The Cell processor is an example of this approach.) Owing to this trend, it seems necessary to assume the distributed memory model, not only to encompass large scale MPP systems, but also to assure scalability at the chip level on future generations of chip multiprocessors.

In fact the DAG-based visual programming environments of the early 90s [34–42] were designed around the PVM communication package for distributed memory systems. On the other hand, the ideas of asynchronous execution and dynamic scheduling are typically associated with shared memory systems. One reason for this is the simplicity of implementing synchronization using straightforward shared memory mechanisms like mutexes (locks). Another is global view of the memory, where processor-to-memory affinity plays a much smaller role. There are, however, no fundamental obstacles in implementing dynamic scheduling in distributed memory environments. Issues that require particular attention are synchronization and data affinity.

Synchronization relates to tracking the progress of the algorithm. It is easily implemented by shared (centralized) progress tables on shared memory systems, an approach, that would create bottlenecks and prevent scaling on distributed memory systems. The problem can be remedied by the use of replicated or distributed data structures for progress tracking. Replicated progress tables have been successfully used for minimizing synchronization overhead on the Cell processor [3].

Processor and data affinity refer to pinning memory regions to processors, serve the purpose of maximizing locality of reference. By creating fixed processorto-memory associations, affinity reduces opportunities for dynamic scheduling, but does not eliminate them completely. Good examples can be found, specifically in the context of dense linear algebra: The *High Performance Linpack* (HPL) benchmark relies on fixed, block-cyclic data distribution, yet benefits from the performance advantages of multiple-depth look-ahead [51]. More recently, experiences with dynamic Cholesky factorization on distributed memory systems were reported by Gustavson [19]. Typically, in dense linear algebra, affinity is either strictly enforced or not enforced at all, while advantages of both approaches could be combined in a "flexible affinity" scheme.

The solutions for the distributed memory model should simply extrapolate to the shared memory model. Shared memory, NUMA systems are likely to benefit from behavior similar to message-passing, where sends and receives are replaced with memory copies (and can sometimes be accelerated with DMA mechanisms). On the other hand, specialization of this model to truly symmetric SMP systems may be as straightforward as replacing the copy operations with aliasing.

Related Topics of High Importance

This section is devoted to topics that are envisioned as integral parts of the development of efficient software for systems based on multicore technology. These include: the use of specialized data structures (*block data layout* - BDL), the technique of iterative refinement for improving numerical accuracy of computations and porting the technology to specialized hardware (such as the Cell processor).

Block Data Layout

The use of a "standard" column-major or block-major data layout in linear algebra algorithms leads to multiple performance problems, causing cache misses at all cache levels and *translation look-aside buffer* (TLB) misses and preventing good utilization of memory banks on architectures with bank memory organization.

The issues have long been identified and addressed by copying data to block layout internally within BLAS. Although this is a valid performance-improving solution, it has to be observed that the copy operation introduces overhead, which can be avoided if data is stored in such a layout to begin with. At the same time, block data layout is a perfect match for the "tile algorithms," which process the input in fixed-size blocks. If input blocks are stored continuously in the memory, a number of performance problems can be eliminated altogether.

With the right choice of block size, *conflict misses* and *capacity misses* can be completely eliminated for a tile operation, leaving only the *compulsory misses* (for all levels of cache and the TLB). Also, owing to the continuous storage of blocks in memory, block data layout provides an opportunity to utilize the bandwidth to the fullest on systems with multiple memory banks. This is, for instance, the case on the Cell processor, where the use of BDL results in exactly the same number of reads/writes issued to all 16 memory banks, which results in more than 90% of utilization of the memory bandwidth [3, 52].

Overall, the use of BDL imposes much less strain on the cache mechanisms and makes the algorithm much less vulnerable to specific shortcomings of a particular cache system, which historically led programmers to unappealing workarounds, like padding of the input matrices.

A software library should, thus, rely internally on non-canonical data representation, which may vary for different architectures and be subject to tuning and autotuning (see Section 3.4.4). It is important to shield the average user from low level mechanisms with convenient interfaces and memory management facilities, as well as provide the advanced user with efficient translation tools.

Iterative Refinement

The motivation for the use of the iterative refinement technique is twofold: Short vector SIMD processing allows us to trade accuracy for computing speed; the new class of "tile algorithms" may, in some cases, deliver slightly lower accuracy than their traditional counterparts.

Until recently, ILP has been exploited mainly through multiple instruction issue (superscalar execution), a costly solution in terms of the circuitry required for dependency tracking, register renaming, etc. In recent years, short vector SIMD processing has rapidly gained popularity. Today it is a mandatory feature of modern processors and is almost certain to be ubiquitous in future generations of multicore processors. Short vector SIMD processing allows us to trade accuracy for performance, which makes mixed-precision algorithms an attractive target, specifically in dense linear algebra, where the accuracy lost in lower precision calculations can often be easily recovered through refinement of the solution [53]. On the first generation Cell processor, where the full (double) precision performance is severely inhibited by arbitrary design choices, the technique was used to speed up calculations by a factor of 10 [3, 52].

The other motivation is the use of the class of "tile algorithms," which, in some important cases, may deliver solutions of slightly lower accuracy. A good example of such an algorithm is LU factorization with partial pivoting, where in the "tile algorithm" the *full partial pivoting* is replaced by *block pairwise pivoting*, which slightly affects the accuracy of the result. In this case iterative refinement allows for quick recovery of the lost accuracy.

Cell Broadband Engine

The Cell processor follows many current trends in modern chip design: It is a multicore processor with simple in-order execution cores, powerful SIMD capabilities and relies on software controlled local memory model. The use of the Cell processor in computational sciences has been severely hindered by the difficulty in programming the chip, which, in most cases, requires manual MIMD parallelization between the cores and manual SIMD vectorization within the cores. Despite the difficulties, the processor proves superior to other multicore chips currently available for important classes of scientific workloads, including problems in dense linear algebra [3, 52], sparse linear algebra [54], Fourier transforms [55], and alike. Although the Cell architecture certainly does not provide answers to all problems facing processor design, it may provide a glimpse into certain aspects of programming future generations of multicores.

Very significant is the fact that the BLAS model fails miserably to deliver performance on the Cell processor. Despite efforts to tune the execution parameters, it proved impossible to exceed 10% of peak of a single Cell chip for a dense linear algebra workload, such as Cholesky factorization, using the Cell BLAS library released in IBM Cell SDK 3.0 (Figure 3). At the same time, the proposed algorithmic approach along with the asynchronous execution model allows us to exploit the architecture to the fullest and achieve 90% of the peak performance on a single Cell processor and also 90% of the peak of a dual-socket Cell blade system with NUMA memory organization. This success can be attributed to the use of the tile algorithm (Section 3.1), fine granularity of parallelization and pipelined mode of execution, and also block data layout. Not without significance is the fact that the iterative refinement technique (Section 3.4.2) allows us to recover the accuracy lost in single precision calculations [3, 52].

Tunability and architecture adaptivity are mandatory features of the proposed software. State of the art dense linear algebra algorithms have a number of tunable execution parameters trading off utilization of different system resources. Specifically, the class of tile algorithms introduces two important parameters: the outer block size and the inner block size. The outer block size trades off parallelization granularity and scheduling flexibility with single core utilization. The inner block size trades off memory load with redundant calculations [1].

In rare cases, these parameters are fixed by architecture constraints. One such example is the Cell processor. Here the inner blocking is mostly predetermined by the SIMD vector size, and the outer blocking is mostly predetermined by local memory size. For most architectures, however, the choices are much less obvious and arbitrary decisions have to be replaced by software auto-tuning capabilities. Auto-tuning has been successfully exploited before in the field of dense linear algebra in projects like ATLAS (Automatically Tuned Linear Algebra Software) [56].

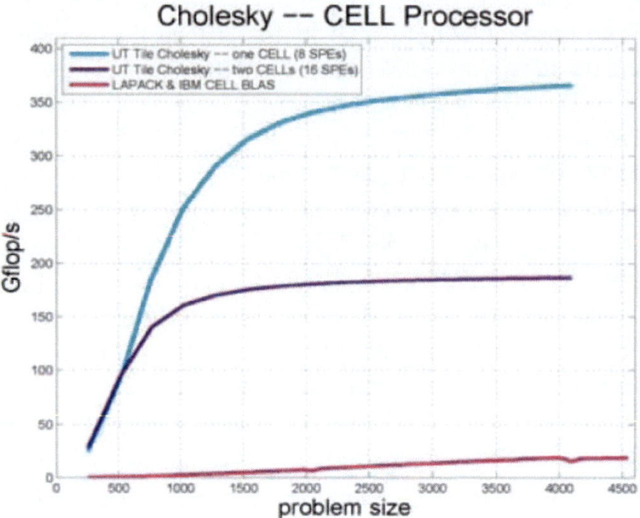

Figure 3. Performance of Cholesky factorization on a dual-socket QS20 Cell blade. Software Tunability and Adaptivity.

Conclusions

The goal of the PLASMA project is to create a new generation of dense linear algebra libraries that achieve the fastest possible time to an accurate solution on massively parallel systems by efficiently using all of the processors that such systems will make available. To work within this design space and leverage the power of million way parallelism, the PLASMA software must combine together new, highly scalable algorithms, a programming and execution model that can exploit massive task parallelism, and flexible memory management facility that can help optimize data locality across a range of different platforms. Advancing to the next stage of growth for computational simulation and modeling will require the solution of basic research problems in Computer Science and Applied Mathematics at the same time as a new paradigm for the development of scientific software is created and promulgated. To make progress on both fronts simultaneously will require a level of sustained, interdisciplinary collaboration among the core research communities.

References

[1] A. Buttari, J. Langou, J. Kurzak, and J. J. Dongarra. Lapack working note 191: A class of parallel tiled linear algebra algorithms for multicore architectures. Technical Report UT-CS-07-600, Electrical Engineering and Computer Sciences Department, University of Tennessee, 2007.

[2] A. Buttari, J. Langou, J. Kurzak, and J. J. Dongarra. Parallel tiled QR factorization for multicore architectures. In *PPAM'07: Seventh International Conference on Parallel Processing and Applied Mathematics*, 2006.

[3] A. Kurzak, J. Buttari and J. J. Dongarra. Solving systems of linear equation on the CELL processor using Cholesky factorization. *IEEE Trans. Parallel Distrib. Syst.* in press.

[4] Excerpts from a conversation with Gordon Moore: Moore's Law. Intel Corporation, 2005.

[5] G. E. Moore. Cramming more components onto integrated circuits. *Electronics*, 38(8), 1965.

[6] K. Asanovic, R. Bodik, B. C. Catanzaro, J. J. Gebis, P. Husbands, K. Keutzer, D. A. Patterson, W. L. Plishker, J. Shalf, S. W. Williams, and K. A. Yelick. The landscape of parallel computing research: A view from Berkeley. Technical Report UCB/EECS-2006-183, Electrical Engineering and Computer Sciences Department, University of California at Berkeley, 2006.

[7] John L. Manferdelli. The many-core inflection point for mass market computer systems. *CTWatch Quarterly*, 3(1), February 2007.

[8] D. Geer. Industry trends: Chip makers turn to multicore processors. *Computer*, 38(5):11–13, 2005.

[9] F. Pollack. New microarchitecture challenges in the coming generations of CMOS process technologies. In 32nd Annual International Sym posium on Microarchitecture, page 2, 1999.

[10] L. S. Blackford, J. Choi, A. Cleary, E. D'Azevedo, J. Demmel, I. Dhillon, J. Dongarra, S. Hammarling, G. Henry, A. Petitet, K. Stanley, D. Walker, and R. C. Whaley. *ScaLAPACK Users' Guide*. SIAM, 1997.

[11] E. Anderson, Z. Bai, C. Bischof, L. S. Blackford, J. W. Demmel, J. Dongarra, J. Du Croz, A. Greenbaum, S. Hammarling, A. McKenney, and D. Sorensen. *LAPACK Users' Guide*. SIAM, 1992.

[12] R. C. Agarwal and F. G. Gustavson. A parallel implementation of matrix multiplication and LU factorization on the IBM 3090. In M. Wright, editor, *Aspects of Computation on Asynchronous Parallel Processors*, pages 217–221, 1988.

[13] P.E. Strazdins. A comparison of lookahead and algorithmic blocking techniques for parallel matrix factorization. *Int. J. Parallel Distrib. Systems Networks*, 4(1):26–35, 2001.

[14] K. Dackland, E. Elmroth, and B. Kågström. A ring-oriented approach for block matrix factorizations on shared and distributed memory architectures. In Sixth SIAM Conference on Parallel Processing for Scientific Computing, pages 330–338, 1993.

[15] K. Dackland, E. Elmroth, B. Kågström, and C. Van Loan. Parallel block matrix factorizations on the shared-memory multiprocessor IBM 3090 VF/ 600J. *Int. J. Supercomput. Appl.*, 6(1):69–97, 1992.

[16] R. C. Agarwal and F. G. Gustavson. Vector and parallel algorithm for Cholesky factorization on IBM 3090. In *SC'89: 1989 ACM/IEEE Conference on Supercomputing*, pages 225–233, 1989.

[17] E. Chan, E. S. Quintana-Ortí, G. Quintana-Ortí, and R. van de Geijn. SuperMatrix out-of-order scheduling of matrix operations for SMP and multi-core architectures. In *SPAA'07: ACM Symposium on Parallelism in Algorithms and Architectures*, pages 116–125, 2007.

[18] J. Kurzak and J. J. Dongarra. Implementing linear algebra routines on multi-core processors with pipelining and a look-ahead. In *PARA'06: Workshop on State-of-the-*

Art in Scientific and Parallel Computing, 2006. Lecture Notes in Computer Science 4699, Springer, 2007.

[19] F. Gustavson, L. Karlsson, and B. Kågström. Distributed SBP Cholesky factorization algorithms with near-optimal scheduling. Technical Report RC24342(W0709-011), IBM, 2007.

[20] H. T. Kung. Memory requirements for balanced computer architectures. In ISCA '86: Proceedings of the 13th annual international symposium on Computer architecture, pages 49–54, Los Alamitos, CA, USA, 1986. IEEE Computer Society Press.

[21] A. Agarwal and M. Levy. The kill rule for multicore. In *ACM/IEEE Design Automation Conference*, pages 750–753, 2007.

[22] GotoBLAS. http://www.tacc.utexas.edu/resources/software/.

[23] N. Park, B. Hong, and V. K. Prasanna. Analysis of memory hierarchy performance of block data layout. In *Proceedings of the International Conference on Parallel Processing*, 2002.

[24] N. Park, B. Hong, and V. K. Prasanna. Tiling, block data layout, and memory hierarchy performance. *IEEE Trans. Parallel and Distrib. Systems*, 14(7):640–654, 2003.

[25] E. L. Yip. FORTRAN Subroutines for Out-of-Core Solutions of Large Complex Linear Systems. Technical Report CR-159142, NASA, November 1979.

[26] E. Quintana-Ortí and R. van de Geijn. Updating an LU factorization with pivoting. Technical Report TR-2006-42, The University of Texas at Austin, Department of Computer Sciences, 2006. FLAME Working Note 21.

[27] B. C. Gunter and R. A. van de Geijn. Parallel out-of-core computation and updating of the QR factorization. *ACM Trans. Math. Softw.*, 31(1):60–78, 2005.

[28] C. Bischof, X. Sun, and B. Lang. Parallel tridiagonalization through twostep band reduction. In *Proceedings of the Scalable High-Performance Computing Conference, 1994. 23-25 May 1994*, pages 23–27.

[29] Michael W. Berry, Jack Dongarra, and Youngbae Kim. A parallel algorithm for the reduction of a nonsymmetric matrix to block upper-hessenberg form. *Parallel Computing*, 21(8):1189–1211, 1995.

[30] PBLAS: Parallel basic linear algebra subprograms. http://www.netlib. org/scalapack/pblas qref.html.

[31] BLACS: Basic linear algebra communication subprograms. http://www.netlib. org/blacs/.

[32] V. S. Sunderam. PVM: A framework for parallel distributed computing. *Concurrency: Pract. Exper.*, 2(4):315–339, 1990.

[33] A. Geist, A. L. Beguelin, J. J. Dongarra, W. Jiang, R. Manchek, and V. Sunderam. *PVM: Parallel Virtual Machine: A Users' Guide and Tutorial for Networked Parallel Computing*. MIT Press, 1994.

[34] A. L. Beguelin. Deterministic Parallel Programming in Phred. PhD thesis, University of Colorado at Boulder, Department of Computer Science, 1991.

[35] Adam Beguelin and Gary Nutt. Visual parallel programming and determinacy: A language specification, an analysis technique, and a programming tool. Technical Report CS-93-166, Carnegie Mellon University, School of Computer Science, 1993.

[36] Ö. Babaoğlu, L. Alvisi, A. Amoroso, R. Davoli, and L. A. Giachini. Paralex: An environment for parallel programming in distributed systems. In *ICS'92: 6th International Conference on Supercomputing*, pages 178–187, 1992.

[37] A. L. Beguelin and J. J. Dongarra. Graphical development tools for network-based concurrent supercomputing. In *SC'91: 1991 ACM/IEEE Conference on Supercomputing*, pages 435–444, 1991.

[38] A. L. Beguelin, J. J. Dongarra, A. Geist, and R. Manchek. HeNCE: A heterogeneous network computing environment. *Scientific Programming*, 3(1):49–60, 1993.

[39] A. L. Beguelin, J. J. Dongarra, A. Geist, and V. Sunderam. Visualization and debugging in a heterogeneous environment. *Computer*, 26(6):88–95, 1993.

[40] A. L. Beguelin, J. J. Dongarra, G. A. Geist, R. Manchek, K. Moore, P. Newton, and V. Sunderam. *HeNCE: A Users' Guide, Version 2.0*. Netlib, June 1994.

[41] J. C. Browne, M. Azam, and S. Sobek. CODE: A unified approach to parallel programming. *Software*, 6(4):10–18, 1989.

[42] P. Newton and J. C. Browne. The CODE 2.0 graphical parallel programming language. In *ICS'92:6th International Conference on Supercomputing*, pages 167–177, 1992.

[43] J. J. Dongarra and D. C. Sorensen. A portable environment for developing parallel FORTRAN programs. *Parallel Comput.*, 5(1–2):175–186, 1987.

[44] OpenMP. http://www.openmp.org/.

[45] Cell Superscalar. http://www.bsc.es/ → Computer Sciences → Programming Models → Cell Superscalar.

[46] SMP Superscalar. http://www.bsc.es/ → Computer Sciences → Programming Models → SMP Superscalar.

[47] M. Cosnard and E. Jeannot. Compact DAG representation and its dynamic scheduling. *J. Parallel Distrib. Comput.*, 58(3):487–514, 1999.

[48] P. Bellens, J. M. Perez, R. M. Badia, and J. Labarta. CellSs: A programming model for the Cell BE architecture. In *SC'06: 2006 ACM/IEEE Conference on Supercomputing*, page 5, 2006.

[49] R. D. Blumofe, C. F. Joerg, B. C. Kuszmaul, C. E. Leiserson, K. H. Randall, and Y. Zhou. Cilk: An efficient multithreaded runtime system. In *PPOPP'95: Fifth ACM SIGPLAN Symposium on Principles and Practice of Parallel Programming*, pages 207–216, 1995.

[50] R. D. Blumofe and C. E. Leiserson. Scheduling multithreaded computations by work stealing. *J. ACM*, 46(5):720–748, 1999.

[51] J. J. Dongarra, P. Luszczek, and A. Petitet. The LINPACK Benchmark: past, present and future. *Concurrency Computat.: Pract. Exper.*, 15:803–820, 2003.

[52] J Kurzak and J. J. Dongarra. Implementation of mixed precision in solving systems of linear equations on the CELL processor. *Concurrency Computat.: Pract. Exper.*, 19(10):1371–1385, 2007.

[53] A. Buttari, J. J. Dongarra, J. Langou, J. Langou, P. Luszczek, and J. Kurzak. Mixed precision iterative refinement techniques for the solution of dense linear systems. *Int. J. High Perform. Comput. Appl.*, 21(4):457–466, 2007.

[54] S. Williams, L. Oliker, R. Vuduc, J. Shalf, K. Yelick, and J. Dennel. Optimization of sparse matrix-vector multiplication on emerging multicore platforms. In *SC'07: 2007 ACM/IEEE Conference on Supercomputing*, 2007.

[55] A. C. Chowg, G. C. Fossum, and D. A. Brokenshire. A programming example: Large FFT on the Cell Broadband Engine, 2005.

[56] ATLAS. http://math-atlas.sourceforge.net/.

In: Cyberinfrastructure Technologies and Applications
Editor: Junwei Cao
ISBN: 978-1-60692-063-3

Chapter 2

SHARING SCIENTIFIC INSTRUMENTS FOR HIGHER EDUCATION AND RESEARCH IN CHINA

Jie Yin[1], Yuexuan Wang[2] and Cheng Wu[3]

[1] National CIMS Engineering and Research Center;
Tsinghua University, Beijing 100084, P. R. China
[2] Institute for Theoretical Computer Science;
Tsinghua University, Beijing 100084, P. R. China
[3] National CIMS Engineering and Research Center;
Tsinghua National Laboratory for Information Science and Technology;
Tsinghua University, Beijing 100084, P. R. China

ABSTRACT

Cyberinfrastructure (CI) for instrument sharing is an infrastructure that aims to facilitate effective resource sharing of expensive scientific instruments, e.g. telescopes and observatories, through a system of grid services. This cyberinfrastructure consists of three components: an instrument pool alliance, instrument pools, and physical instruments. When a user submits an experiment to the CI environment,the instrument pool alliance is responsible to allocate instruments in related instrument pools to conduct the experiment. After the experiment is finished and results are returned, the user will appraise performance of corresponding services.

In this chapter, fuzzy random scheduling algorithms are proposed in instrument pools when a job is submitted to one of instruments within a pool. The randomness lies in the probability of which instrument be chosen for an experiment and the fuzziness origins from vagueness of users' feedback opinions on experimental results. Users' feedback information is utilized to improve overall quality of service (QoS) of an instrument CI. Several algorithms are provided to increase utilization of instruments providing higher QoS and decrease utilization of those with poor QoS. This is demonstrated in details using quantitative simulation results included in this chapter.

[1] E-mail address: yinjie05@mails.tsinghua.edu.cn
[2] E-mail address: wangyuexuan@tsinghua.edu.cn
[3] E-mail address: wuc@tsinghua.edu.cn

INTRODUCTION

With the development of scientific research, more and more research activities require multidisciplinary collaboration. There are some large scale research tasks that need the coordination of resources from different management domains, for example, computational resources, storage resources, especially instrument resources. In this situation, how to organize and coordinate geographically distributed instruments is of very importance. How to connect expensive instruments and provide public sharing access of them, which can increase the utilization ratio of these instruments and eliminate the inconveniences brought by geographical distances, are very challenging issues.

Cyberinfrastructure for Instrument Sharing

Current situation in China is that there are some instruments, like microscope, mass spectrograph, Raman spectra, high resolution X-ray diffract meter etc, in some university or research institutes. On one hand, there are some users want to carry out their experiments but do not have corresponding instruments. What they can do is to contact the universities or institutes that have the instruments they require and reserve them for a proper time and duration. When time approaches, they go to the place where the instruments are located and conduct their experiments. This process of using geographically distributed instruments is time and cost consuming. On the other hand, there are some universities and institutes that have low utilization of their expensive instruments. The high maintenance cost became a burden for them. A new technology and management method is urgently required to address this issue.

Grid technologies [19, 20] can be adopted to address corss-domain resource sharing issues. It connects all resources and information distributed on the web into a loosely organized infrastructure, which can provide powerful computational ability, large amount of storages, transparency access, and information aggregation and sharing. In order to share scientific instruments among national universities in China, Ministry of Education of China funded a project, China Education and Resource System (CERS) [15, 31]. In CERS, grid technologies are used to connect some expensive instruments geographically distributed in universities all over the country. With advanced computing and information technologies, CERS aims at high level instruments sharing and coordinated working and an infrastructure to enable such activities, which is know as Cyberinfrastructure (CI) [16, 26] for instrument sharing [1-10, 31, 33]. The instrument CI runs on the basis of economic rules. It encourages instrument owners to provide their resources for sharing and charges from users for maintenance purposes.

Research Background

Several aspects of related research background on Cyberinfrastructure are included in this section, including CI architecture, resource scheduling, information management, etc.

CI Architecture

The architecture of the instrument Cyberinfrastructure is proposed in [1], as shown in figure 1. The architecture is divided into eight layers, including the portal, application, simulation and modeling, middleware, resource encapsulation, protocol, resource and network layers. Basic functions of these layers are described below.

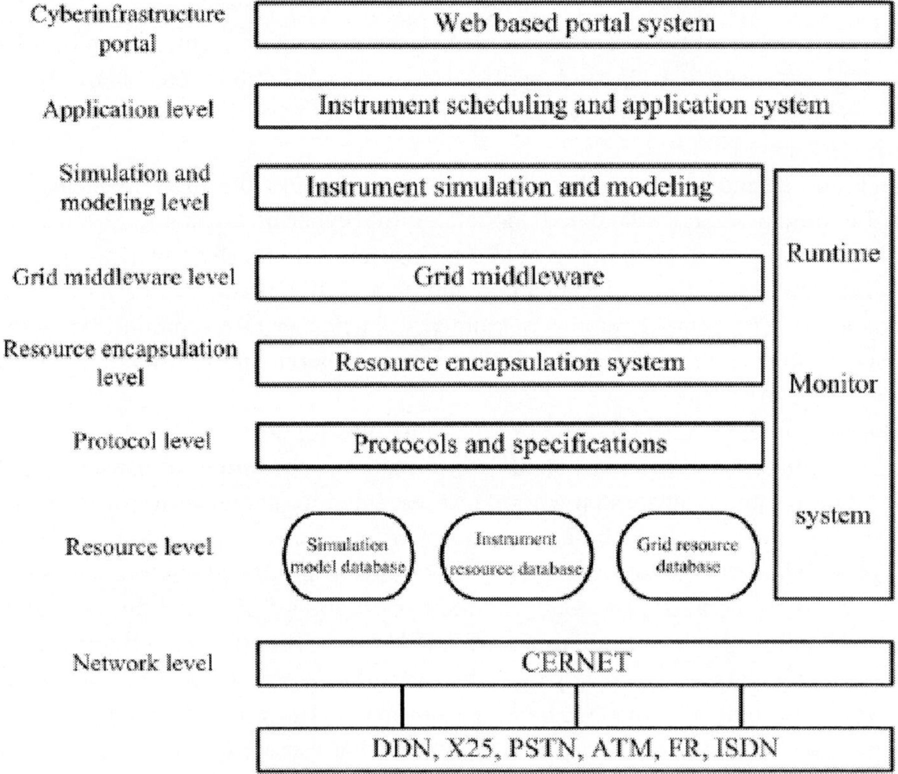

Figure 1. CI architecture.

The bottom layer is based on China Education and Research Network (CERNET) [11] that provides a network infrastructure to enable access of resources in the instrument CI and remote access ability to geographically distributed instruments.

The second and third layers from bottom, which are resource and protocol layers respectively, provide some key technical supports for the construction and operation of instruments, including instrument model database, instrument resource database, basic protocols and related specifications.

Instrument model database provides virtual mapping of remote instruments during the runtime. Instrument resource database provides information about all instruments shared in CI, their state information, operation and training information etc. Basic protocols refer to protocols that should be complied by all related staffs for the CI construction. Basic network protocol, multi agent protocols and related technical standards and specifications belong to basic protocols.

The fourth layer from bottom is resource encapsulation. In this layer, the local resources are encapsulated into global resources that accessible in the CI. The heterogeneity of different kind of instruments is smoothed by grid technology and it provides a transparent way for users to access. The management system in resource encapsulation level provides control and management of encapsulated resources and real time monitor for the resources, which provides the basis for resource location and scheduling optimization.

Grid middleware level provides basic functions for instrument sharing and service coordination using grid techniques. These functions include remote manipulation and management for instruments, coordinated allocation of instruments, storage access, information security, quality of service, advance reservation of instruments, agent transaction, appraisal of instrument sharing and application of open funding etc. They also include language, compiler, library, application program interface and developing environment for the cyberinfrastructure.

Simulation and modeling level is the level that simulates the distributed model, which includes the simulator and visualized model, of the cyberinfrastructure. Using visualized method, it can design and modify the model of instruments and their workflow process and provide visualized output like verification and model files. Simulation models have the similar functions with related remote instruments. They can telecommunicate with remote instruments through cyberinfrastructure. When users operate the simulation models, they operate the real instruments.

Application level provides following functions, like application of instruments shared in the cyberinfrastructure, collection of statistic information, scheduling of instrument resources, scheduling of jobs, approval for application of open funding, appraisal for instrument sharing and management for costs. Scheduling of instrument resources and scheduling of jobs are the two vital parts. The principle for resource scheduling is to minimize the total costs, maximize the profit, near distance first, fast response time first, high QoS first and high utilization first. The job scheduling can also manipulated by human through user interface. The job scheduling includes parse of job requests, choose suitable resources according to the jobs, processing of jobs and tasks management, accept or reject to process user jobs according to current work load in the cyberinfrastructure, monitoring remote instruments and running of jobs.

Cyberinfrastructure portal provides uniform and secure user interface on web for users, which can provide a uniform interface for users to access related services taking into no account of his location and identity, which is realized by role based information agents. These information agents record tastes of all users, like interface style, interesting contents that every user caring about, notification of some subscribed contents, sending and receiving of instructions, filtration and searching of information etc.

Resource Scheduling

The essence of resource scheduling is to distribute m independent jobs on n available resources, which may be heterogeneous. The goal of scheduling is to minimize the span time of jobs and at the same time makes utilization ratio of all available resources high. In the cyberinfrastructure, scheduling methods are used to optimize the utilization of resources. In [32], the process of how to schedule an instrument when resource scheduling model receives this request is discussed.

- The user submit his experiment request to Grid Resource Allocation Manager (GRAM);

- GRAM notifies GRIS (Grid Resource Information Service);
- GRIS searches for the instrument or instruments needed, then submit the location information to GIIS (Grid Index Information Service);
- The instrument agent is notified;
- The instrument agent negotiates with instrument pool and select a suitable instrument;
- The scheduling result is returned;
- Notify the users when the remote task finished and resource scheduling model finished this job scheduling.

Resource scheduling issues for clusters and grids has been discussed for many years. Especially, using historical QoS data to improve scheduling performance has been proved to be very effective. In a parallel and distributed computing environment, QoS data can be defined easily using quantitative values, e.g., job execution time [28], queue waiting time [29], data transfer time [18], CPU workloads [32], which can be modeled and analyzed using performance prediction technologies [30] and utilized to improve resource scheduling performance. However, it is difficult for users to characterize instrument performance quantitatively since various criteria (e.g. time, cost and precision) may play different roles in different experiments. In general, users can only provide an overall impression of instrument QoS. The fuzzy random theory is adopted here, which is suitable and straightforward when applied to the scheduling scenarios involved in an instrument cyberinfrastructure, though not necessarily providing the best scheduling solution. A similar work using fuzzy methods for grid scheduling can be found in [17], but the exact model and algorithms are different.

Information Management

In the cyberinfrastructure, the design of grid components should be adjustable according to different application requirements. Information management system plays an important role in the system and it is the base for resource scheduling and monitoring, like MDS4 in Globus and Matchmaker in Condor. It is a better way to improve the function of information management system by setting up classification mechanism. According to the information management problems in the cyberinfrastructure, [5] combined MDS4 in Globus Toolkits4 with UDDI registration center in Web Service and proposed a two-level information management system. The classification mechanism is imported into the proposed system, which increased the searching efficiency when there is much resource information presented.

UDDI and MDS can realize the management of information independently but also has their inefficiency. UDDI can locate a specific type of instrument resource services quickly but can not provide support for the dynamic information in these services. MDS4 is a bit different. It can reflect the availability and change of resource states well but has no classification mechanism. So when the number of services of instrument resource increased and the hierarchy of information service increased, the service searching process will take quite a long time. So the two-level information management mechanism that takes both advantages of UDDI and MDS will enhance the ability for information management in the cyberinfrastructure.

To be accordance with the two-level, which are UDDI level and MDS level, in the cyberinfrastructure, there are two phases in the release and discovery of resource service information. In UDDI level, the information about service description and type of resources is managed here and in MDS, basic index service is imported and detailed instance information about resource service is managed. Information about instrument resources needs to be released when the resources join the cyberinfrastructure at the first time and provide services there. First the instruments need to find a suitable classification they belong to according to manual of information management system and provide related static information about these resources. If they are new instruments and not in the index of the cyberinfrastructure, the system administrator will create related service information data for them in UDDI registration center, and then with the help of system administrator the runtime instance of resource service will to pointed to a specific index service and register there.

Related Work

There are several projects most related to the work described in this chapter and detailed introduction are given below.

NEES

The first project focus on the cyberinfrastructure is NEES [12], which was set up in 1999 and got support from National Science Foundation. The object of this project is to accelerate the research of earthquake engineering through the cyberinfrastructure and lower the damages caused by earthquake. One key function of NEESGrid is the integration of instruments, which integrates many earthquake detectors, like all kinds of sensors, remote control instruments, remote monitoring instruments etc, into the cyberinfrastructure for instruments. This cyberinfrastructure facilitates researchers doing remote design, execution and monitoring earthquake simulation experiments. NEESGrid is not only an infrastructure for network interconnection in experiment instruments and monitoring instruments, it also provides a platform for all earthquake engineering researchers in US to work coordinately, to share these expensive scientific and engineering instruments and research products, which is supposed to be a new problem solving pattern for difficult and complex problems.

XPort

XPort [13] is another scientific project supported by US government. The object of this project is to provide remote operations for geographically distributed scientific instruments. Under the support of grid technologies, the remote operations towards several expensive X-ray crystallizers are realized. The manipulation and operation of these instruments, the acquisition, filtration and analysis of data are also specified. Users only need to post the crystals they want to research to the location where a near crystallizer located, they can gain the interior image of their specimens in their own laboratories. XPort greatly simplified the analysis process for the structure of huge molecule crystal. It not only enable enables the users to use remote expensive instruments, thus increases the utilization ratio of these instruments, but also provides a coordinated research platform to facilitate team research, which integrates some former isolated research groups and boosts the development of some interdisciplinary research.

CancerGrid

CancerGrid project [14] is a medical instruments testbed supported by Cambridge University UK and West England Cancer Web. This project started in 2005 and currently has provided cancer related medical services for more than 1.6 million persons. The system can support real time transportation, storage and searching of radiation images and multi-part video conferences. It can also provide remote access to physical therapy simulation computational program to provide assistant clinical diagnoses by data mining in the patient case records. With the support of CancerGrid, patients who lived locally distributed can get diagnosis and therapy from authorized experts.

CERS

From 2004, Tsinghua University had cooperated with some other universities in China to do research on the field of the cyberinfrastructure. As a sub project of CERS, we invited the participation of more than thirty universities to share their expensive instruments. It provides following functions, like entrance application, instrument information service, instruments reservation, job scheduling, cost management, assessment and evaluation and statistical analyse. Our work is under the support of this sub project.

SCHEDULING INTRUMENT SHARING

This work focuses on scheduling remote access of scientific instruments with consideration of quality of service (QoS) issues. A layered model of instrument pools is introduced. In our previous work the role of human was not taken into account and there was no QoS feedback mechanism to reflect whether users are satisfied with experimental results. In this chapter the feedback information regarding instrument QoS is considered to be a fuzzy variable with one of the following linguistic values, *terrible*, *bad*, *normal*, *good* and *excellent*. The probability whether an instrument could be chosen for a job is dynamically adjusted according to users' QoS feedback information. As a result, utilization of instruments providing higher QoS according to users' feedback is increased so that QoS of an instrument cyberinfrastructure as a whole is dramatically improved. This is quantitatively illustrated using detailed modeling and simulation results included in this section.

Scientific Instrument Sharing

As shown in figure 2, in an instrument cyberinfrastructure, similar instruments are organized into an instrument pool and different instrument pools constitute an instrument pool alliance. When a user wants to do experiment via the instrument cyberinfrastructure, he submits the job to the instrument pool alliance, which analyses the job and verifies whether it can be accomplished with existing pools within it. If the job can be fulfilled, the instrument pool alliance will submit it to the required instrument pools by order of the job's inherent requirements. When an instrument pool receives a job, it will find an available instrument to do it.

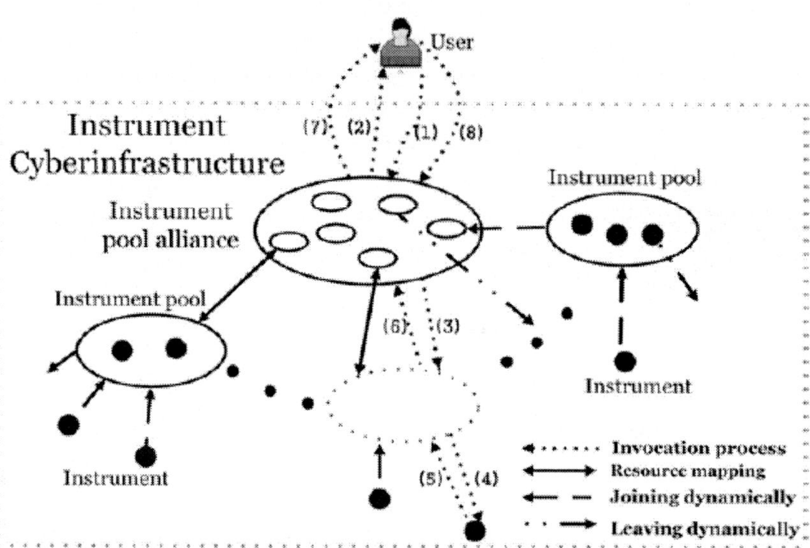

Figure 2. The process of invoking a service in the cyberinfrastructure.

Every instrument in figure 2 belongs to a certain instrument pool and can join and leave the pool dynamically. All instrument pools have their images in the instrument pool alliance and can also join and leave the pool alliance dynamically. When a user wants to do an experiment and submits it to the instrument pool alliance in Step 1, the instrument pool alliance will check whether the instrument cyberinfrastructure has the required instruments needed to fulfill the experiment. If not all resources needed are presented, the pool alliance will reply the user with refusal information in Step 2. Otherwise the alliance will decompose the experiment into parts and submit the related parts to corresponding pools in Step 3. All the related pools will find suitable resources and submit job parts to chosen instruments in Step 4. In Step 5, chosen instruments return results of the experiment to pools after the experiment was done and the pools return results to the pool alliance in Step 6. The pool alliance composes all middle results and returns a final result to the user in Step 7. In Step 8, the user feed back his opinion about the experimental result, which is important to improve QoS of the instrument cyberinfrastructure as discussed later.

Fuzzy Random Scheduling

As we mentioned before, instrument QoS can be hardly described using explicit parameters. In this section, we introduce a fuzzy random theory to characterize users' feedback QoS information.

Fuzzy Random Theory

The fuzzy random theory is an emerging field in uncertain theory, a branch of modern mathematics. It takes two aspects of uncertain factors, randomness and fuzziness, respectively, into account and has attracted many research interests. Some key concepts of the fuzzy random theory are given in this section. The detailed introduction can refer to [24].

A fuzzy random variable is a measurable function from a probability space to the set of fuzzy variables. In other words, a fuzzy random variable is a random variable taking fuzzy values. The notion of fuzzy random variable was first introduced by Kwakernaak in [22] and [23]. This concept was developed in [21], [25] and [27] by different requirements of measurability. Definition of fuzzy random variable is as follows [25]:

A fuzzy random variable is a function ξ from a probability space ($\boldsymbol{\Omega}$, $\boldsymbol{A}$, $\boldsymbol{Pr}$) to the set of fuzzy variables such that $Cr\{\xi(\omega)\in \boldsymbol{B}\}$ is a measurable function of ω for any Borel set $\boldsymbol{B}$ of $\boldsymbol{R}$, real number domain. $\boldsymbol{\Omega}$ is a nonempty set, $\boldsymbol{A}$ is algebra over $\boldsymbol{\Omega}$ and $\boldsymbol{Pr}$ is probability measure. Cr is credit of a fuzzy variable, which is similar to probability of a random variable. Definition of the expected value of a fuzzy random variable ξ introduced in [25] is as follows:

$$E[\xi]=\int_{0}^{+\infty}\Pr\{\omega\in\Omega\mid E[\xi(\omega)]\geq r\}dr-\int_{-\infty}^{0}\Pr\{\omega\in\Omega\mid E[\xi(\omega)]\leq r\}dr \tag{1}$$

providing that at least one of the two integrals is finite.

From the definition above, expected value of a fuzzy random variable is a scalar value. In Equation (1), $\xi(\omega)$ is a fuzzy variable and E in the left of the equation is the expectation of a fuzzy random variable, while E on the right is the expected value of a fuzzy variable. In most real world instances, the expectation calculation of a fuzzy random variable can be simplified.

Scheduling Models

The fuzzy random scheduling model refers to the schedule process of Step 4 in figure 2, which is an essential step in an instrument cyberinfrastructure for resource sharing. The scheduling model described in this work take users' feedback information into account and try to satisfy user requirements better.

Consider an instrument pool with N instruments in it, as shown in figure 3.

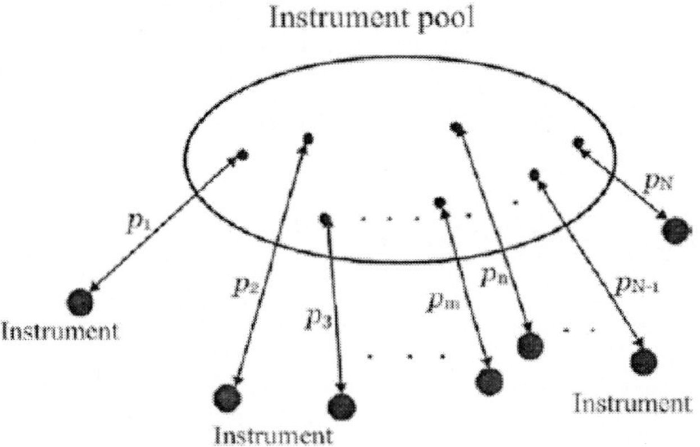

Figure 3. Job scheduling in an instrument pool.

When a new experiment is submitted to an instrument pool, the probability that the experiment runs on each instrument is p_i ($i \in [1, N]$). It is obvious that the following equation holds:

$$\sum_{i=1}^{N} p_i = 1 \tag{2}$$

when an experiment is submitted to any chosen instrument, there are many factors which have influence on users' appraisals, for example the cost of experiment this instrument charges for, the execution time and waiting time, whether the result from this instrument is reliable and whether the precision of the instrument can satisfy the experiment requirement. All these factors differ with different instruments and can be looked as a virtual parameter of the instrument. In this work, this parameter is named as QoS of instrument and denoted by q, and q_i means the QoS of the *i*th instrument in an instrument pool according to a specific experiment. The QoS of the same instrument will be different when the users' constrains changed. The pool adjusts the probability p_i according to the user appraisal, Q, to the experiment after he received his result from the instrument cyberinfrastructure. Both variables q and Q are fuzzy variables because a user can not depict how he satisfied with a result accurately. Only vague linguistic values like *terrible*, *bad*, *normal*, *good* and *excellent* can express his appraisal towards the result from the instrument cyberinfrastructure.

When a user submits a job with detailed experiment specifications to an instrument cyberinfrastructure, the instrument pool alliance will pass this job to corresponding instrument pools. If the experiment is submitted to the *i*th instrument in instrument pool, q_i has the value as one of the following linguistic values, *very bad*, *bad*, *normal*, *good* and *very good*. In most cases, a q_i with *very good* value has a large probability to receive *excellent* value of Q , *good* to *good*, *normal* to *normal*, *bad* to *bad* and *very bad* to *terrible* of Q. Because the value of q to a specific experiment is not known by instrument pool and can only be reflected by the user appraisal towards the total process of the experiment, the instrument pool will adjust p_i to make the instrument with *good* or *very good* appraisal higher utilization ratio to satisfy users. In some urgent experiments, users may attach more importance on time constrain. In such case instrument with shorter job execution time and waiting time will more satisfied the users. While in some experiments, users may care more about costs.

This fuzzy random scheduling model is a close loop model, which takes the user response into account and is believed to be able to provide higher instrument QoS.

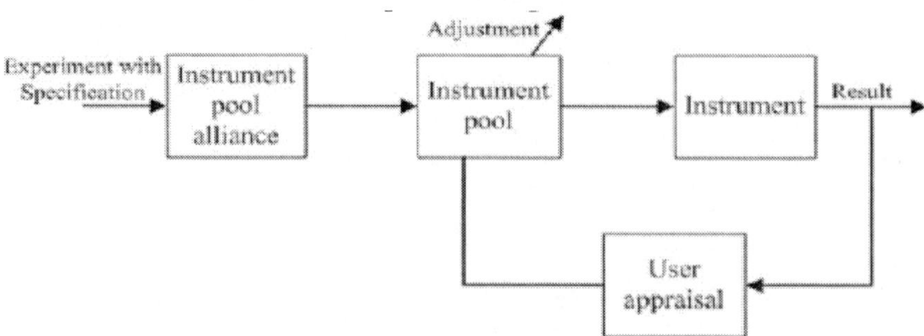

Figure 4. System block of the cyberinfrastructure.

Many good scheduling strategies and algorithms can be designed on the basis of the fuzzy random model introduced above. Most importantly, user QoS feedback information is used in this model thus it complies with the user intentions better.

Scheduling Algorithms

The adjustment of p_i from users' appraisals is described in this section. In this work, the algorithm to adjust p_i is proportional to the expected value of fuzzy random variable $preq_i$, which is the prediction of the fuzzy value q_i, as shown in Equation (3). The reason why the $preq_i$ is used in Equation (3) instead of q_i is that the instrument pool has no information of q_i and has to predict what the value it is through users' appraisals.

$$p_i = E[preq_i] / \sum_{i=1}^{N} E[preq_i] \tag{3}$$

In the following examples, the membership function of the fuzzy variable q_i is shown in figure 5.

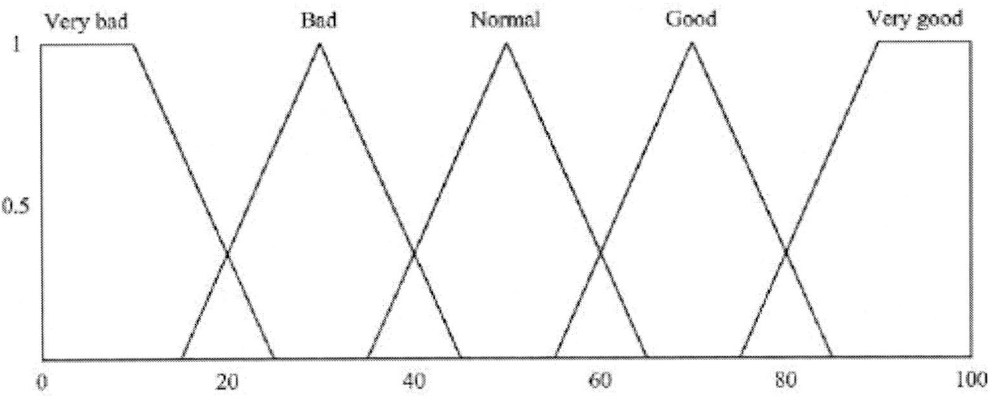

Figure 5. The QoS membership function.

The distribution of $preq_i$ is as Equation (4). In Equation (4), $prep^k{}_i$ ($1 \le k \le 5$) means the probability that $prep_i$ equals to kth value in equation (4). The initial values of $prep^k{}_i$ are the same and equal to 0.2. Because $preq_i$ is a fuzzy random variable, we can calculate its expected value using Equation (1).

$$preq_i = \begin{cases} \text{"Verygood"} & \text{with probability } prep_i^1 \\ \text{"Good"} & \text{with probability } prep_i^2 \\ \text{"Normal"} & \text{with probability } prep_i^3 \\ \text{"Bad"} & \text{with probability } prep_i^4 \\ \text{"VeryBad"} & \text{with probability } prep_i^5 \end{cases} \tag{4}$$

In this example the expectation of $preq_i$ can be simplified to Equation (5).

$$E[preq_i] = \sum_{k=1}^{5} prep_i^k \times c_i \tag{5}$$

in which c_i is defined to be the center of membership function and in this case, they are 17.5, 30, 50, 70 and 82.5 respectively. It should be noted that when the membership function changed, the expected value of $preq_i$ will also be different.

According to the users' feedback information, the $prep^k{}_i$ will be adjusted by Algorithm 1.

***Algorithm 1*:**

switch (*appraisal*)
{
case "*very good*":
for (*i*=1; *i*<=5; *i*++)
{ $prep^k{}_i = prep^k{}_i * (1 - 4 * increment)$; }
$prep^1{}_i - prep^1{}_i + 4 * increment$;
break;
case "*good*":
for (*i*=1; *i*<=5; *i*++)
{ $prep^k{}_i = prep^k{}_i * (1 - 2 * increment)$; }
$prep^2{}_i = prep^2{}_i + 2 * increment$;
break;
case "*normal*":
for (*i*=1; *i*<=5; *i*++)
{ $prep^k{}_i = prep^k{}_i * (1 - increment)$; }
$prep^3{}_i = prep^3{}_i + increment$;
break;
case "*bad*":
for (*i*=1; *i*<=5; *i*++)
{ $prep^k{}_i = prep^k{}_i * (1 - 2 * increment)$;}
$prep^4{}_i = prep^4{}_i + 2 * increment$;
break;
case "*very bad*":
for (*i*=1; *i*<=5; *i*++)
{ $prep^k{}_i = prep^k{}_i * (1 - 4 * increment)$; }
$prep^5{}_i = prep^5{}_i + 4 * increment$;
break;
}

In the above algorithm, the instrument that can satisfy users will have a higher probability to be used according to Equations (3) and (5). Parameter *increment* is a constant number. If a new instrument joins into the instrument pool, the following algorithm works.

***Algorithm 2*:**

$N = N + 1$;
$p_N = 1 / N$;
for ($i = 1; i < N$; *i*++)
{ $p_i = p_i * (N - 1) / N$;}

In Algorithm 2, any new instrument joining into an instrument pool will have the average probability to be used. *N* is the existing number of instruments in a pool.

When an instrument wants to leave the pool, the probabilities are adjusted according to Algorithm 3. In Algorithm 3, the *k*th instrument in an instrument pool is supposed to leave the pool.

Algorithm 3:

```
totalP=0 ;
p_k=0 ;
for ( i=1; i<=N; i++ )
  { totalP = p_i + totalP ; }
for ( i=1; i<=N; i++ )
  { p_i = p_i / p ; }
for (i = k; i<N; i++)
  { p_i = p_i+1 ; }
N = N–1;
```

Algorithm 3 only allows the instrument without any experiment running on it at that time to leave. Any instrument with job running on it is not permitted to leave. If it leaves by some inevitable reasons, the pool will record the instrument as unstable and it will have trouble when next time it wants to join the pool.

Performance Evaluation

In this section three case studies are given to illustrate the fuzzy random scheduling model and algorithms introduced in Section 3.2.3. The programming language of the simulation environment is Java.

Case Study I

A simple experiment, which requires only one instrument, is submitted to the pool alliance. An instrument pool with N instruments, which can run the experiment, is chosen by the pool alliance. Every instrument has the same initial probability to run the experiment. In this example N equals to 50, and 10 of them have *very good* QoS and may receive users' feedback value of *excellent*, 10 *good*, 10 *normal*, 10 *bad* and 10 *terrible*. Figure 6 is the result when QoS feedback information is used to adjust probabilities of instruments in 100,000 such experiments. The vertical axis represents the number of jobs and the horizontal axis represents users' feedback information in terms of vague value. It is also the same in figure 7, 9, 10, 11 and 12. For the purpose of comparison, the result without probability adjustment is also given in Figure 6. The parameter *increment* is a constant and in this example the values are 0.02% and 2%, respectively.

As shown in Figure 6, when feedback information from users' appraisals is considered, those instruments which can not satisfy users well will have fewer chances to be used. If the owners of these instruments want to have more chances for their instruments to be used, they should improve the QoS of their instruments, like decreasing the price their instruments charge for or shortening the execution time of their instruments.

When the probability adjustment strategy is improved on the basis of Equation (3), better results can be obtained and the occurrence of *excellent* experiments will increased.

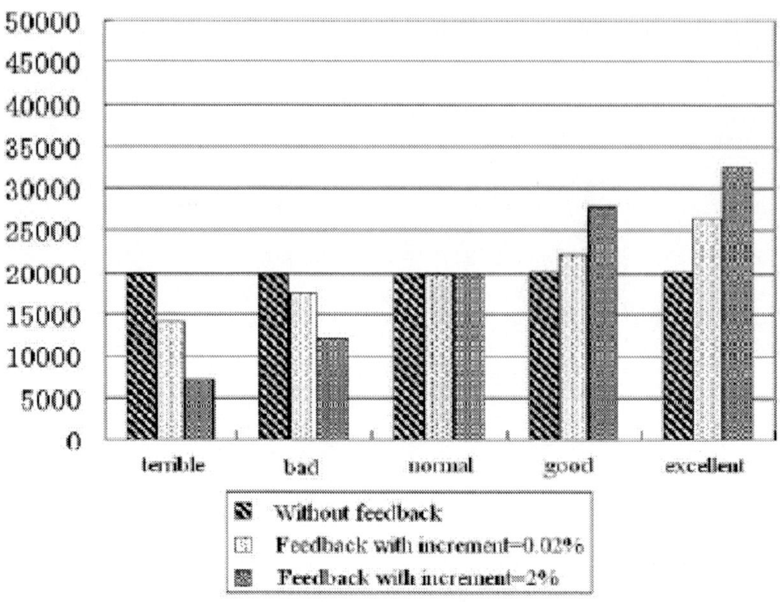

Figure 6. Results of user appraisals for 100,000 experiments.

There are more complicated scenarios for remote instrument access. For example, an experiment may involve multiple instruments, which is discussed in the case study II. Also serving more tasks will somehow decrease QoS levels of instruments, which is not considered in this case. For example, if task arrival rate is high enough to exceed processing capability of an instrument, responding delays will decrease QoS levels of user feedback information. This is discussed in the case study III using detailed simulation results.

Case Study II

In this example, two instruments in two different instrument pools are required to complete an experiment. The numbers of instruments in the two pools are N_1 and N_2, respectively. Every instrument in each pool has the same initial probability to be used. The number of instruments with different QoS values in each pool is the same. The final QoS value of an experiment is $q_i^1 \Lambda q_j^2$ when the *i*th instrument in one pool works coordinately with the *j*th instrument in another pool. This means the appraisal to the overall experiment is the worse one of the two instruments. In this example N_1 and N_2 are both 50. The user feedback information will have the same impact on the two instruments used.

Figure 7 includes simulation results when feedback information is used to adjust probabilities of both instrument pools. When feedback information is used, less *bad* or *terrible* experiments appeared. In comparison with Figure 5, no more *excellent* experiments are achieved and there is no obvious QoS improvement in this case. This is caused by that the two instruments are coupled in one experiment and the user can only provide feedback information on the whole experiment instead of each instrument.

Case Study III

In this example 100,000 similar experiment requests are submitted to the pool alliance and an instrument pool will be chose as the execution pool of these experiment requests.

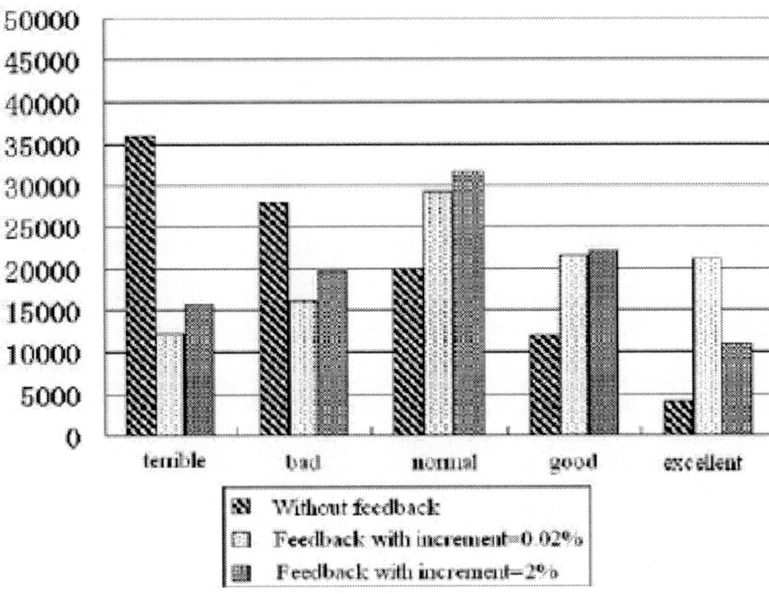

Figure 7. Results of appraisals for 100,000 experiments involving 2 instruments.

Similar to the case study I, there are 50 instruments in the chosen pool. Different from example 1, job execution times are taken into account. The execution time of the instruments with *very good* QoS complies with an exponential distribution and the expected value of the distribution is E_1, E_2 for *good*, E_3 for *normal*, E_4 for *bad* and E_5 for *very bad*. For the purpose of illustration and simulation the five expected values from E_1 to E_5 in this example are 1/250, 1/180, 1/150, 1/120 and 1/100, respectively. The request arrival time is supposed to be a Poisson distribution with λ, where λ is the average arrival rate in a Poisson distribution. In this case study, two situations are considered.

If experiment requests come beyond execution capabilities of an instrument pool, a queue is unavoidable. In this situation, instruments with high QoS feedback are chosen first and those with poor QoS feedback next. In the example, λ_1 and λ_2 are given according to this case. λ_1=5000 results in request arrivals far beyond a pool capability and λ_2=1000 corresponds to a situation that the request arrival is only slightly beyond a pool capability. The other situation is that experiment requests are within the capability of all instruments in a pool. Corresponding λ values are λ_3=600 and λ_4=100. The following simulation results are obtained using the flow chart described in Figure 8.

One thing we should bear in mind is that too long responding time, including waiting time in a queue and execution time on instruments, will degrade users' appraisals towards the results they got. With consideration of this situation, additional rules are applied.

- If $T_1 < RT < T_2$, the appraisal will degrade by one level.
- If $T_2 < RT$, the appraisal will degrade by two levels.

In above rules, *RT* represents the total responding time to an experiment request. T_1 and T_2 are two time limits that users can bear. We suppose that T_1 and T_2 are about ten to twenty times of execution time, thus T_1 =10 and T_2 =20 in this example.

Effects of these rules are also shown in table 1, which describes relationships between users' appraisals and the responding time. For example, a *very good* experiment in a user impression could be downgraded to be *good* if responding time is longer than T_1 and *normal* if responding beyond T_2.

Table 1. RT and corresponding user appraisal

Level / RT	Very good	Good	Normal	Bad	Very bad
$< T_1$	Very good	Good	Normal	Bad	Very bad
$[T_1, T_2]$	Good	Norm al	Bad	Very bad	Very bad
$> T_2$	Normal	Bad	Very bad	Very bad	Very bad

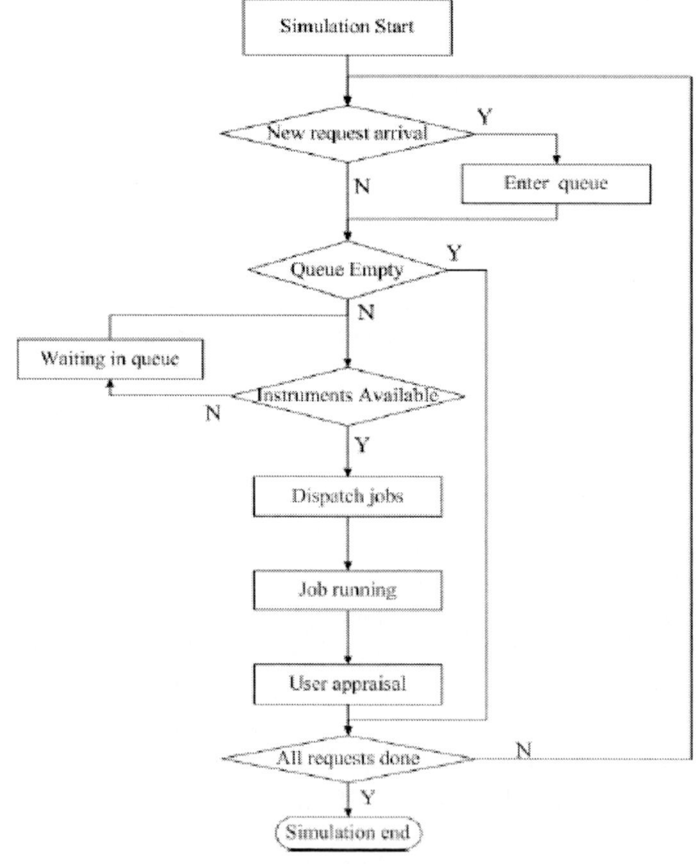

Figure 8. The flow chart of simulations.

In Figures 9 to 12, simulation results of the case study III are illustrated. In each figure, simulation results with probability adjustments algorithms described in Section 3.2.3 and those without probability adjustments are all given for the purpose of comparison. As shown in Figures 9, when request arrival speed is far beyond a pool's processing capability, the probability adjustment algorithm does not work well to provide users with more *excellent* service, since *bad* services have to be utilized anyway. Also when requests arrive too fast and have to wait in a queue, a longer responding time will downgrade users' appraisals even if an *excellent* service is supposed to be provided. The only way to still ensure high QoS for users is to let more similar instruments join the pool to increase the pool's processing capability. The situation is improved when request arrival speed is lower in Figure 10.

As shown in Figures 11 and 12, arrival requests are within a pool's processing capability, more satisfactory appraisals will achieve through the adjustment of probability in an instrument pool. Since a queue seldom appears in these situations, requests do not have to be served with *bad* instruments and downgrade rules are not often applied. These results are conformed to those achieved in the case study I.

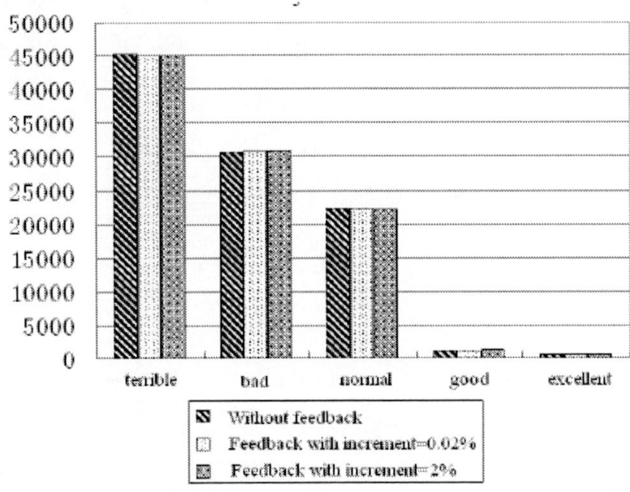

Figure 9. Results of user appraisals under λ_1.

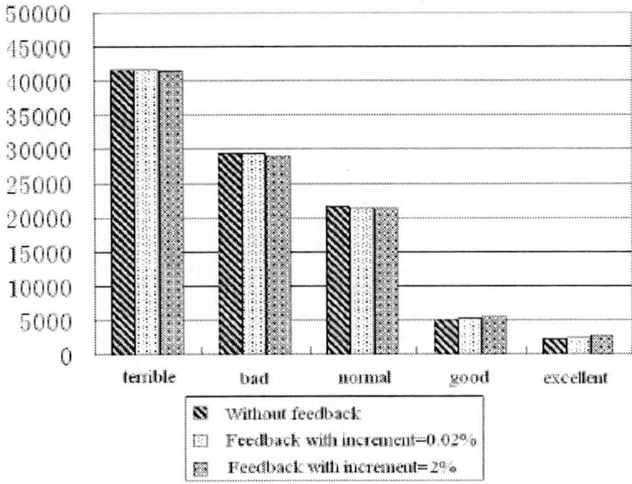

Figure 10. Results of user appraisals under λ_2.

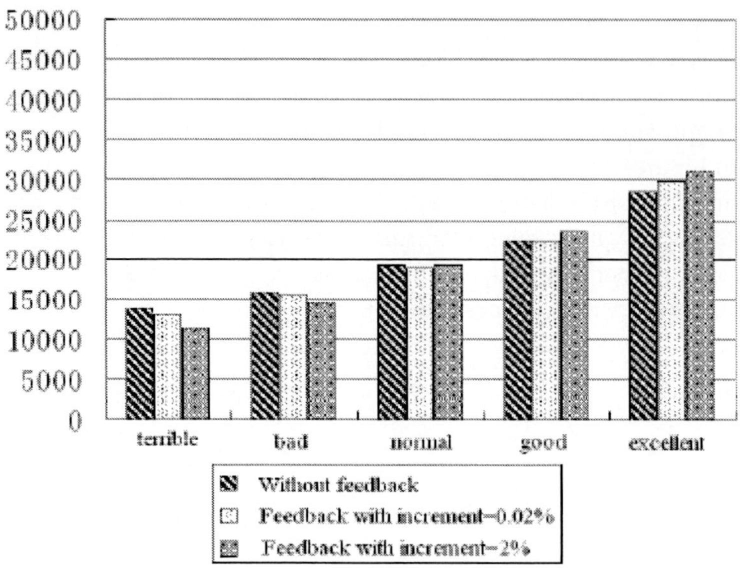

Figure 11. Results of user appraisals under λ_3.

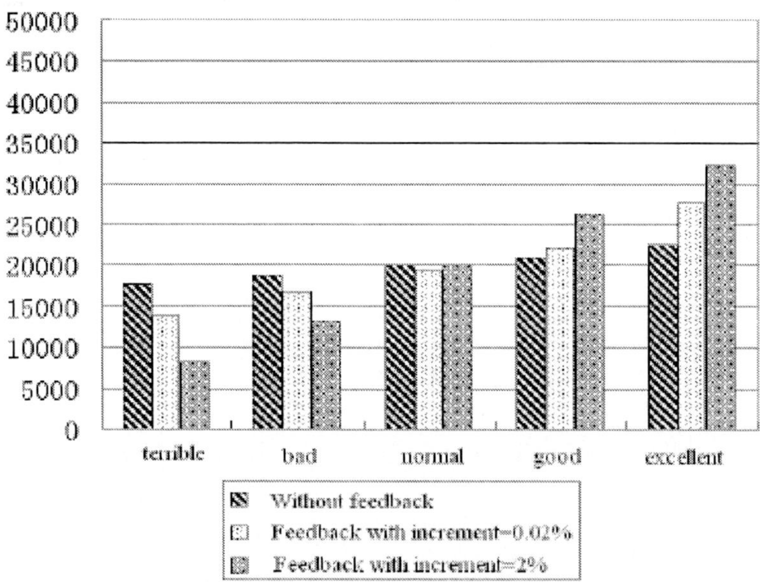

Figure 12. Results of user appraisals under λ_4.

Summary

The contribution of this section lies in the proposal of a fuzzy random scheduling model, which takes the users' QoS feedback information into account to provide more satisfactory services for users in an instrument CI. The QoS appraisals from users can not be represented in an accurate and quantitative way, since there are many factors in instrument QoS that have effects on users' appraisals. In many real world scenarios, users' feedback information is fuzzy and the fuzzy random model is suitable and straightforward when applied to the scheduling scenarios described in this work.

The algorithms provided in this work to increase the utilization probability of some instruments with higher QoS and decrease usage of those with lower QoS, is proved to be effective in a CI environment for scientific instrument sharing when pool capability is beyond experiment requests. In situations when request arrival speed is far beyond processing capability of an instrument pool, algorithms supposed to improve instrument QoS do not work, since long queuing time downgrades users' appraisals and instruments with low QoS feedback have to be used anyway.

CONCLUSION

In this chapter, we introduced the CI for education and research, especially for instrument sharing in China. We introduce some research projects being developed in the world, currently hot research field and topics. We focus our effort on scheduling issues in the CI using fuzzy random theory. Simulation results show that our algorithms can improve user satisfactions.

Ongoing work is focused on applying these proposed scheduling algorithms into the CI testbed being developed at Tsinghua University using grid software and technologies. When applying the work described in this chapter into a real world situation, additional issues have to be considered besides resource management and scheduling. Ongoing work includes an information service providing detailed equipment and experiment data, a workflow enactor to manage experiments involving multiple equipments, a mechanism to ensure reliable and coherent of different user appraisals, management of provenance information and its storage and a layered security mechanism for authentication and authorization of remote equipment access.

ACKNOWLEDGEMENT

This work is supported by Ministry of Education of China under the 211/15 project "National University Instrument and Resource Sharing Systems".

REFERENCES

[1] Wang Yuexuan, Wu Cheng and Hu Xixiang, "Study of instrument grid based on simulation and modeling", *Computer Integrated Manufacturing Systems*, 10(9), pp1031-1035, (2004)

[2] Wang Yuexuan, Wu Cheng and Ni Wangcheng, "Study on instrument grid service chain sharing technology and method", *Journal of Huazhong University of Science and Technology* (Nature Science Edition), 33, pp15-17, (2005)

[3] Yuan Li-xiang, Wang Wen-yong and Luo Guang-chun, "Study on device grid technology", *Computer Applications*, 25(12), (2014-2015), (2005)

[4] Qu Zheng-wei, Zhao Wei and Huang Song-ling, "The instrument and instrument grid for resource sharing and collaboration", *Electrical Measurement and Instrumentation*, 144(494), pp1-5, (2007)

[5] Jiang Jian-jun, Liu Ji-guang, Xiao Zhi etc, "A UDDI-Based Two-Level Information Management System for Instrument Grid", *Computer Engineering and Science*, 28(8), pp86-89, (2006)

[6] Yuexuan Wang, Cheng Wu, "Study on Instrument Interoperation Chain Model in Grid Environment", *Lecture Notes in Computer Science*, 3758, pp588-595, (2005)

[7] Wang Yuexuan, Wu Cheng, Xu Ke, "Study on π-Calculus Based Instrument Grid Service Chain Model", *Lecture Notes in Computer Science*, 3779, pp40-47, (2005)

[8] Yuexuan Wang, Lianchen Liu, Cheng Wu, "Research on instrument resource scheduling in grids", T*he Third International Conference on Grid and Cooperative Computing*, 2004, pp. 927-930.

[9] Yuexuan Wang, Lianchen Liu, Cheng Wu, "Research on Instrument Grid Platform for Resource Sharing", *World Engineers Convention (WEC)*, 2004, pp. 148-151.

[10] Wang Yuexuan, Liu Lianchen, Hu, Xixiang etc, "The study on simulation grid technology for instrument resource sharing system", *Proceedings of the World Congress on Intelligent Control and Automation*, 4, 2004, pp. 3235-3239.

[11] http://www.edu.cn/

[12] http://www.nees.org

[13] http://www.cs.indianan.edu/ngi

[14] http://www.escience.cam.ac.uk/projects/telemed.html

[15] http://www.cers.edu.cn

[16] D. E. Atkins, K. K. Droegemeier, S. I. Feldman, H. Garcia-Molina, M. L. Klein, D. G. Messerschmitt, P. Messina, et. al., "Revolutionizing Science and Engineering through Cyberinfrastructure", National Science Foundation Blue – Ribbon Advisory Panel on Cyberinfrastructure, January 2003.

[17] J. Cao, S. A. Jarvis, S. Saini and G. R. Nudd, "GridFlow: Workflow Management for Grid Computing", in *Proceedings of 3rd IEEE/ACM International Symposium on Cluster Computing and the Grid*, Tokyo, Japan, 2003, pp. 198-205.

[18] M. Faerman, A. Su, R. Wolski and F. Berman, "Adaptive Performance Prediction for Distributed Data-Intensive Applications", in *Proceedings of ACM/IEEE Supercomputing Conference*, 1999.

[19] I. Foster and C. Kesselman, The Grid: Blueprint for a New Computing Infrastructure, Morgan-Kaufmann, 1998.

[20] I. Foster, C. Kesselman and S. Tuecke, "The Anatomy of the Grid: Enabling Scalable Virtual Organizations", *International Journal of Supercomputer Applications*, 15(3), 2001.

[21] R. Kruse and K. D. Meyer, Statistics with Vague Data, D. Reidel Publishing Company, Dordrecht, 1987.

[22] H. Kwakernaak, "Fuzzy Random Variables-I. Definitions and Theorems", *Information Sciences*, 15, pp1-29, (1978).

[23] H. Kwakernaak, "Fuzzy Random Variables-II. Algorithms and Examples for the Discrete Case", *Information Sciences*, 17, pp253-278, (1979)

[24] B. Liu and J. Peng, *A Course in Uncertainty Theory*, Tsinghua University Press, Beijing, 2005.

[25] Y. K. Liu and B. Liu, "Fuzzy Random Variables: a Scalar Expected Value Operator", *Fuzzy Optimization and Decision Making*, 2(2), pp143-160, (2003)

[26] NSF Cyberinfrastructure Council, NSF's Cyberinfrastructure Vision for 21st Century Discovery, Version 5.0, January 20, 2006.

[27] M. L. Puri and D. Ralescu, "Fuzzy Random Variables", *Journal of Mathematical Analysis and Applications*, 114, pp409-422, (1986)

[28] W. Smith, V. Taylor and I. Foster, "Predicting Application Run Times Using Historical Information", *Job Scheduling Strategies for Parallel Processing*, LNCS 1459, Springer Verlag, pp122-142, (1998)

[29] W. Smith, V. Taylor and I. Foster, "Using Run-Time Predictions to Estimate Queue Wait Times and Improve Scheduler Performance", *Job Scheduling Strategies for Parallel Processing*, LNCS 1659, Springer Verlag, pp.202-219, (1999)

[30] D. P. Spooner, S. A. Jarvis, J. Cao, S. Saini and G. R. Nudd, "Local Grid Scheduling Techniques Using Performance Prediction", *IEE Proceedings – Computers and Digital Techniques*, 150(2), pp87-96, (2003).

[31] Y. Wang and C. Wu, "A Study on Education Resource Sharing Grid", *International Journal of Information Technology, Special Issue on Grid Computing I*, 11(3), pp73-80, (2005).

[32] L. Yang, J. M. Schopf and I. Foster, "Conservative Scheduling: Using Predicted Variance to Improve Scheduling Decisions in Dynamic Environments", *Proceedings of ACM/IEEE Supercomputing Conference*, 2003.

[33] J. Yin, J. Cao, Y. X. Wang etc, "Scheduling Remote Access to Scientific Instruments in Cyberinfrastructure for Education and Research", *International Symposium on Cluster Computing and the Grid*, May 2007.

In: Cyberinfrastructure Technologies and Applications
Editor: Junwei Cao
ISBN: 978-1-60692-063-3

Chapter 3

AN INTEROPERABLE INFORMATION SERVICE SOLUTION FOR GRIDS

Anand Padmanabhan[1], Eric Shook[2], Yan Liu[3] and Shaowen Wang[4]

[1] CyberInfrastructure and Geospatial Information Laboratory (CIGI)
National Center for Supercomputing Application (NCSA)
University of Illinois at Urbana-Champaign

[2] CyberInfrastructure and Geospatial Information Laboratory (CIGI)
University of Illinois at Urbana-Champaign

[3, 4] CyberInfrastructure and Geospatial Information Laboratory (CIGI)
National Center for Supercomputing Application (NCSA)
University of Illinois at Urbana-Champaign

ABSTRACT

Information services are a critical piece of Grid infrastructure, they collect, aggregate, and organize sources of Grid resource information, and provide the information to applications to schedule computational tasks, and enable Virtual Organizations (VO) to share distributed computing resources. To be successful an information provider needs to be able to aggregate diverse information from numerous sources, publish information into different Grid monitoring systems, function under different Grid deployments, and be adaptable to different information schemas. Addressing these challenges would provide a platform for an interoperable information service.

In this chapter, we present the Modular Information Provider (MIP) that address the aforementioned challenges and eliminates the shortcomings of the existing Grid information providers. MIP adopts a XML-based data model to enable robust schema

[1] E-mail address: apadmana@uiuc.edu
[2] E-mail address: eshook2@uiuc.edu
[3] E-mail address: yanliu@uiuc.edu
[4] E-mail address: shaowen@uiuc.edu

validation as well as to support the latest generation of service-oriented Grid monitoring software.

MIP provides a critical capability for production Grid environments focusing on achieving Grid interoperability and manageability. To validate this we conduct experiments on the Open Science Grid and Grid Australia. These experiments demonstrate MIP's Grid-independent approach is successful and can easily adapt to different software stacks and the heterogeneous configuration of several sites on a Grid. We use GISolve, a TeraGrid science gateway as an example to demonstrate how MIP is useful from the perspective of Grid application. GISolve provides a geospatial problem solving platform that hides the complexity of Grid information services based on MIP, which enables GISolve applications to access multiple Grids seamlessly. We illustrate how MIP is integrated within GISolve so that GISolve applications can benefit from the interoperable information service MIP provides.

INTRODUCTION

Grid technologies enable Virtual Organizations (VO) to share distributed computing resources in a coordinated manner [10]. Information services are a critical component of the Grid infrastructure providing information which can be used by applications to schedule computational tasks, detect and recover from faults, and predict performance [9]. Recent active research in Grid monitoring has covered a broad scope of research topics [11], which include Grid monitoring architectures [35], monitoring information modeling [5], query methods for Grid information services [3], performance study of monitoring for distributed systems [38], and information providers [20].

A Grid information provider serves as an interface to both static and dynamic information and plays a vital role in collecting, aggregating, and organizing sources of Grid resource information. Grid applications depend on the published information from such a provider and thus require a stable and dependable platform. Developing an information provider poses significant challenges, because the information provider 1) must collect and aggregate diverse information from numerous sources (e.g., Ganglia [12]); 2) should be capable of publishing information into various information schemas and different Grid monitoring systems (e.g., BDII (Berkeley Database Information Index) [4], MDS4 (Monitoring and Discovery Services) [20]); and 3) needs to function under diverse Grid deployments and environments. An information provider that addresses these challenges would provide a platform for an interoperable information service that Grid applications can depend on.

In this chapter, we present the Modular Information Provider (MIP) that eliminates the shortcomings of existing Grid information providers. The modular design in MIP facilitates the separation of Grid-specific information from the Grid-independent framework through the use of modules, configuration files, and packaging utilities. MIP adopts a XML-based data model to enable robust schema validation as well as to support the latest generation of service-oriented Grid monitoring software. The MIP software toolkit provides a critical capability that could be leveraged in production Grid environments such as the Open Science Grid, EGEE, and TeraGrid in order to achieve Grid interoperability and manageability. We also describe several scenarios of MIP deployment.

Challenges

Diverse information sources pose several challenges for information providers. Evolving interfaces and varying data formats complicate information collection. Each interface needs to be individually managed and the unique information must be parsed and organized before being processed. Information from sources distributed across multiple systems must be combined and managed.

Information providers must enable error-free information handling while maintaining stability. Deployment needs to be simple and straightforward to lower the possibility of erroneous configuration. The information produced must be verified for correctness to guarantee accuracy. Machine impact during information collection should be minimized to maintain scalability while supporting large amounts of complex information. Stability must be sustained to insure information availability.

Information providers are responsible for enabling Grid-interoperability, a key requirement for many Grids. This typically requires the ability to produce information for multiple schemas (e.g. CIM [5], GLUE [16]) in different languages (e.g. XML, ClassAd [7]) and interface with a wide-variety of Grid information services.

The primary challenges faced by the information providers can be summarized as follows:

- Efficiently collect, aggregate, and manage information from a large number of Grid information sources such as Grid and cluster monitoring tools (e.g., MonALISA [23] and INCA [18], local operating system information, Grid configuration data, and cluster resource management tools (e.g., Torque [36] and Condor [34]);
- Adapt to multiple different information schemas;
- Automate the process of configuring and executing Grid information providers;
- Resolve conflicts arising from identical information being published by multiple information sources;
- Assure the correctness of the information published within large scale production Grid environments; and
- Achieve stability, scalability, and interoperability across Grid environments.

To address these challenges, the Modular Information Provider successfully supplies monitoring information from local and distributed sources, primarily with the contribution of the following three components: Modules, Information Producers, and the framework. The framework is composed of a Collector, Module Handler, and Integrator Service. Modules and Information Producers use the framework for interaction and communication. Modules collect local information while the Integrator manages distributed information. MIP modules collect information across several sources such as Ganglia [12] and Condor and provide crucial information on the status and availability of resources. The integrator enables collection of resource information from these geographically distributed sources. Information Producers format and output information based on specific schemas and languages such as GLUE and Condor ClassAd. The Modular Information Provider (MIP) approach also utilizes a Grid-independent framework to support re-usability. The Grid-independent framework contains several components that utilize an XML data-model to support multiple information modeling schemas and languages. Additional MIP components interact with the framework to support the formatted collection and output of information. Experiments were conducted on the Open Science Grid [25] and Grid Australia (formerly known as Australian Partnership for Advanced Computing (APAC)) [2] successfully demonstrated that MIP's Grid-independent approach can easily adapt to different software stacks and the heterogeneous configuration of several sites on a Grid

To demonstrate the usefulness of MIP from the application communities' perspective, we use GISolve [14] as a use-case to evaluate it. GISolve provides a suitable environment to demonstrate MIP's ability to seamlessly integrate within a TeraGrid [33] science gateway. GISolve applications can benefit from Grid information provided by MIP, while not being exposed to the complexities of the information services. GISolve is a TeraGrid Science Gateway toolkit for GIScience that provides user-friendly capabilities for performing geographic information analysis using computational Grids, and help non-technical users directly benefit from accessing cyberinfrastructure capabilities. GISolve aggregates the enormous computational resources that are needed to store and manage geographic information that is collected for wide variety of application domains (e.g., environment science, transportation, and public health), and to conduct computationally intensive geographic information analysis. In order to deliver these computational resources, GISolve would be required to know the status of the Grid resources to make appropriate job allocation decisions. MIP aims to satisfy this requirement from the application community.

The rest of the chapter is organized as follows. In section 2 we will review the current landscape of Information Services deployed on production Grids. Section 3 will provide the details of MIP's architecture, and data model. Section 4 will evaluate MIP, based on deployment experiences on OSG and APAC Grid and a case-study of its usage by GISolve, a TeraGrid GIScience Gateway. We will conclude in section 5 with a conclusion and pointers to future work.

Current Information Services Landscape

In this section we will try to overview the current information services landscape. Most production grids currently use LDAP based information services, such as MDS2 (the Monitoring and Discovery Service from Globus Toolkit 2 (GT2), the pre-web services version of the Globus Toolkit) or BDII (used by LCG [19]). However the GT4 provides resource information service using MDS4 which uses a Web Services framework. MDS4 and Web Services will be the logical direction in the future, but the information providers designed to capture and publish resource information based on Web Services technology are not well developed. MIP hopes to fill this niche.

The Monitoring and Discovery Services

The Monitoring and Discovery Service (MDS), a component of the Globus Toolkit, serves as the standard tool for publishing and accessing Grid resource information. MDS provides functionality to allow the aggregation of information from multiple resources on the Grid site. It aggregates information in a hierarchical manner, e.g., an instance of MDS may collect all the information about Grid resources at a site, while another instance could collect information from all the site level MDSs to provide a Grid level view of the available resources. MDS aims to provide a single standard interface and schema for accessing the heterogeneous resource information within the Grid, while providing flexibility to publish additional information, by allowing system administrators to customize existing providers or write their own resource information providers. MDS can be configured to use the Grid Security Infrastructure (GSI) to allow only authorized access.

There are two versions of the MDS provided by the Globus Toolkit: MDS2 and MDS4. MDS2 [21] is the pre-web service version of the information service. The information is published in an LDAP Data Interchange Format (LDIF) format using an OpenLDAP [24] server. MDS2 forms a hierarchical structure composed of the Grid Index Information Service (GIIS), Grid Resource Information Service (GRIS) and Information Providers. An information provider, according to MDS2, is a program that queries the information sources and publishes information according to a schema understood by GRIS. The GRIS advertises information about the resource, aggregating information from one or more information providers. GIIS acts as an index service that collects information from one or more GRIS or a lower level GIIS.

MDS4 is the Web Service (WS) implementation of MDS. This version of MDS includes WSRF implementations of the Index Service, a Trigger Service, WebMDS and the underlying framework, the Aggregator Framework [27]. The Index Service collects data from Information Providers and provides a mechanism to query that data. The Trigger Service collects data from multiple sources and provides the functionality that performs actions when certain conditions are satisfied based on collected data. The common Aggregation Framework infrastructure provides shared interfaces and mechanisms for working with data sources. The WebMDS provides a simple XSLT-transform based visual web-based interface to the data.

Though originally the MDS4 could only use a built-in schema to publish information, recently developed Resource Property Provider framework [26] addresses this issue. The RPProvider Framework is an extensible software component that can be used to dynamically generate XML values for one or more WSRF Resource Properties in any given GT4 Java WSRF-Core compatible service [37]. This package allows MDS4 to publish information using any schema.

Berkeley Database Information Index

The Berkeley Database Information Index [4], like MDS2, also uses a LDAP server and provides an alternative to MDS2. The BDII was developed within the WLCG to replace the MDS2, since the later has been shown to be unstable when it has to index information from a large number of sites or process a large number of queries. BDII employs two or more LDAP servers, to perform update and queries compared to only one with MDS2, there by providing better scalability. It can be used to replace both the GRIS and GIIS components.

Generic Information Provider

The Generic Information Provider [13] is deployed on several large production grids (WLCG, OSG) and has proven to be a reliable means of collecting local resource information. GIP's design has focused primarily on meeting the needs of a specific Grid environment. While this is advantageous to a single Grid environment, significant work is required to customize GIP to support multiple Grid environments. For example GIP is designed to support a single schema not multiple schemas, which is needed to support multiple Grid information services. Also currently there is no software available to convert GIP output to XML format as needed by MDS4. Further development of this component would be challenging due to the inherent differences between the LDAP and XML implementation of the Glue schema, which the production Grids rely on.

MDS4 Information Providers

The Globus Toolkit 4 has built-in Information Providers which can provide some basic information about the batch system, memory size, Operating System (OS) name, OS version, etc [20]. These information providers can use PBS, Ganglia, Hawkeye [17], and Nagios [22] as their information sources. The issue with the built-in providers is that they all use the GLUE 1.1 schema which is at least one generation old. Most production Grids use a later version of Glue schema, for e.g. OSG uses Glue schema 1.3 in its latest production release.

Other Grid Monitoring Systems

Grid monitoring systems such as Inca [18] and MonALISA [23] support distributed information collection, but differ in the communication method. MonALISA uses a distributed agent-based model to communicate the information. Inca on the other hand utilizes a flat hierarchy composed of several components such as reporters, agents, and depots; while MIP is designed to utilize a multi-level hierarchy through the use of Integrators.

MonALISA and Inca also require multiple layers or components to establish an effective monitoring system where as MIP utilizes a single component for distributed information collection. This design feature reduces complexity and simplifies distributed information collection. Additionally the Integrator is an optional component, so it can be disabled if distributed information collection is not needed.

Modular Information Provider (MIP)

The MIP software provides is a critical but missing capability in production Grid environments such as the Open Science Grid, EGEE, and TeraGrid. MIP focuses on achieving Grid interoperability and manageability, and is designed to interface multiple information sources and information systems such as Hawkeye and MDS4 (figure 1).

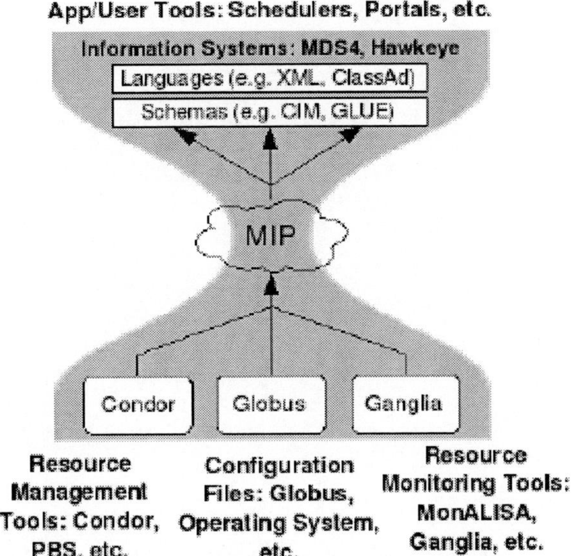

Figure 1. MIP in the context of Grid information protocol hourglass model (After [27]).

MIP Architecture and Components

The Modular Information Provider (MIP) leverages a modular design to separate Grid-specific components from those which are Grid-independent. Specifically, MIP utilizes a Grid-independent framework that is easily customized using Modules and Information Producers. The MIP framework uses a XML data-model to process and transfer information from Modules to Information Producers.

MIP consists of several components including: Modules, a Module Handler, a Collector, an Integrator Service, and Information Producers. Modules, the basic building blocks of MIP, are invoked by the Module Handler. The Integrator Service is used to receive distributed information forming an information hierarchy. The Collector accepts information from the Module Handler and the Integrator Service, it then forwards the information to Information Producers or other Integrator Services. Information Producers format the information to a specific schema using a particular language. The information can then be used by a wide variety of systems (e.g. MDS4, Hawkeye). Figure 2 illustrate the MIP architecture and the relationship between the components.

The unique nature of each Grid environment requires a different combination of Modules and Information Producers. Modules collect information from various sources and these sources vary from one Grid environment to another. Therefore the combinations of Modules must also change in each environment. Information Producers format information to a specific schema and language. Grids may have different schema and language requirements, so they also require different Information Producers. These components have been separated from the framework, because they must be changed or customized for each Grid. The framework however will remain unchanged to simplify deployment and reduce duplicated effort. Each component is described below to provide further detail.

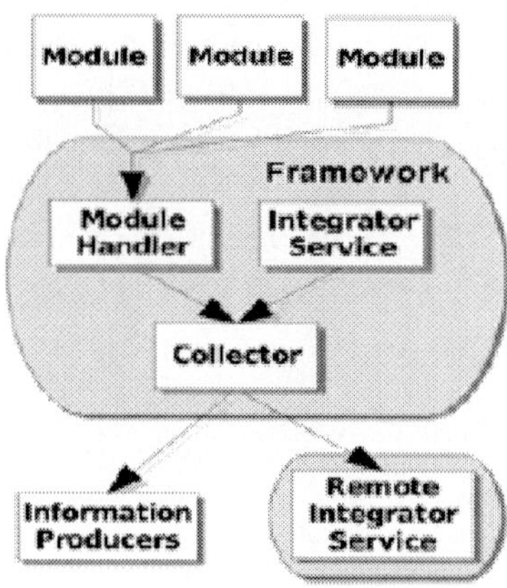

Figure 2. MIP architecture.

Modules

To manage multiple information sources MIP employs Modules. A Module is an executable file that accepts certain command-line parameters and outputs information based on a XML data-model. Modules are easy to create and maintain and dramatically simplify information collection. This simplification lowers the barrier to adoption for site administrators, because they can easily change, remove, or add new Modules to properly describe the status of the site, with a site being a set of computational resources that is owned by a single entity, such as a university. Site-level customizations greatly improve the accuracy of information compared to previous information providers which typically offer little support for site-specific customizations. Furthermore, Modules can be shared and distributed across sites and Grids, supporting re-usability and reducing duplicated effort.

Modules reduce complexity and simplify information collection, but are restricted to the interfaces provided by the information sources. Unfortunately, many sources lack support for remote interaction or information collection. This constrains Modules to collecting local information.

Module Handler

The Module Handler coordinates and manages the execution of Modules based on optional filter controls and priorities. Filter controls limit the scope of information collection to pertinent information, increase efficiency and reduce machine overhead. This is accomplished by not executing filtered Modules based on the filter controls.

Priorities determine the execution order of Modules. Site and Grid administrators have the ability to change the priority of each Module. Priorities are used to resolve potentially conflicting information from similar sources [32]. Once the Modules are executed they return their collected information, which is then forwarded from the Module Handler to the Collector.

Integrator Service

The MIP framework includes the Integrator Service to extend the functionality of MIP beyond local information collection. The Integrator is a service that receives information from remote MIP instances, enabling distributed information collection.

Distributing potentially sensitive information, such as software versions or Grid accounting information, raises security concerns. To address such concerns the Integrator supports Secure Socket Layer (SSL) data transfers. SSL encryption enables MIP to securely transfer sensitive data across a network. Additionally, the Integrator can be configured to accept information from a select number of machines. This security feature reduces the likelihood of publishing false information from malicious machines.

MIP Integrators enable hierarchical distributed information collection. Grids and sites can use several Integrators at various levels to establish an information hierarchy. Other Grid tools and services can be easily incorporated into this hierarchy. Grid information services, such as MDS Aggregator Framework [1], can interface with Information Producers. Grid monitoring systems, such as Inca and MonALISA, can provide information to Modules. MIP and the Integrator connect these two pieces to establish a robust information hierarchy.

Resources behind a restrictive firewall or in an internal network, (such as within a cluster) represent a problem for collecting information. Information cannot be collected from these resources using a pull model, without modification to the network infrastructure (i.e. port forwarding, opening ports). To support information collection for these resources the Integrator uses a push model.

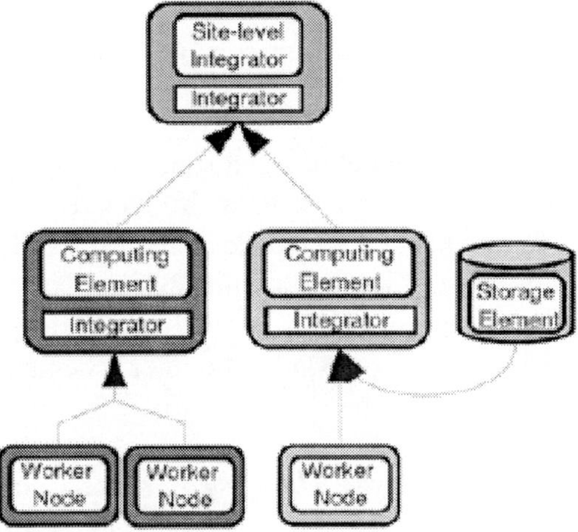

Figure 3. An integrator hierarchy example.

Figure 3 shows an example configuration of a MIP Integrator hierarchy. The site-level Integrator (top of diagram) is feeding information to an MDS4 instance. In this example the MDS Aggregator Framework is interfacing with MIP as the Grid information service. A deep hierarchy is not necessarily required; all nodes could be configured to push information directly to the site-level Integrator Service. This example shows the flexibility of the Integrator Service within the MIP framework, which ensures a scalable design as more MIP instances are added.

Collector

The Collector gathers information from the Module Handler and Integrator Service. Once the Collector gathers the information it presents it to a combination of Integrator Services or Information Producers.

All components in the framework, including the Collector, transmit information using XML. This standards-based language provides maximum flexibility for framework components. XML offers several advantages, such as: easy schema-validation, ability to handle very large amounts of information, and it provides rich semantics to support schema translations.

Information Producers

Information Producers accept XML and output information in a particular schema based on a specific language. The implementation of an Information Producer is dependent on the specific schema and language. To allow maximum flexibility, MIP places few restrictions on a Producer.

Unfortunately thc complex nature of converting information from one language or schema to another can make Information Producers cumbersome to write. Fortunately, many Grids share a common or similar schema or utilize the same Grid information service, in effect, limiting the number of Information Producers that must to be written. MIP attempts to minimize the complexity by using XML, which provides semantics for schema conversion. Next we will discuss the XML data model employed by MIP.

MIP Data Format

XML Data Model

MIP adopts a XML-based data model to support multiple schemas and languages because 1) XML-based data model describes both the data and its structure in the same document; 2) XML components can be combined and separated without loss of information; 3) XPath and XQuery provide a straightforward interface to full-text and structural queries; 4) XML can easily handle a large amount of information as well as efficiently validate information; 5) XML has rich semantics to support its translation to other schema implementations.

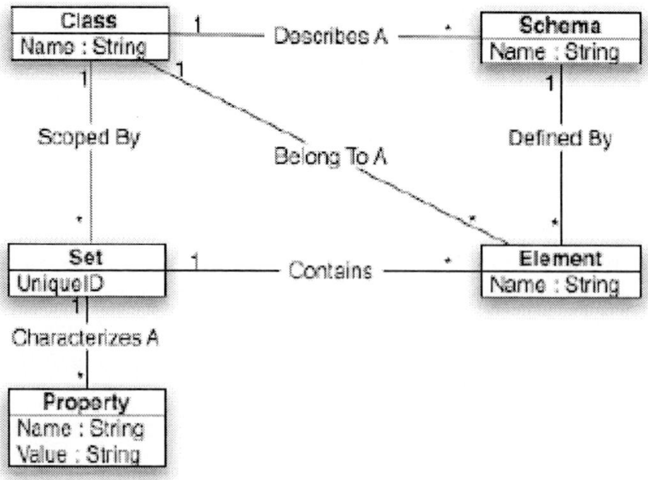

Figure 4. MIP data model.

The MIP data model can be described by the meta-schema in figure 4. This meta-schema is loosely derived from the CIM meta-schema [6] and shares some of its terminology. This meta-schema permits MIP to describe and relate Classes of objects. To reduce the scope of the description, MIP uses multiple Schemas to define Elements already known to be useful. These Schemas can easily be extended if needed. A Set is used to add meta-data to a group of Elements. Properties are used to characterize a Set. This meta-schema is sufficiently generic to support existing diverse schemas and languages, such as CIM, XML, and ClassAd as well flexible enough to support future schemas and languages.

Schema and Languages

The aforementioned MIP meta-schema is capable of structuring information into MIP's XML-based format. This format will be based on a series of Classes, some example Classes are: OperatingSystem, Computer, and Storage. An example below provides concrete evidence supporting MIPs ability to handle multiple schemas and languages using this format.

```
<OperatingSystem>
<Set UniqueID="OSuniqueID">
<Elements>
<Name>Scientific Linux</Name>
<Version>4.3</Version>
<Release>Beryllium</Release>
<Type>Linux</Type>
<Description>Scientific Linux is a ...</Description>
</Elements>
<Properties>
<Hostname>mip.cigi.uiuc.edu</Hostname>
<Role>WorkerNode</Role>
</Properties>
</Set>
</OperatingSystem>
```

This example has one Set within the OperatingSystem Class with Elements of Name, Version, and Description. The two Properties: Hostname and Role, characterize the example Set. This small amount of information is sufficient to describe the Operating System components in both GLUE and CIM. More specifically, to convert this data format to GLUE Schema, an Information Producer knows this Set will describe the Operating System of a worker node, or SubCluster in GLUE Schema terminology, with hostname 'mip.cigi.uiuc.edu' based on the Role Property and Class name. It will use this meta-data to insert this Operating System information within the SubCluster section of the GLUE Schema. This example works with GLUE schema, however CIM is much more complex. The CIM Schema requires the Operating System to be related to the File System and Computer components using Association and Composition Aggregation relationships. The CIM Information Producer will know this Set describing the Operating System of a particular Computer based on the Hostname Property and Class name. With this, the Information Producer can relate all the Sets sharing the same Hostname. This includes Sets from OperatingSystem, Computer, and Storage Classes, where the Storage Class holds the File System information as needed by the CIM schema. The CIM Information Producer, aware of CIM's relationship requirements, will then make the necessary Association and Composition Aggregation relationships between the information from the various Sets to meet CIM's Schema requirements. The CIM example, albeit more complex, also is supported by MIP's flexible data-model. Further demonstrating MIP's data-model flexibility, this example will convert this information to another language, ClassAd. To convert this information to another language such as ClassAd all one has to do is take the XML "<Attribute>Value</Attribute>" pairs from each Element and place them into ClassAd "Attribute=Value" pairs with the Class name in front of each pair. This example would look like:

```
OperatingSystemName=Scientific Linux
OperatingSystemVersion=4.3
OperatingSystemRelease=Beryllium
OperatingSystemType=Linux
OperatingSystemDescription=Scientific Linux is a ...
```

The data-model can be extended to contain many more Properties, but as the previous examples have demonstrated it is simple and straightforward to establish a robust XML data-model using MIP's meta-schema.

An Example of MIP Deployment

In order to better understand the MIP framework and how it provides the functionality, let us consider a simple example; a site is running a PBS batch system that can be accessed through a Globus gatekeeper. Any application that potentially wants to use the resource is interested in knowing the status of the gatekeeper and the batch system. For example a resource user, such as GISolve, is interested in the number of free CPUs to the batch system, the queue time, the maximum wall clock time available to the potentially submitted jobs, the priority of the job when submitted, and many others. To make an informed decision for job scheduling GISolve must collect similar information from all the resources.

The Grid site administrators may be interested in sharing cluster information using Globus MDS4. In this example MIP will be installed on all head-nodes that are running the PBS server as well as on the machine hosting Globus MDS4. A MIP *pbs* Module installed on the head-node will query the status of queues and collect relevant information such as node status and job information. The Module Handler will receive the information from the *pbs* Module. Then the Collector will gather the information from the Module Handler and forward the information to the MIP Integrator on the Globus MDS4 machine. The Integrator of the MDS4 MIP will gather the information and forward it to the Information Producer. The Information Producer will format the information for Globus MDS4 which will read the information from all PBS clusters where it can be seen by Grid users and applications. This will answer the questions we mentioned earlier, enabling Grid applications to make informed job scheduling decisions.

Grid Interoperability

Interoperability is an essential requirement for many Grids and the information provider framework is a key component supporting this requirement. MIP has several techniques to support interoperability across Grids

Through the use of MIP Framework, Modules, and Information Providers, MIP has the flexibility to support multiple Grids with minimal customization. MIP Framework provides a stable foundation upon which to customize MIP for a particular Grid. Modules are used to customize the collection of information for the particular Grid environment. Producers can interface with the MIP framework to format information in a specific language or schema based on the Grids' needs. This combination of customizable components and stable framework provides a robust solution toward supporting multiple Grids.

Interoperability is easily supported, as shown in the two scenarios described below. If two or more Grids use the MIP information provider framework, information exchange is simple. The Grids will each establish a Grid-level MIP Aggregator that will share data using the MIP XML data model. Alternatively, if one or more Grids use a different information provider, named ProviderX, then interoperability can be achieved in one of two ways. If ProviderX supports multiple schemas, it can use MIP's XML data model to interact with MIP to share information. The second technique will use an Information Provider to send data in ProviderX's schema or language and will retrieve ProviderX's data using a MIP Module. The first technique is, in general, much cleaner, but since not all information providers support multiple schemas or languages, MIP provides the second technique to accommodate older generation or closed specification technologies.

Grids utilizing MIP's information provider framework can also share and exchange modules. This can simplify the need to support multiple information sources and reduce the need to "reinvent the wheel" for each Grid.

EVALUATING MIP

In this section we will provide experimental evidence of how MIP achieves its design goals. For this we will use two case studies: a) Deployment of current Grid Infrastructures: we will share the experiences of deploying MIP on Open Science Grid (OSG) and Australian Partnership for Advanced Computing (APAC) Grid; b) Integration with GISolve, a TeraGrid Science Gateway.

Case Study on Production Grid Infrastructures

Open Science Grid (OSG)

A case study was conducted on a single site of the Open Science Grid [25]. In this study, MIP was deployed on both Production and Integration Testbed (ITB) resources. A single site-level Integrator Service collected the information and an Information Producer formatted the information according to Glue Schema 1.2 [16]. MDS4, the Grid information service read this information and published it using WebMDS [20]. The case study fully leveraged the capabilities from MDS4 in particular, Resource Property Providers.

Each resource used a different configuration. The Production resource, used in the study, consisted of a combined Compute/Storage Element (head-node) and 2 Worker Nodes. A large storage system on the head-node provided shared storage to the Worker Nodes using a Network File System (NFS). This resource used the Condor batch system [8] as its job manager and the latest version of the OSG software stack. The Integration Testbed resource consisted of a Compute Element (head-node), Storage Element, and a single Worker Node. The Storage Element provided shared storage to the nodes using NFS. This resource used Torque [36] as its job manager and the latest version of the OSG software stack for the ITB.

MIP was deployed on the ITB resources by a non-privileged user, (had privileged access to PBS batch system). Modules such as *pbs* (*torque*), *cluster*, and *computingelement* were configured to collect information on the head-node. The Storage Element used Modules such as storagearea to collect storage information. The Worker Node used Modules such as subcluster, which collects architecture specific information.

Next, MIP was deployed on a Production resource. This installation was similar to the ITB resource with the following exceptions:

- A *condor* Module was used instead of *pbs* (*torque*).
- The head-node also had storage Modules installed.
- Head-node Modules slightly modified because of software version differences

All machines in the study, including the worker nodes, were configured to push information directly to the site-level Integrator Service, resulting in a flat hierarchy. This study showed that MIP can be deployed on top of OSG infrastructure with minimal effort and it could publish resource information consistent with Glue Schema.

Grid Australia

A more extensive deployment was conducted on Grid Australia (Formerly known as The Australian Partnership for Advanced Computing (APAC) grid) [2]. In this study, approximately 7 sites deployed MIP. The remaining sites plan to deploy MIP as additional customized Modules become available. This study is particularly important because it demonstrates the benefit of MIPs Grid-independent design. The framework remains unchanged between the two studies even though the Grid environments varied drastically.

The APAC Grid utilizes Virtual Machine (VM) technologies for their Gateway Machines. The Gateway Machines offer several uniform interfaces for Grid users, one such interface is MDS4. Each participating site installed a site-level MIP Integrator on the MDS4 VM. Similar to OSG, APAC used Resource Property Providers to interface MIP with MDS4.

After establishing the site-level Integrator, MIP was installed on each cluster head-node. The APAC deployment required a different set of Modules compared to OSG. APAC was able to reuse some OSG Modules, while others needed to be customized, and several new Modules needed to be created. Differences in job managers, software stacks, software versions, and operating systems all contributed to the need for differing Modules.

The deployment on APAC provides evidence that MIP addressed several challenges. In particular, we moved the framework from one Grid environment to another without changes, demonstrating MIPs Grid-independent design. The quick customization made by APAC members' shows MIPs modular design is simple, flexible, and easy-to-use. MIP has shown to be scalable on the APAC grid environment.

APACs use of MIP demonstrates it can reliably collect local and distributed information in a production Grid environment.

GISolve: A TeraGrid GIScience Gateway

GISolve is a Grid-based problem solving environment for computationally and data intensive geospatial information analysis. Geospatial information analysis is widely used to support scientific investigations and decision-making in a variety of application domains (e.g., environment science, transportation, public health, and business). Methods of geospatial information analysis often require the use of computationally intensive search, simulation, optimization, and statistical methods, particularly when they are applied to large, realistic problems. As massive increases in the quantity of geographic data are readily observable, these methods must also be able to access these data to address problems that range across a broad spectrum of spatial and temporal scales. GISolve addresses these challenges by exploiting Grid computing resources and developing a Grid-based geo-middleware toolkit for large-scale geospatial information analysis [28].

GISolve is deployed and used as a U.S. National Science Foundation TeraGrid GIScience Gateway. It provides a set of geospatial science-specific tools, applications, data collections, and services that are integrated via the GISolve web portal. GISolve portal provides user-friendly capabilities for performing geographic information analysis using computational Grids, and helps non-technical users directly benefit from accessing Grid computing capabilities. Currently, three types of geospatial information analyses are supported: Bayesian geostatistical modeling, detection of local spatial clustering, and inverse distance weighted interpolation.

As a geo-middleware, GISolve consists of four major components to facilitate above analyses: a domain decomposition module, a task scheduling module, a geographic data access module, and an information broker [29]. Given a geospatial information analysis problem, the domain decomposition module is used to divide the problem into several sub-problems each of which can be processed on a single Grid computing element. The task scheduling module interacts with the information broker to discover dynamically available Grid resources and employ a proper scheduling algorithm to optimally allocate Grid resources for sub-problem execution. The geographic data access module is used to manage the transfer, replication, and manipulation of geographic data on the Grid. Data access functionalities are requested by the domain decomposition and task scheduling modules, and are implemented through protocols and services for data access on the Grid (e.g., Globus GridFTP). The information broker leverages underlying Grid information services to provide: 1) Grid resource information for the task scheduling module; 2) geospatial data and computing provenance information. All components are accessible interactively via web or machine-readable via web services.

GISolve has been developed to manage Grid complexity within a service-oriented architecture that allows scalable integration between geographic information analysis and basic Grid services (e.g., security and accounting, data management, job execution, and information services). Geo-middleware components are implemented as Grid services that interact with OGSA Grid services. There are three types of GISolve Grid services: application-specific services, OGSA services, and interaction-specific services [31].

Information broker is an interaction-specific service that plays a key role in GISolve to provides Grid and GISolve analysis status information to application-specific services via standard web service interface (for service invocation) and user interface (for portal users). It acts as both information provider and information consumer.

GISolve Service Information Publication

GISolve services are implemented and deployed in a service-oriented framework which employs OGSA service model for application and higher level Grid service development. OGSA is an extension of web service standards for Grid computing. An important extension of OGSA is the WSRF (Web Service Resource Framework) specification which defines stateful information support in Grid services. Stateful information support is critical to support Grid workflows which define a Grid application process as dependent service interactions. In GISolve, each service has an embedded WSRF interface to access service data which is the status information holder for a particular Grid service. GISolve interacts with MIP to publish service data of a particular service instance deployed on a Grid site. MIP collects service data information from all the deployed GISolve services and presents them as service information for further retrieval, aggregation, and search. Between GISolve services and MIP, GLUE schema [16] is used to define service description data format. Information published to MIP includes service handle, service status, service WSDL definition, owner, service start time, and service-specific data. Figure 5 is an example of GISolve service information definition in WSDL format.

```
<definitions name="DCService" targetNamespace=http://uiuc.edu/cigi/namespaces/2005/07/gisolve/DC/GID>
... ...

        <import location="serviceinfoTypes.wsdl"
    namespace="http://uiuc.edu/cigi/namespaces/2005/07/gisolve/serviceinfotypes"/>
  <types>
    <xsd:schema...>
      <!-- Service Information: GLUE Schema Service format -->
      <xsd:element name="ServiceInfoRP" type="serviceinfotypes:serviceinfoType"/>
      <xsd:element name="DCResourceProperties">
        <xsd:complexType>
          <xsd:sequence>
            <xsd:element ref="tns:ServiceInfoRP" minOccurs="1" maxOccurs="1"/>
          </xsd:sequence>
        </xsd:complexType>
      </xsd:element>
    </xsd:schema>
</types>
</definitions>
```

Figure 5. An example of GISolve service information publish.

Grid Information Service Integration in GISolve

The information broker module in GISolve collects Grid and GISolve service information to GISolve application-specific services to support resource discovery, selection, and scheduling, data movement, job monitoring, and geospatial experiment provenance. Currently, GISolve supports information retrieval from the following three Grid information providers:

- MIP: GISolve uses MIP to discover, select, and monitor Grid resources and services used in a geospatial information analysis process in GISolve.
- MAGGIS: MAGGIS (Multi-Agent System Architecture for End-User Level Grid Monitoring Using Geographic Information Systems) [30] is a multi-agent framework for Grid information services. GISolve integrates MAGGIS client agent for Grid computing element information monitoring.
- Grid service deployment: GISolve directly accesses service data on their service deployment sites for third-party Grid services and GISolve services not aggregated in MIP.

CONCLUSION

In this chapter we presented an information provider solution, the Modular Information Provider that utilized a Grid-independent framework. The Modular Information Provider (MIP) addressed the challenges associated with local and distributed information collection utilizing components such as Modules and the Integrator Service.

We demonstrated MIP's ability to interface with multiple Grid information services without sacrificing simplicity or ease-of-use. The deployment of MIP was shown to scale across multiple Grid infrastructures in experiments on the Open Science Grid and Australian Grid. The experiments demonstrated the capabilities of MIP, particularly the flexibility of Modules used in local information collection, the Integrator Service's ability to support distributed information collection, and the benefits of a Grid-independent design. These experiments can be extended in future to extensively test MIP's ability to support interoperability by using information provided by MIP to submit computing jobs between Grids. Further we demonstrated that MIP is used to publish service oriented data from GISolve as well as to gather resource information about Grid sites which is used by GISolve. Thus GISolve serves both as producer as well as consumer of MIP information.

MIP's Grid-independent design minimized the amount of work needed to adapt MIP to a new Grid environment, thus reducing the significant cost associated with developing, deploying, and supporting information provider solutions. MIP represents an important stepping stone to achieving interoperation between Grids. The MIP software framework will help eliminate current error-prone processes of deploying and managing Grid information providers and greatly improve the efficiency of developing individual Grid information provider. MIP achieves a lightweight and customizable architecture that facilitates the interaction between information sources and Grid information systems. Enhancing the MIP data model and extending its framework to support new features and capability would be good directions for further research.

Acknowledgments

This research was done using resources provided by the Open Science Grid, which is supported by the National Science Foundation and the U.S. Department of Energy's Office of Science. To conduct the case study with GISolve, this research used resources from TeraGrid and NCSA. The authors would like to thank the participating sites and individuals from the Australian Grid.

References

[1] Aggregator. (2007) Information Services - Aggregator Framework. http://www.globus.org/ toolkit/docs/4.0/info/aggregator.
[2] APAC. (2007) Grid Australia (APAC National Grid), http://grid.apac.edu.au/.
[3] Andrzejak, A., and Xu, Z. (2002) Scalable, efficient range queries for grid information services. In Proceedings of Second International Conference on Peer-to-Peer Computing (P2P 2002).
[4] BDII. (2007) Berkeley Database Information Index. https://twiki.cern.ch/twiki//bin/view/EGEE/BDII.
[5] CIM. (2007) DMTF - Common Information Model (CIM). www.dmtf.org/standards/cim/.
[6] CIM-Metadata. (2007) DMTF - Common Information Model (CIM) Metadata. http://www.wbemsolutions.com/tutorials/CIM/metaschema.html.
[7] ClassAd. (2007) Condor Classified Advertisements. http://www.cs.wisc.edu/condor/classad/.
[8] Condor. (2007) Condor project. http://www.cs.wisc.edu/condor/.
[9] Czajkowski, K., Fitzgerald, S., Foster, I., and Kesselman, C. (2002) Grid Information Services for Distributed Resource Sharing. In Proceedings of the Tenth IEEE International Symposium on High-Performance Distributed Computing (HPDC-10). Washington, DC, USA.
[10] Foster, I., Kesselman, C., and Tuecke, S. (2001) The Anatomy of the Grid: Enabling Scalable Virtual Organizations. International Journal of Supercomputer Applications, 15 (3): 200-222.
[11] Foster, I., and Kesselman, C. (2004) The Grid 2: Blueprint for a New Computing Infrastructure. Morgan Kaufmann, 2 edition.
[12] Ganglia. (2007). Ganglia Monitoring System. http://ganglia.sourceforge.net/.
[13] GIP. (2007) Generic Information Provider. https://twiki.cern.ch/twiki//bin/view/EGEE/GIP.
[14] GISolve. (2007) GISolve: TeraGrid GIScience Gateway. http://www.gisolve.org/.
[15] Globus. (2007). Globus Toolkit. http://www.globus.org/toolkit/.
[16] Glue. (2007) Glue Schema. http://glueschema.forge.cnaf.infn.it/.
[17] Hawkeye. (2007) Hawkeye: A Monitoring and Management Tool for Distributed Systems. http://www.cs.wisc.edu/condor/hawkeye/.
[18] Inca. (2007) INCA: user-level grid monitoring. http://inca.sdsc.edu/.

[19] LCG. (2007) Large Hadron Collider (LHC) Computing Project (LCG). http://lcg.web.cern.ch/LCG/.
[20] MDS. (2007) GT Information Services: Monitoring and Discovery System (MDS). http://www.globus.org/toolkit/mds/.
[21] MDS2. (2007) Monitoring and Discovery Service in GT2. http://www.globus.org/toolkit/mds/#mds_gt2.
[22] Nagios. (2007) Nagios: A service and network monitoring program. http://www.nagios.org/.
[23] Newman, H. B., Legrand, I.C., Galvez, P., Voicu, R., and Cirstoiu, C. (2003) MonALISA: A Distributed Monitoring Service Architecture. CHEP 2003, La Jola, California.
[24] OpenLDAP. (2007). Open Source Lightweight Directory Access Protocol Server. http://www.openldap.org/.
[25] OSG. (2007) Open Science Grid. http://opensciencegrid.org/.
[26] RP-Provider. (2007) Resource Property Provider Framework and Information Providers for MDS4. http://www.globus.org/toolkit/docs/4.0/info/usefulrp.
[27] Schopf, J., D'Arcy, M., Miller, N., Pearlman, L., Foster, I., and Kesselman, C. (2005) Monitoring and Discovery in a Web Services Framework: Functionality and Performance of the Globus Toolkit's MDS4. Argonne National Laboratory Tech Report ANL/MCS-P1248-0405.
[28] Wang, S., Armstrong, M. P., and Bennett, D. A. (2002) Conceptual Basics of Middleware Design to Support Grid Computing of Geographic Information, Proceedings of 2nd International Conference on Geographic Information Science, Boulder, CO, USA, September, 25 - 28, 2002.
[29] Wang, S. (2003) Grid-based Geo-middleware. Paper presented at The Association of American Geographers 99th Annual Meeting, New Orleans, LA, USA, March 2003.
[30] Wang, S., Padmanabhan, A., Liu, Y., Briggs, R., Ni, J., He, T., Knosp, B. M., Onel, Y. (2004) A Multi-Agent System Architecture for End-User Level Grid Monitoring Using Geographic Information Systems (MAGGIS): Architecture and Implementation. In the conference proceeding of 2nd International Workshop of Grid and Cooperative Computing (GCC2003)), Lecture Notes in Computer Science, LNCS vol 3032, pp. 536-543.
[31] Wang, S., Armstrong, M. P., Ni, J., Liu, Y. (2005) GISolve: A Grid-based problem solving environment for computationally intensive geographic information analysis. In: Proceedings of the 14th International Symposium on High Performance Distributed Computing (HPDC-14) – Challenges of Large Applications in Distributed Environments (CLADE) Workshop, Research Triangle Park, NC, USA, July 24, 2005.
[32] Wang, S., Shook, E., Padmanabhan, A., Briggs, R., Pearlman, L. (2006) Developing a modular information provider to support interoperable Grid information services. In: Proceedings of Grid and Cooperative Computing - GCC 2006: The Fifth International Conference, IEEE Computer Society, pp. 448-453.
[33] TeraGrid. (2007) TeraGrid, http://teragrid.org/.
[34] Thain, D., Tannenbaum, T., and Livny, M. (2005) Distributed Computing in Practice: The Condor Experience. Concurrency and Computation: Practice and Experience, Vol. 17, No. 2-4, pages 323-356, February-April, 2005.

[35] Tierney, B., Aydt, R., Gunter, D., Smith, W., Taylor, V., Wolski, R., Swany, M. (2002) A Grid Monitoring Architecture. Global Grid Forum (GGF). http://www-didc.lbl.gov/GGF-PERF/GMA-WG/papers/GWD-GP-16-2.pdf.

[36] Torque. (2007) TORQUE Resource Manager. http://www.clusterresources.com/pages/products/torque-resource-manager.php.

[37] UsefulRP. (2007) GT 4.2-drafts MDS4 UsefulRP. http://www.globus.org/toolkit/docs/development/4.2-drafts/info/usefulrp/index.html.

[38] Zhang, X., Freschl, J. L., Schopf, J. M. (2003) A performance study of monitoring and information services for distributed systems. In Proceedings of 12th IEEE International Symposium on High Performance Distributed Computing.

In: Cyberinfrastructure Technologies and Applications
Editor: Junwei Cao
ISBN: 978-1-60692-063-3

Chapter 4

PERFORMANCE-ORIENTED WORKLOAD MANAGEMENT FOR MULTICLUSTERS AND GRIDS

Ligang He[1] and Stephen A. Jarvis[2]
Department of Computer Science, University of Warwick
Coventry, CV4 7AL, UK

ABSTRACT

This chapter addresses the dynamic scheduling of parallel jobs with QoS demands (soft-deadlines) in multiclusters and grids. Three performance metrics (over-deadline, makespan and idle-time) are combined with variable weights to evaluate the scheduling performance. These three metrics are used for measuring the extent of jobs' QoS demands compliance, resource throughput and resource utilization, respectively. Therefore, clusters situated in different administrative organizations can utilize different weight combinations to represent their different performance requirements. Two levels of performance optimisations are applied in the multicluster. At the multicluster level, a scheduler, (which we call MUSCLE), allocates parallel jobs with high packing potential to the same cluster; MUSCLE also takes the jobs' QoS requirements into account and employs a heuristic to allocate suitable workloads to each cluster to balance the performance. At the local cluster level, a workload manager, called TITAN, utilizes a genetic algorithm to further improve the scheduling performance of the jobs sent by MUSCLE. The extensive experimental studies are conducted to verify the effectiveness of the scheduling mechanism in MUSCLE; the results show that comparing with the traditional workload allocation policies in distributed systems (Dynamic Least Load and Weighted Random), the comprehensive scheduling performance (in terms of over-deadline, makespan and idle-time) of parallel jobs is significantly improved and well balanced across the multicluster.

---

[1] E-mail address: liganghe@dcs.warwick.ac.uk
[2] E-mail address: saj@dcs.warwick.ac.uk

INTRODUCTION

Clusters are increasingly being interconnected to create multicluster or grid computing architectures. These constituent clusters may be located within a single organization or across different administrative organizations [1, 2, 7]. Work management and scheduling is a key issue in grid computing. Parallel jobs constitute a typical workload type which is often encountered in the scheduling scenario. Parallel jobs can be classified into two categories: rigid and moldable [12]. Rigid parallel jobs are run in the user-specified number of computers while moldable jobs can be run in the variable number of computers, which is often determined at run-time [12].

Qualities of service (QoS) are often requested by the jobs submitted to a grid system [6][8]. An example of this is jobs with user-defined deadlines [2]. The QoS of a job is satisfied if it finishes before the specified deadline while the QoS decreases as the excess (of completion over deadline) increases. Therefore, over-deadline can be used to measure the extent to which the QoS demands of a set of jobs are satisfied. Over-deadline is defined as the sum of excess time of each job's finish time over its deadline.

The scheduling of parallel jobs has been extensively studied in a single cluster environment [9, 10, 12, 14, 16]. In such a scenario, backfilling is a popular solution [10, 14] and resource utilization is a commonly used system-oriented metric [12]. The backfilling approach runs small jobs before larger jobs that arrive earlier to utilize otherwise idle computers aiming to achieve higher resource utilization. Idle-time in a resource can be viewed as the metric for measuring resource utilization [12]. However, backfilling is applied to a single cluster and submitted jobs typically have no QoS requirements. A new mechanism is necessary to address the dynamic scheduling of parallel jobs with QoS request in multiclusters or grids.

An additional goal in job scheduling in computational grid environments is high resource throughput. In these scenarios, makespan is a commonly used metric to measure resource throughput [8]. Makespan is defined as the duration between the start time of the first job and the finish time of the last executed job.

In the multicluster architecture assumed in this chapter, the constituent clusters may be located in different administrative organizations and as a result be managed with different performance criteria. In this scheduling work, we combine three metrics (over-deadline, makespan and idle-time) with additional variable weights; this allows the resources in different locations to represent different performance scenarios.

It may be the case that maximising user-oriented performance metrics (e.g., achieving a low over-deadline) conflicts with the goal of maximising system-oriented metrics (e.g. ,obtaining low idle time or short makespans) [8, 12]. To some extent, if parallel jobs can be packed tightly, the over-deadline can also be reduced. However, overemphasizing the packing of parallel jobs may compromise performance improvements in terms of over-deadline. It is therefore necessary to find the schedule that offers high comprehensive performance according to the performance requirements of each individually managed resource.

In this chapter, the multicluster architecture is equipped with two levels of performance optimisation. A MUlticluster SCheduling LEver (MUSCLE) is developed at the multicluster-level to allocate jobs submitted to the multicluster to constituent clusters. MUSCLE is able to allocate parallel jobs with high packing potential (i.e., they can be packed more tightly) to the same cluster. It also takes the QoS demands of jobs into account and exploits a heuristic to allocate suitable workload to each cluster. As a result, the comprehensive scheduling performance (in terms of over-deadline, makespan and idle-time) of parallel jobs is significantly improved and well balanced across the multicluster. When MUSCLE makes scheduling decisions to distribute jobs to individual clusters, it also determines a seed schedule for the jobs allocated to each cluster. These seed schedules are sent to the corresponding clusters where a local workload manager (TITAN [13]) uses a genetic algorithm to transform the schedule into one that improves the (local) comprehensive performance.

In grid systems, the global scheduler usually has no control over the local schedulers. This presents difficulties when designing effective schedulers for such an environment. The two-level optimisation architecture developed in this chapter overcomes this difficulty. MUSCLE has no control over the operations of the TITAN scheduler at each local cluster. There is also no communication between global-level decision making (using MUSCLE) and local-level optimisation (using TITAN).

Another challenge in grid scheduling is that jobs can be submitted from different locations and so a grid scheduler generally lacks a complete view of the jobs in the grid. In this chapter we preserve realistic usage patterns in that users are allowed to submit jobs to the multicluster through MUSCLE or through the TITANs of each local cluster. If a job is submitted through TITAN, another (existing) grid component called A4 (Agile Architecture and Autonomous Agents) [3, 5] prompts MUSCLE as to the submission and provides summary metadata (e.g., job size, arriving time) for the job. A4 is responsible for resource advertisement and discovery as well as for transferring jobs (or job metadata) among schedulers or resources if necessary. This chapter focuses on presenting the scheduling mechanism found in MUSCLE. The detailed design of the A4 component is addressed in [3, 5].

A grid is a dynamic system with resources that change over time. The genetic algorithm used in TITAN is an evolutionary process and is therefore able to deal with changes in the resources in its domain (such as the addition/deletion/failure of processing nodes). MUSCLE can also realize these changes via A4 so as to adjust its allocation operations accordingly.

The rest of this chapter is organized as follows. The system and workload model is introduced in Section 2. Section 3 briefly presents the genetic algorithm used in TITAN. The design of MUSLCE is proposed in Section 4. Section 5 presents the experimental results. Finally, Section 6 concludes the chapter.

SYSTEM AND WORKLOAD MODEL

The multicluster architecture assumed in this chapter (shown in figure 1) consists of n clusters, C_1, C_2, …, C_n; where a cluster C_i ($1 \leq i \leq n$) consists of m_i homogeneous computers (i.e., the size of cluster C_i is m_i), each with a service rate of u_i. There are two scheduling levels in the architecture; MUSCLE acts as the global scheduler for the multicluster while TITAN schedules the jobs sent by MUSCLE within each local cluster. MUSCLE and TITAN are interconnected through the A4 system. Users can submit parallel jobs to the multicluster through MUSCLE or through TITAN. If a job is submitted via TITAN, MUSCLE is made aware of it through metadata sent by the A4 agents. Also, if the resources (computers) in the multicluster change (addition or deletion), MUSCLE is also made aware of this via A4. Hence, the multicluster is a virtual organization from the perspective of users. The submitted jobs will be judiciously allocated by MUSCLE to suitable clusters for the further scheduling and execution. The PACE (Performance Analysis and Characterisation Environment) toolkit [11, 14] is incorporated into the architecture to provide execution time predictions for the submitted jobs. This performance prediction information is used to support the scheduling decisions made by MUSCLE and TITAN.

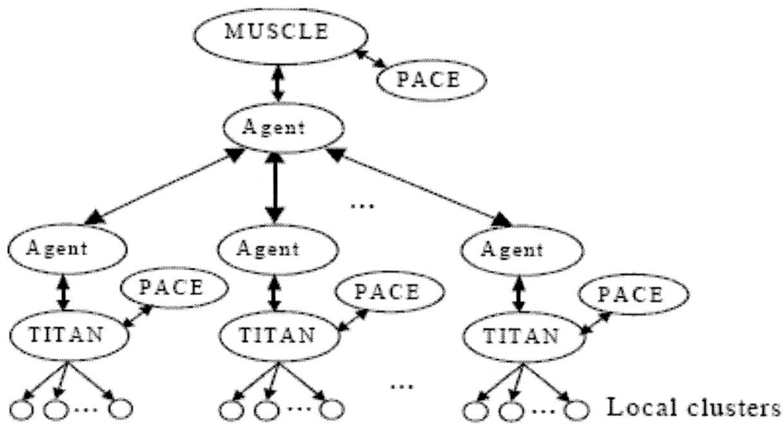

Figure 1. The multicluster architecture.

Parallel jobs considered in this chapter are rigid. A parallel job, denoted by J_i, has the following attributes:

- Job size (i.e., the requested number of computers to be run), denoted by s_i;
- Arrival time, denoted by a_i;
- Execution time in cluster C_j ($1 \leq j \leq n$) (predicted by PACE), denoted by et_{ij}; Soft-deadline (QoS), denoted by d_i.

GENETIC ALGORITHM

This section briefly describes the genetic algorithm used in the TITAN [13]. A two dimensional coding scheme is developed to represent a schedule of parallel jobs in a cluster. Each column in the coding specifies the allocation of the processing nodes to a parallel job, while the order of these columns in the coding is also the order in which the corresponding jobs are to be executed. An example is given in figure 2 and illustrates the coding for a schedule as well as the execution of the corresponding jobs.

Three metrics (over-deadline, makespan and idle-time) are combined with the weights to form a comprehensive performance metric (denoted by CP), which is used to evaluate a schedule. The CP is defined in Eq.1, where Γ, ω and θ are makespan, idle-time and over-

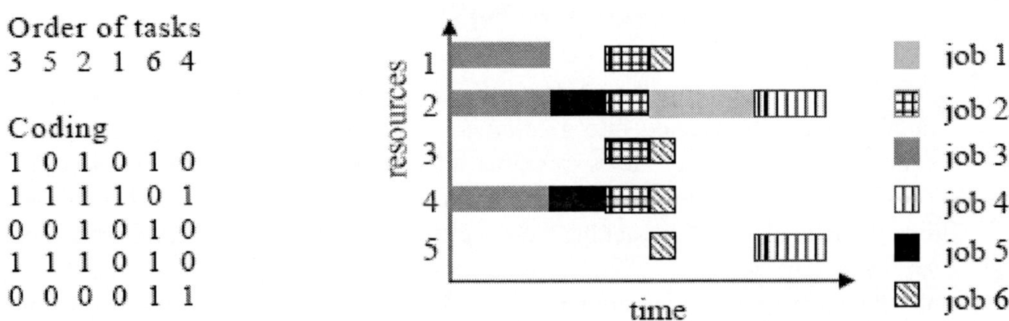

Figure 2. A case study for the coding scheme of a schedule as well as the corresponding execution on computers (hosts).

deadline, respectively; and W^i, W^m and W^o are their weights. For a given weight combination, the lower the value of CP, the better the comprehensive performance.

$$CP = \frac{W^i\Gamma + W^m\omega + W^o\theta}{W^i + W^m + W^o} \tag{1}$$

A genetic algorithm is used in TITAN to find a schedule with a low CP. The algorithm first generates a set of initial schedules (one of these is the seed schedule sent by MUSCLE and others are generated randomly). The performance of a schedule, evaluated by the CP, is normalized to a fitness function using a dynamic scaling technique which is shown in Eq.2, where $CP^{\max}$ and $CP^{\min}$ represent the best and the worst performance in the schedule set; and CP^k is the performance of the k-th schedule.

$$f_k = \frac{CP^{\max} - CP^k}{CP^{\max} - CP^{\min}} \tag{2}$$

In the genetic algorithm, the value of the fitness function of a schedule determines the probability that the schedule is selected to create the next generation. *Crossover* and *mutation* operations are performed to generate the next generation of the current schedule set. The procedure continues until the performance improvement between two generations of schedule sets is less than a predefined threshold. The crossover operation selects two schedules each time from the current schedule set, cuts the schedules at a random location, merges the head portion of one schedule with the tail portion of the other and then reorders the schedules to produce two legitimate children. The mutation operation consists of two parts; one is to exchange the execution order of two randomly selected jobs while the other involves randomly adjusting the allocation of jobs to host computers. The probabilities of performing crossover and mutation are predetermined.

DESIGN OF MUSCLE

The main operations performed by MUSCLE are as follows. First, MUSCLE determines which parallel jobs can be packed into a *computer space* with a given size (i.e., available computers of a given number). It is possible that in a set of parallel jobs, different compositions of jobs can be packed into a given computer space. These possible compositions of parallel jobs for packing into a given computer space are organized into a *composition table*. The *i*-th row of the table corresponds to the possible compositions of parallel jobs for packing into a computer space with size *i*. After the composition table is constructed, MUSCLE searches the table for suitable parallel jobs when it wants to allocate jobs into an available computer space in a cluster. When the computer space is available in multiple clusters, MUSCLE orders the processing of the space using a heuristic.

Organizing Parallel Jobs

Suppose the maximum cluster size in the multicluster is m_{MAX} and that at some time point, *p* parallel jobs, J_1, J_2, …, J_p are in the queue in the multicluster (they are queued in MUSCLE or TITAN). Algorithm 1 outlines the steps for constructing the composition table. The *p* jobs are filled into suitable rows in the table. When trying to fill job J_i (with size s_i) into the *j*-th row (i.e., packing the job into the computer space with size *j*), the algorithm checks if there exists such a composition in the $(j\text{-}s_i)^{th}$ row that no job in the composition has appeared in the *j*-th row. If such a composition exists, it indicates that J_i and the jobs in the composition can be packed into the computer space with size *j*. Hence, J_i and these jobs are filled into this row.

The composition table has m_{MAX} rows. The *r*-th row ($1 \le r \le m_{MAX}$) corresponds to a computer space with size *r* and it contains the compositions of jobs for packing into the computer space. In this row, the total size of all jobs in every composition is *r*. A composition in the composition table is located by a pair (*r*, *l*) and the composition is denoted by *cps*(*r*, *l*), where *r* is the row number which the composition lies in and *l* is the position subscript of the first job of the composition in the row. A job is not allowed to appear more than once in the same row. This rule is to guarantee that a job will not be allocated to different computer space.

Algorithm 1. Constructing the composition table

1. ***for*** *each parallel job* J_i *($1 \le i \le p$) to be scheduled* ***do***
2. ***for*** *each* j *satisfying* $1 \le j \le m_{MAX}$ ***do***
3. ***if*** $s_i = j$
4. Append J_i *to the tail of the* j*-th row of the table;*
5. ***if*** $s_i < j$
6. $r \leftarrow j - s_i$;
7. ***if*** *the* r*-th row in the table is not NULL*
8. ***if*** *there is such a composition of parallel jobs in the* r*-th row, in which no job is in the* j*-th row of the table;*

9. Append J_i *as well as the parallel jobs in the composition obtained in the* r*-th row to the tail of the* j*-th row;*

Table 1. The working example of all case studies in this chapter

Clusters	C_1	C_2				
Size	4	6				
Service rate of computers	1	1				
Jobs	J_1	J_2	J_3	J_4	J_5	J_6
Job size	2	1	4	3	2	1
Execution time	2	4	4	6	2	4
Slack	6	8	14	12	4	8

There are two for-loops in Algorithm 1. Step 8 searches a row for the qualified composition. In the worst case, the time taken by Step 8 is O(p). Hence, the worst-case time complexity of Algorithm 1 is O($p^2 m_{MAX}$). Algorithm 1 is illustrated by the following case study. The cluster setting and the parameters of the jobs in queue are listed in table 1, which are used as the working example for all case studies in this chapter unless otherwise stated. Table 2 shows the composition table after filling J_1-J_6.

Table 2. This is a sample table

1	J_2	J_5	
2	J_1	J_5, J_2	J_6
3	J_2, J_1	J_4	J_6, J_5
4	J_3	J_4, J_2	J_6, J_1
5	J_3, J_2	J_4, J_1	
6	J_3, J_1	J_6, J_4, J_2	

Searching the Composition Table

Job allocation proceeds as follows. First, MUSCLE finds the earliest available computer space in each cluster. MUSCLE then searches the composition table for suitable jobs to allocate to this space. If multiple clusters offer available computer space, a heuristic is used to determine the cluster which should be processed first. After jobs are allocated to the computer space in a cluster, the new earliest available computer space in that cluster is updated. The heuristic is then called once again to determine the cluster to which MUSCLE should allocate further jobs. The heuristic aims to balance the CP (see Eq.1) among the constituent clusters in the multicluster. This subsection demonstrates how the composition tables are searched for parallel jobs to allocate to a given computer space in a cluster. In subsection 4.3, a heuristic is proposed to determine which cluster in the multicluster MUSCLE should process first.

The procedure of allocating jobs to a computer space with size r in a cluster proceeds as follows (Algorithm 3). First, it searches the composition table from the r-th row up to the first row to obtain the first row which is not null. Then, in this row, the algorithm selects the composition in which the number of jobs having been allocated is the least. If the number is zero (i.e., no jobs in the composition have been allocated), then these jobs are allocated to the computer space. If a job, J_i, whose size is s_i, in the composition has been allocated, a function (Algorithm 2) is called to search the s_i-th row for alternative jobs for J_i. The function is called recursively if a composition cannot be found in the s_i-th row in which no job in it is allocated. The recursive call ceases when there is only one composition in a searched row (i.e., there are no alternative jobs) or when the composition consisting of unallocated jobs is found. If the algorithm fails to allocate jobs to the computer space with size r, it continues by trying to identify jobs to allocate to the computer space with size r=r-1. The searching procedure continues until r reaches 0, which means that no jobs can be allocated to the given computer space. After the jobs to allocate to a computer space have been determined, the schedule of these jobs can also be calculated (Step 11 in Algorithm 3).

Algorithm 2 outlines the function of selecting alternative jobs for allocation in a given composition, $cps(r, l)$. Algorithm 3 outlines the allocation of jobs to the given composition in a cluster. A computer space is represented by $cs(t, r, cp)$, where t is the time when the space is available, r is the size of the space and cp is the computer number that the space starts from. After Algorithm 3 is finished, array q contains the jobs allocated to the computer space.

As can be seen from the above description, Algorithm 3 always attempts to identify the jobs that maximally occupies the given computer space. By doing so, the number of computers left idle is minimized. Consequently, the jobs sent to a cluster are tightly packed.

Algorithm 2. Getting the alternative jobs for the allocated jobs in a composition cps(r, l)

Input: the position of the composition in the composition table (r, l); an array, q (used to contain alternative jobs).

Output: if succeed, return 1; otherwise, return 0; (array q contains partial alternative jobs).

1. lc←l; *noway*←0;
2. **while** lc does not reach the end of the composition and *noway* equals 0
3. Get the job J_i pointed by lc;
4. **if** J_i has not been allocated
5. Append J_i to array q;
6. **else if** there is only one composition in the s_i-th row of the composition table (s_i is the size of J_i), which consists of J_i itself
7. *noway*←1;
8. **else**
9. In the s_i-th row (except the composition consisting of J_i itself), get such a composition $cps(s_i, l)$ in which the number of allocated jobs is minimum (if more than one compositions have the same minimum number, select the one first found);
10. **if** the number of allocated jobs in $cps(s_i, l)$ is 0
11. Append the jobs in the composition $cps(s_i, l)$ to array q;
12. **else**
13. Call Algorithm 2 with (s_i, l) and array q as the inputs;
14. **if** its output is 0
15. *noway*←1;
16. lc←lc+1;
17. **end while**;
18. **if** *noway* equals 0
19. **return** 1;

20. **else**
21. **return** 0;

Algorithm 3. Getting the alternative jobs for the allocated jobs in a composition cps(r, l)

Input: the computer space (*t*, *r*, *cp*); an array, *q* (used to contain jobs allocated to the computer space).

1. *rc*←*r*;
2. **while** the *rc*-th row is NULL
3. *rc*←*rc*−1;
4. Get such a composition of jobs in which the number of allocated jobs is minimum in the *rc*-th row (if more than one compositions have the same minimum number, select the one first found; suppose the composition obtained is $cps(r_0, l)$);
5. **if** the number of allocated jobs in the composition is 0
6. Put the jobs in the composition to array *q*;
7. **else**
8. Call Algorithm 2 with (r_0, l) and array *q* as inputs;
9. **if** Algorithm 3 returns 0
10. *rc*←*rc*−1; Go to Step 2;
11. The starting time of these found jobs is *t*; these jobs are allocated to the computers in the order they are in array *q*, starting from computer *cp*.

The time complexity of Algorithm 3 (including Algorithm 2) is based on the number of jobs that are allocated. The best-case time complexity is O(1) while the worst-case time complexity is $O(p^2 n_{MAX})$, since the worst case involves searching the whole composition table.

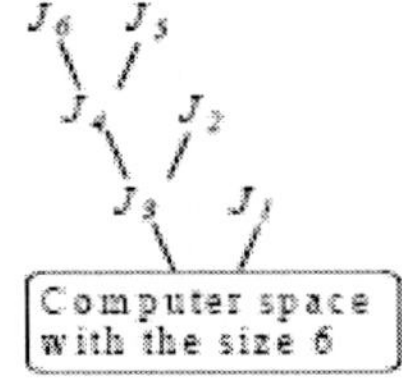

Figure 3. The case study for Algorithms 2 and 3.

A case study is given below to illustrate Algorithm 2 and 3. The constructed composition table is table 2. Suppose job J_3 and J_4 have been allocated. Algorithm 3 tries to find the jobs to allocate to a computer space *cs*(0, 6, 1), i.e., the computer space is available at time 0; its size is 6 and it starts from the computer numbered 1. In the sixth row of the table, there are two compositions $\{J_3, J_1\}$ and $\{J_6, J_4, J_2\}$. The number of allocated jobs in both compositions is 1. The first composition is selected; this is $\{J_3, J_1\}$. Since J_3 has been allocated, Algorithm 2 is called to search the forth row (4 is the size of J_3) for alternative jobs for J_3. All compositions, except $\{J_3\}$ that contains itself, are considered in this row. The number of allocated jobs in the remaining compositions is 1 (J_1 is also regarded as the allocated job since it has been used in the selected composition, $\{J_3, J_1\}$). The composition $\{J_4, J_2\}$ is selected. Since J_4 has been allocated, Algorithm 2 is called again (in Step 13) to search the third row for the alternative jobs for J_4. The composition $\{J_6, J_5\}$ is now selected. The recursive call of Algorithm 2 ceases as neither J_6 nor J_5 are allocated.

As shown in Step 11 in Algorithm 3, the starting times of these jobs $\{J_6, J_5, J_2, J_1\}$ are 0. J_6 is allocated to computer 1-2, J_5 to computer 3, J_2 to computer 4 and J_1 to computer 5-6. Therefore, the scheduling of these jobs is determined. After that, the earliest available computer space in the cluster is updated. The pre-scheduling of these jobs forms part of the seed schedule sent to the TITANs at the local clusters.

Employing a Heuristic to Balance the Performance

In each local cluster, TITAN uses a genetic algorithm to adjust the seed schedule sent by MUSCLE, aiming to further improve the CP. Although MUSCLE has no control over the detailed operations of the genetic algorithm, it analyses the objective factors influencing the performance and allocates different levels of workloads to each cluster through a heuristic so that the CP can be well balanced among the constituent clusters.

The fundamental attributes of the workload (parallel jobs) allocated to a cluster include:

- The number of jobs, denoted by *jn*;
- The sum of execution times of all jobs, denoted by *et_sum*;
- Total slacks (relating to the QoS of jobs), denoted by *slk_sum* (the slack of a job is its deadline minus its execution time and its arrival time);
- Total job sizes, denoted by *size_sum*.

The fundamental attribute of a cluster is the number of computers (the attribute of the service rate of computers has been reflected in the execution times of jobs).

The scheduling performance achieved in a cluster is determined by the resultant forces of these fundamental attributes. In addition to this, the packing potential for the parallel jobs allocated to a cluster is also considered a critical factor. This problem however is solved by Algorithm 3, which always selects the jobs which can maximally occupy a given computer space so that the jobs sent to a cluster can be packed tightly.

Table 3. Workload allocation trends by comparing the current attribute values

Attribute	Definition	Value	Workload allocation trend
jn	The number of jobs	Low	More
et_sum	The sum of execution times	Low	More
slk_sum	The sum of slacks	Low	Less
size_sum	The sum of job sizes	Low	More
m_i	The cluster size	Low	Less

If separately comparing the values of these attributes among clusters, the workload allocation should follow the following trend, listed in table 3, where “Low” in the “Value” field means that the value of the attribute in a cluster is lower than that in other clusters; “More” means more workloads should be allocated to the corresponding cluster. These attributes are integrated to form a new metric (denoted by ε), shown in Eq.3, where the attributes whose value are “More” in the “Workload allocation trend” fieldis are placed in the numerator while others are placed in the denominator. Let ε be 0 if *slk_sum* is 0 (no jobs are allocated).

$$\varepsilon = \frac{jn \times et\_sum \times size\_sum}{slk\_sum \times m_i} \qquad (3)$$

when multiple clusters can offer available computer space, the cluster with the smallest ε is given the highest priority and will be allocated the jobs. Such an integration of these attributes is consistent with the allocation trends in table 3. When more than one cluster has the same value of ε, the cluster with the greatest size is given the highest priority. If multiple clusters are still of the same size, a cluster is selected randomly. The complete scheduling procedure of MUSCLE is outlined in Algorithm 4.

Algorithm 4. The complete scheduling procedure of MUSCLE

1. **while** the current makespans of all clusters are greater than a predefined valve
2. Waiting;
3. Collecting the current jobs in queue in the multicluster;
4. Call Algorithm 1 to construct the composition table for the clusters whose current makespans are less than the predefined valve;
5. Get the earliest available computer space in each cluster;
6. **do**
7. Calculate ε using Eq.3 for each cluster and get the cluster with the minimal ε, C_i;
8. Call Algorithm 3 to allocate jobs to the earliest available computer space in C_i;
9. Update the earliest available computer space in C_i;
10. Update the current value of workload attributes in each cluster;
11. **while** not all jobs have been allocated;
12. Go to Step 1;

The time of Algorithm 4 is mainly taken by Step 4 and Step 8 in the do-while loop. Their time complexities have been analysed in subsection 4.1 and 4.2. A case study is presented below to illustrate Algorithm 4. The working example has been listed in table 1. The corresponding composition table is table 2. Figure 4 shows the evolution of the job scheduling in C_1 and C_2 (figure 4.1 and 4.2 are for C_1; and figure 4.3 and 4.4 for C_2).

Table 4 shows the evolution of the values of the workload attributes, ε and earliest available computer space (the *cs* field in table 4). Initially, there is no workload in both clusters. The earliest available computer spaces in C_1 and C_2 are (0, 4, 1) and (0, 6, 1), respectively. C_2 has the higher priority. Therefore, in this round MUSCLE calls Algorithm 2 to search table 2 for the jobs to allocate to the space (0, 6, 1). Consequently, J_3 and J_1 are found and allocated to the space. The schedule of these jobs in C_2 is shown in figure 4.3. After the allocation, MUSCLE obtains the new earliest available computer space in C_2, which is (2, 2, 5). MUSCLE also updates the values of the workload attributes in C_2 (i.e., *jn*, *et_sum*, *slk_sum* and *size_sum*; their current values are shown in Column “Round 2” of table 4) to get the new value of ε. Since ε for C_1 is less than that for C_2 at this time, MUSCLE search the composition table for jobs to allocate to the earliest available computer space in C_1, which is (0, 4, 1). J_4 and J_2 are obtained and their schedule in C_1 is shown in figure 4.1. Similarly, the earliest available computer space and ε in C_1 are updated. MUSCLE compares the current values of ε in these two clusters to select a computer space. The procedure continues until all jobs are scheduled.

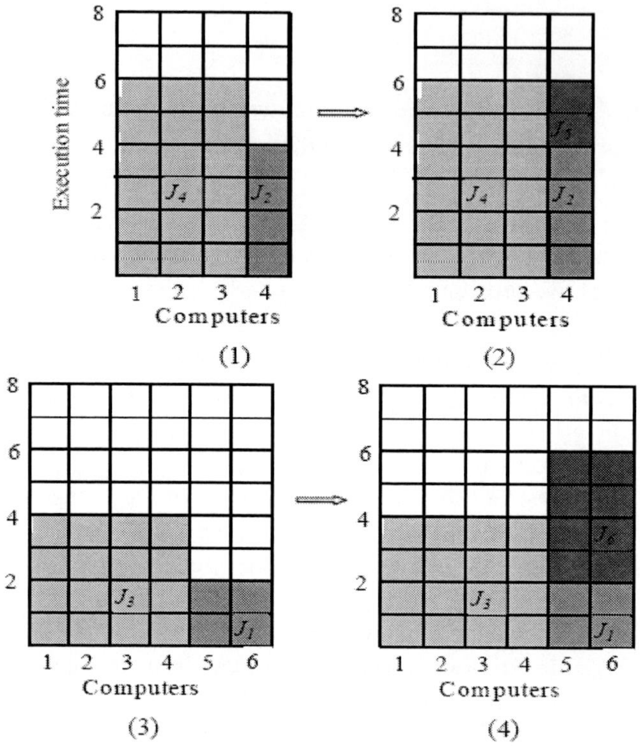

Figure 4. Evolution of job scheduling in Cluster C_1 and C_2; (1) and (2) are for C_1; (3) and (4) for C_2.

Table 4. Evolution of the values of the workloads attributes, ε and earliest available computer space (cs) during the job allocation

		Round 1	Round 2	Round 3	Round 4	Round 5
C_1	*jn*	0	0	2	2	3
	et_sum	0	0	10	10	12
	slk_sum	0	0	20	20	24
	size_sum	0	0	4	4	5
	m_i	4	4	4	4	4
	ε	0	0	1	1	1.88
	cs	(0, 4, 1)	(0, 4, 1)	(4, 1, 4)	(4, 1, 4)	(6, 4, 1)
C_2	*jn*	0	2	2	3	3
	et_sum	0	6	6	12	12
	slk_sum	0	20	20	28	28
	size_sum	0	6	6	8	8
	m_i	6	6	6	6	6
	ε	0*	0.6	0.6	1.71	1.71
	cs	(0, 6, 1)	(2, 2, 5)	(2, 2, 5)	(4, 4, 1)	(4, 4, 1)

After Algorithm 4 is finished, the seed schedules in both clusters are determined, which are shown in figure 4.2 and 4.4. It can be seen that these jobs have been packed tightly. These seed schedules will be sent to TITANs situated in C_1 and C_2 for further performance improvement.

EXPERIMENTAL STUDIES

A simulator is developed to evaluate the performance of the scheduling mechanism in MUSCLE. The presented experimental results focus on showing the performance advantages of the scheduling mechanism in MUSCLE over the scheduling policies frequently used in distributed systems. Weighted Random (WRAND) and Dynamic Least Load (DLL) policies are two selected representatives. The advantages of TITAN over other typical cluster-level schedulers have been presented in [13].

In the experiments, the generated parallel jobs are submitted to the multicluster. MUSCLE, DLL or WRAND are used as the multicluster-level scheduling policies respectively while in all cases TITAN is used as the cluster-level scheduler in each local cluster. The combination of the weights for the over-deadline, the makespan and the idle-time is denoted by (W^o, W^m, W^i).

The workloads in the experiments are generated as follows. 20,000 parallel jobs are generated; the submissions of parallel jobs follow Poisson process with the average arrival rate λ; job size is generated following a uniform distribution in [*MIN_S*, *MAX_S*], and the execution times of jobs in cluster C_1 follow a bounded Pareto distribution, shown below, where e_l and e_u are the lower and upper limit of the execution time x.

$$f(x) = \frac{\alpha e_l^{\alpha}}{1-(e_l / e_u)^{\alpha}} x^{-\alpha-1}$$

It is shown in [15] that the distribution can better represent the realistic workload model than the exponential distribution. If a job is scheduled to another cluster consisting of computers with the service rate u_j, the execution time is determined through multiplying by the ratio between u_1 and u_j (i.e., u_1/u_j). A job's deadline, d_i is determined by Eq.4, where dr is the deadline ratio. dr follows the uniform distribution in [*MIN_DR*, *MAX_DR*]. The range is used to measure the deadline range.

$$d_i = \max\{et_{ij}\} \times (1+dr) \qquad (4)$$

Table 5. the Multicluster setting

Clusters	C_1	C_2	C_3	C_4
Size	20	16	12	10
Service rate ratio	1.0	1.2	1.4	1.6

The multicluster in the experiments uses the setting in table 5 unless otherwise stated.

Dynamic Least-Load (DLL) is a scheduling policy extensively used in heterogeneous systems [7, 15]. The workload in cluster C_i, denoted as L_i, is computed by Eq.5. When a parallel job is submitted to the multicluster, the DLL policy schedules the job to the cluster whose workload is the least and whose size is greater than the size of the job.

$$L_i = \frac{\sum_{J_i \in WQ_i} e_i u_i}{m_i} \quad WQ_i \text{ is the set of jobs allocated to cluster } C_i \tag{5}$$

Weighted Random (WRAND) scheduling policy is another frequently used scheduling policy in distributed systems [7, 15]. In the policy, the probability that a job is scheduled to a resource is proportional to its processing capability. The processing capability of a cluster is determined by the sum of the processing capability of all constituent computers. When a job arrives at the multicluster system, the WRAND policy first picks out all clusters whose sizes are greater than the job size (suppose these clusters are C_{i1}, C_{i2}, ..., C_{ij}), and then schedules the job to a cluster C_{ik} ($1 \le k \le j$) satisfying the probability below.

$$\Pr = \frac{m_{ik} u_{ik}}{\sum_{k=1}^{j} m_{ik} u_{ik}}$$

The performance metrics evaluated in the experiments are the *mean comprehensive performance* (MCP) and *performance balance factor* (PB). Each time MUSCLE sends seed schedules to TITANs in individual clusters C_i ($1 \le i \le n$). TITANs further improve the performance in terms of the CP. When the performance improvement between two generations of schedule sets is less than a threshold, the CP performance for cluster C_i, denoted by CP_i, is recorded. The MCP is the average of CP_i ($1 \le i \le n$), calculated by Eq.6, where p_i is the number of jobs allocated to C_i. The procedure continues until all generated jobs are processed. Each point in the performance curves is plotted as the average of the MCP obtained each time.

$$MCP = \sum_{i=1}^{n} CP_i \times \frac{p_i}{p} \tag{6}$$

The PB is defined as the standard deviation of CP_i, shown in Eq.7.

$$PB = \sqrt{\frac{1}{n-1} \sum_{i=1}^{n} (CP_i - MCP)^2} \tag{7}$$

Workload Levels

Figure 5.a and figure 5.b demonstrate the performance difference among MUSCLE, DLL and WRAND scheduling policies under different workload levels. The performance is evaluated in terms of mean comprehensive performance (MCP) and performance balance factor (PB). The workload level is measured by the average arrival rate, by which the average

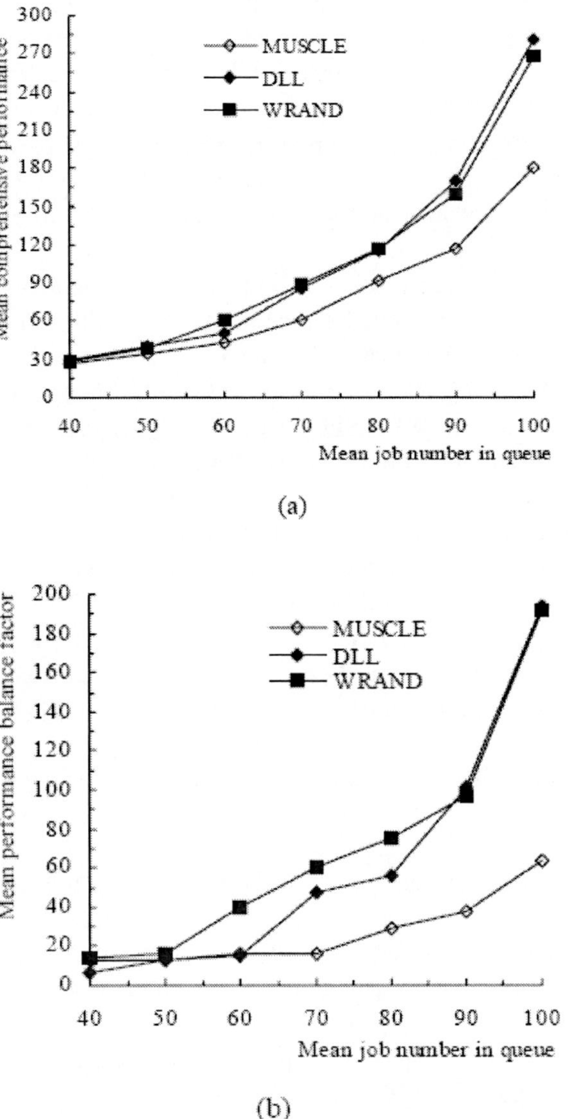

Figure 5. The comparison among MUSCLE, DLL and WRAND under different workload levels in terms of (a) mean comprehensive performance (MCP), and (b) performance balance factor (PB); (W^o, W^m, W^i)=(4, 3, 1); e_l/e_u=5/100; MIN_S/MAX_S=1/10; MIN_DR/MAX_DR=0/5; The cluster size and service rate are same as those in table 5.

job number in the queue in the multicluster is also determined. The x-axis of these figures is labelled by the mean job number in queue.

It can be seen from figure 5.a that MUSCLE outperforms DLL and WRAND under all workload levels. This is because that the jobs are packed tightly in the seed schedules sent by MUSCLE to individual clusters. Therefore, the further improvement in each cluster is based on an excellent "seed". However, the jobs sent by DLL or WRAND to each cluster are random in terms of whether they can be packed tightly. This reduces the possibility of achieving a high MCP in the multicluster.

A further observation from figure 5.a is that the advantage of MUSCLE over other policies becomes increasingly pronounced as the workload increases. When the mean job number in queue is 40, MUSCLE outperforms DLL in terms of the MCP by 12.1% and outperforms WRAND by 8.4%. When the mean job number is 100, the performance advantages are up to 56.7% and 48.9% comparing with DLL and WRAND respectively. This is because that when the job number in queue increases, MUSCLE can gather more information regarding parallel jobs and as a result make better allocation decisions among clusters.

As can be observed from figure 5.b, when the mean job number in queue is less than 60 (the workload is low), the performance balance factor (PB) achieved by MUSCLE is slightly worse than that by DLL. However, when the mean job number is greater than 60, MUSCLE significantly outperforms DLL. When the mean job number in queue is 100, the advantages of MUSCLE over DLL and the WRAND are up to 202.7% and 199.2%, respectively. This can be explained as follows. DLL allocates the jobs to clusters according to the ratios of the workloads in the clusters to their processing capabilities, which is beneficial to obtaining the balanced resource throughput and resource utilization. When the workload is low, a small number of jobs miss their deadlines and the MCP is mainly caused by makespan (resource throughput) and idle time (resource utilization). Therefore, DLL shows more balanced MCP performance (though this does not mean DLL achieves higher overall MCP performance). However, as the workload increases further, more jobs miss their deadlines. The DLL policy ignores the QoS demands of these jobs. In contrast, MUSCLE takes the QoS demands into account so that the MCP performance remains balanced among the clusters when the over-deadline gradually becomes a significant portion in the MCP performance.

Deadlines

Figure 6 compares the performance of MUSCLE, DLL and WRAND under different deadline ranges, which is measured by the value range of *dr* in Eq.4. Figure 6.a and b demonstrate the results in terms of the MCP while figure 6.c and d are for the results in terms of the PB. In figure 6.a and c, the mean job number in queue is 40 while the number is 100 in figure 6.b and d.

A general observation from figure 6.a and b is that the MUSLCE performs better than DLL and WRAND in terms of MCP. Further observations show that the advantages are different under different combinations of workload levels and deadlines. When the workload is low and the deadline is loose (this is the case in figure 6.a where the mean job number in queue is 40 and the deadline range is [0, 5]), the advantage is low. This is because that in this case, the over-deadline, makespan and idle-time are all low. Hence, the potential of MUSCLE cannot be fully released. When the deadlines become more urgent relative to the workload (this is the case in figure 6.a where the mean job number in queue is 40 and the deadline ranges are [0, 1] and [0, 3], also the case in figure 6.b where the number in queue is 100 and the deadline parameter is [0, 5]), the advantage of MUSCLE over DLL and WRAND becomes increasingly prominent. However, when the deadlines become even shorter relative to the workload (this is case in figure 6.b where the mean job number in queue is 100 while the deadline range is less than [0, 5]), the advantage of MUSCLE diminishes. This is because that the workload is saturated relative to the urgent deadlines and the finish times of many jobs exceed their deadlines by a large amount. This performance deterioration is due to the overloading and the scheduling policies do little in this situation.

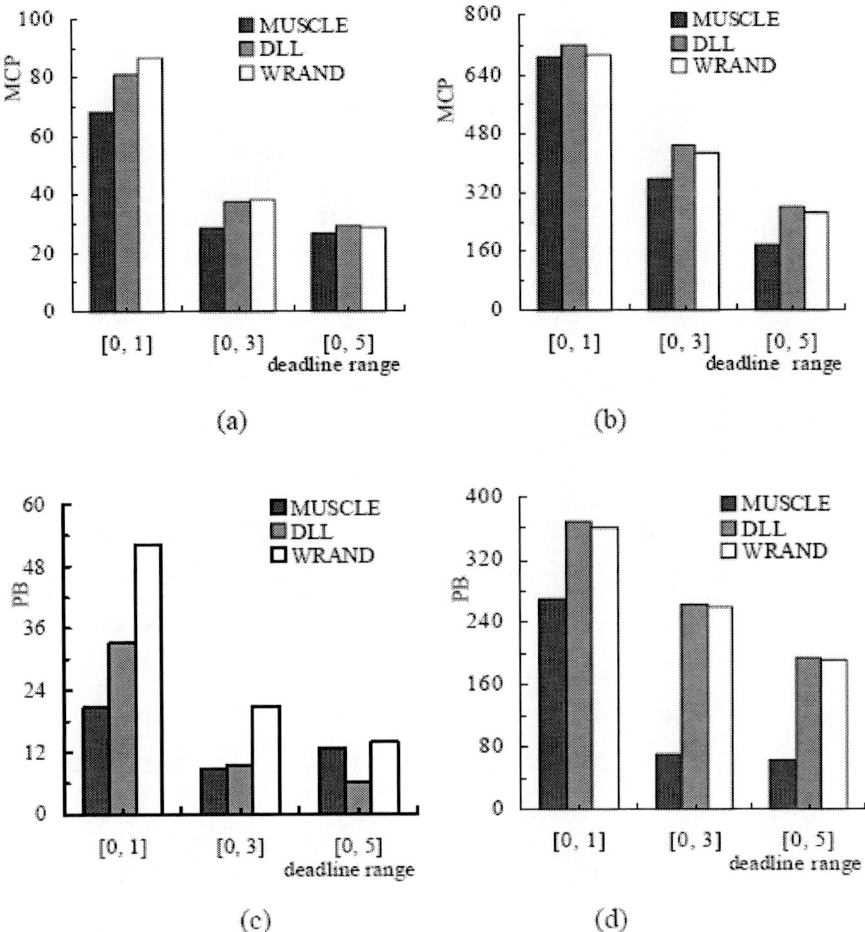

Figure 6. Performance comparison among MUSCLE, the DLL and the RAND under different deadline ranges; (a) and (b) are for the comparison in terms of the MCP; the mean job number in queue is 40 in (a) and 100 in (b); (c) and (d) are for the comparison in terms of the PB; the mean job number in queue is 40 in (c) and 100 in (d); (Wo, Wm, Wi)=(4, 3, 1); el/eu=5/100; MIN_S/MAX_S=1/10.

As for the performance in terms of the PB (shown in figure 6.c and d), MUSCLE significantly outperforms DLL and WRAND under all workloads and deadline ranges (except when the workloads are low and the deadlines are loose). This is because that comparing with the DLL and the WRAND policies, MUSCLE is able to distribute the workload evenly among the multicluster by taking the QoS demands of the jobs into account.

Different Performance Goals

In the multicluster architecture, different clusters can use different weight combinations to cater to their performance goals. Figure 7 compares the performance of MUSCLE, DLL and WRAND when the constituent clusters in the multicluster utilise the different weights, shown in table 6, to get the MCP. Figure 7.a, b and c show the performance comparison in terms of the MCP while figure 7.d, e and f are for the comparison in terms of the PB. The experiments are conducted under the low, medium and heavy workload levels (the mean job number in queue is 40, 100 and 160, respectively).

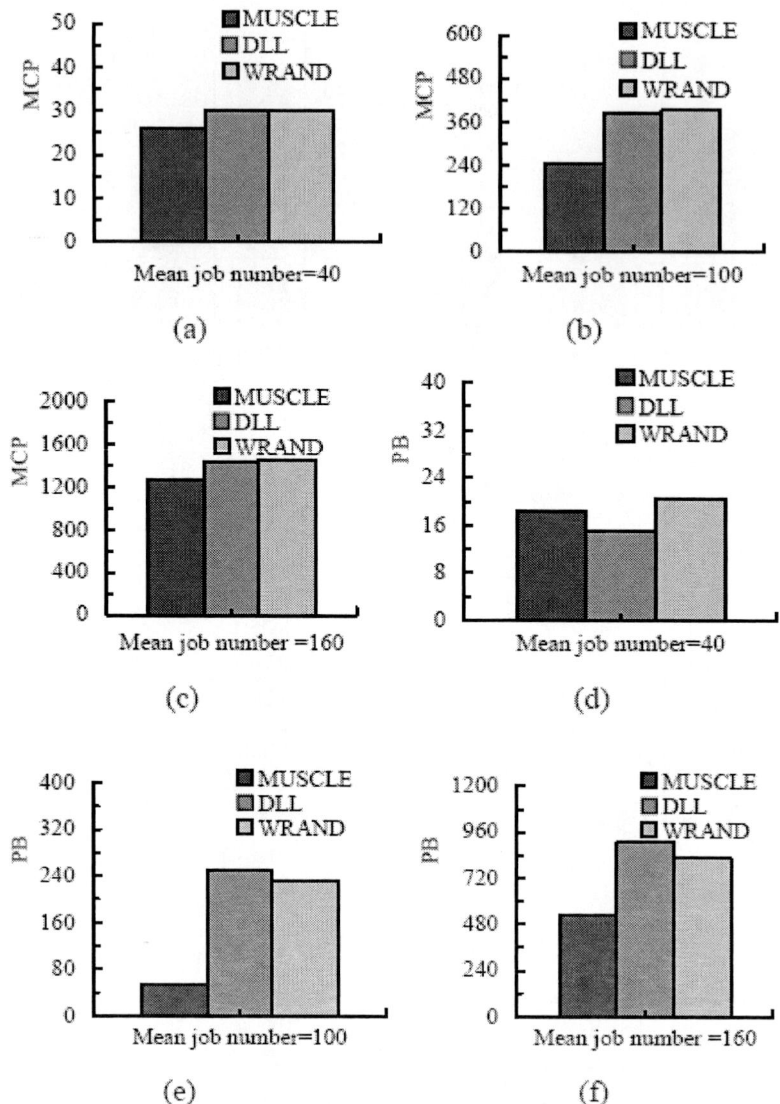

Figure 7. Performance comparison of MUSCLE, the DLL and WRAND in terms of MCP and PB; e_l/e_u=5/100; MIN_S/MAX_S=1/10; MIN_DR/MAX_DR=0/5.

Table 6. The weight combinations in figure 7

	C_1	C_2	C_3	C_4
(W^o, W^m, W^t)	(12, 3, 1)	(8, 3, 1)	(4, 3, 1)	(1, 3, 1)

As can be observed from figure 7.a, b and c, the performance of MUSCLE is superior to other policies in all cases in terms of the MCP. Furthermore, the advantage of MUSCLE over DLL and WRAND is different under different workload levels. When the mean job number in queue is 40, the advantage of MUSCLE over DLL is 15.6% while the advantage increases up to 57.5%, when the job number is 100. When the workload increases further and the mean job number in queue is 160, the advantage of MUSCLE over DLL decreases to 13.7%. These results are consistent with the experimental results demonstrated in Subsection 5.2.

The results for performance comparison in terms of the PB among MUSCLE, DLL and WRAND demonstrate the similar pattern as that shown in figure 6.c and d. When the mean job number in queue is 40, the performance of MUSCLE in terms of the PB is worse than that of DLL (MUSCLE still outperforms DLL in terms of the MCP though). Under other workloads, MUSCLE performs significantly better than DLL and WRAND. These are also consistent with the results in figure 5.b as well as figure 6.c and d.

Conclusion

A multicluster-level scheduler, called MUSCLE, is described in this chapter for the scheduling of parallel jobs with QoS demands in multiclusters, in which the constituent clusters may be located in different administrative organizations. Three metrics (over-deadline, makespan and idle-time) are combined with variable weights to evaluate the scheduling performance. MUSCLE is able to allocate jobs with high packing potential to the same cluster and further utilizes a heuristic to control the workload distribution among the clusters. Extensive experimental studies are carried out to verify the performance advantages of MUSCLE. The results show that compared with the traditional scheduling policies in distributed systems, the comprehensive performance (in terms of over-deadline, makespan and idle-time) is significantly improved and the jobs are well balanced across the multicluster.

References

[1] M. Barreto, R. Avila, and P. Navaux, "The MultiCluster model to the integrated use of multiple workstation clusters," *Proc. of the 3rd Workshop on Personal Computerbased Networks of Workstations* (2000) pp71–80

[2] R. Buyya and M. Baker, "Emerging Technologies for Multicluster/Grid Computing," *Proceedings of the 2001 IEEE International Conference on Cluster Computing*, (2001)

[3] J. Cao, D. J. Kerbyson, and G. R. Nudd, "Performance Evaluation of an Agent-Based Resource Management Infrastructure for Grid Computing," *Proc. of 1st IEEE/ACM International Symposium on Cluster Computing and the Grid*, (2001)

[4] J. Cao, D. J. Kerbyson, E. Papaefstathiou, and G. R. Nudd, "Performance Modeling of Parallel and Distributed Computing Using PACE," *Proceedings of 19th IEEE Intl Performance, Computing, and Communications Conference*, (2000)

[5] J. Cao, D. P Spooner, S. A Jarvis, G. R Nudd, "Grid load balancing using intelligent agents," *Future Generation Computer Systems, Special Issue on Intelligent Grid Environment: Principles and Applications*, 21, 1, (2005) pp135-149

[6] A. Dogan, F. Özgüner, "On QoS-Based Scheduling of a Meta-Task with Multiple QoS Demands in Heterogeneous Computing," *International Parallel and Distributed Processing Symposium* (2002)

[7] L. He, S. A. Jarvis, D. P. Spooner, G. R. Nudd, "Optimising static workload allocation in multiclusters," *Proceedings of 18th IEEE International Parallel and Distributed Processing Symposium* (2004)

[8] X. He, X. Sun, and G. Laszewski, "QoS Guided Min-Min Heuristic for Grid Task Scheduling," *Journal of Computer Science and Technology, Special Issue on Grid Computing*, 18, 4 (2003)

[9] B. G. Lawson and E. Smirni, "Multiple-queue Backfilling Scheduling with Priorities and Reservations for Parallel Systems," In the 8th *Job Scheduling Strategies for Parallel Processing* (2002)

[10] A. W. Mu'alem and D. G. Feitelson, "Utilization, predictability, workloads, and user runtime estimates in scheduling the IBM SP2 with backfilling," *IEEE Trans. Parallel and Distributed Syst.* 12, 6 (2001) pp 529-543

[11] G. R. Nudd, D. J. Kerbyson, E. Papaefstathiou, J. S. Harper, S. C. Perry and D. V. Wilcox, "PACE: A Toolset for the Performance Prediction of Parallel and Distributed Systems," *Intl Journal of High Performance Computing Applications* (1999)

[12] E. Shmueli and D. G. Feitelson, "Backfilling with lookahead to optimize the performance of parallel job scheduling," *Workshop on Job Scheduling Strategies for Parallel Processing* (2003) pp228-251

[13] D. P Spooner, S. A Jarvis, J Cao, S Saini, G. R Nudd, "Local Grid Scheduling Techniques using Performance Prediction," *IEE Proc. Comp. Digit. Tech.*, 150, 2 (2003) pp87-96

[14] D. Talby and D. G. Feitelson, "Supporting priorities and improving utilization of the IBM SP2 scheduler using slack-based backfilling," In *13th Intl. Parallel Processing Symp* (1999) pp513-517

[15] X.Y. Tang, S.T. Chanson, "Optimizing static job scheduling in a network of heterogeneous computers," *29th Intl Conference on Parallel Processing* (2000)

[16] W. Ward, C. Mahood, and J. West, "Scheduling Jobs on Parallel Systems Using a Relaxed Backfill Strategy," In the 8th *Job Scheduling Strategies for Parallel Processing* (2002).

In: Cyberinfrastructure Technologies and Applications
Editor: Junwei Cao
ISBN: 978-1-60692-063-3

Chapter 5

VIRTUALIZING SCIENTIFIC APPLICATIONS AND DATA SOURCES AS GRID SERVICES

Siegfried Benkner[1], Gerhard Engelbrecht[2]
Martin Köhler[3] and Alexander Wöhrer[4]
Institute of Scientific Computing, University of Vienna, Austria

ABSTRACT

Service-oriented Grid computing promises to change the way scientists will tackle future research challenges by offering advanced data and application services, providing transparent access to distributed heterogeneous data sources and to high-end computing facilities for performing computationally demanding and data-intensive modeling, simulation and analysis tasks. In this article we describe the Vienna Grid Environment (VGE), a service-oriented Grid infrastructure based on standard Web Services technologies for virtualizing scientific applications and data sources as Grid services that hide the details of the underlying software and hardware infrastructure. The VGE service provision framework adopts a component-based approach which supports the configuration of application and data services from a set of basic service components providing capabilities like job or query execution, data transfers, QoS negotiation, data staging, and error recovery. VGE relies on a business-oriented model to Grid computing based on a flexible QoS infrastructure, dynamic negotiation of service-level agreements, and on-demand access to Grid services. VGE has been developed and utilized in the context of several European projects for the realization of Grid infrastructures within medical and bio-medical application domains.

[1] E-mail address: sigi@par.univie.ac.at
[2] E-mail address: gerry@par.univie.ac.at
[3] E-mail address: koehler@par.univie.ac.at
[4] E-mail address: woehrer@par.univie.ac.at

INTRODUCTION

Grid computing infrastructures promise to offer seamless access to globally distributed IT resources including clusters, supercomputers, high-speed networks, scientific applications, data and information repositories as well as scientific instruments and sensors. Grid computing technologies cover a wide spectrum ranging from computational Grids, to Data Grids to Collaborative Grid environments. Computational Grids, for example the NSF's TeraGrid in the US, focus on the provision of superior computational capabilities beyond the limits of single computing systems by dynamically aggregating the power of a large number of individual computers in order to provide a platform for advanced high-performance and/or high-throughput applications. Data Grids, as for example the European Data Grid managed by CERN, focus more on managing and sharing of vast quantities of data for a globally distributed scientific community. Collaborative Grids aim at establishing a virtual environment which enables geographically dispersed individuals or groups of people to cooperate within virtual laboratories or to control and manage remote equipment, sensors, and instruments.

The evolution of Grid technologies over the last years was characterized by a shift towards the adoption of a service-oriented paradigm and the increasing utilization of commercial Web Services technologies as base technologies. The Open Grid Services Architecture (OGSA) [11], proposed by the Open Grid Forum, now relies on a service-oriented architecture and on standard Web service technologies. The Web Services Resource Framework [60], initiated by the Grid community and standardized by OASIS, represented a significant step for bringing Grid technologies and Web Services together. Service-oriented architectures facilitate the virtualization of heterogeneous IT resources by providing abstract interfaces that hide the implementation details of services and their underlying physical infrastructure and execution environment [19].

In this article we describe the Vienna Grid Environment (VGE), a service-oriented Grid infrastructure for virtualizing scientific applications and data sources as Grid services that provide a common set of generic interfaces, hiding the details of the underlying software and hardware infrastructure. VGE adopts a service-oriented architecture based on standard Web Services and Grid technologies and offers an infrastructure for the provision, deployment and management of application and data services. Application services may be accessed and configured on demand subject to a client's Quality of Service (QoS) requirements. VGE services are composed of a set of basic service components providing capabilities for data movement, job and query execution, monitoring, and error recovery. Resources virtualized by the VGE infrastructure include compute-intensive scientific applications and their associated compute resources (clusters, supercomputers, computational Grids), as well as scientific data and information resources, e.g. for long term persistence of research results and their connected input data. The VGE middleware offers a generic service provision framework that supports the provision and deployment of application and data services. By leveraging an intuitive graphical user interface, compute-intensive simulation and analysis applications available on clusters or other HPC hardware may be automatically exposed as Grid Web Services to be securely accessed by remote clients over the Internet. In a similar way, data services may be provided in order to enable transparent access to and integration of various information sources including relational data bases, XML data bases and flat files. For the construction of client side applications that interact with application and data services, a high-level client API hides the complexity of dealing with remote data and application services from the client application developer.

As opposed to a model of cost free sharing of Grid resources, VGE assumes a business-oriented model to Grid computing where clients are willing to pay for services, provided the required QoS levels can be offered. Service providers expose parallel simulation applications running on clusters as QoS-enabled application services, capable of dynamically negotiating with clients guarantees on service response time and price. Grid clients are able to choose from several service providers before agreeing to select a specific service. The associated VGE QoS infrastructure relies on a reservation based approach coupled with application specific performance models, advanced reservation mechanisms, and client-driven negotiation of service level agreements (SLAs).

VGE data services facilitate access to and integration of heterogeneous data sources. Data services support the same access patterns, transfer protocols and security mechanisms as application services and are built upon OGSA-DAI [16], the de-facto standard for data access and integration in Grid environments. Going beyond OGSA-DAI, VGE data mediation services offer transparent access to multiple data sources via a virtual global schema.

The VGE application service infrastructure has been initially developed in the context of the European GEMSS project [2][6] for Grid-enabling advanced medical simulation applications to improve pre-operative planning and near real-time surgery support. Currently, components of VGE are utilized and further developed in the EU project @neurIST [1], which aims at realizing an advanced service-oriented Grid infrastructure for the management of all processes linked to research, diagnosis and treatment of complex, multi-factorial diseases encompassing data repositories, computational analysis services and information systems handling multi-scale, multi-modal information at distributed sites.

The remainder of this chapter is structured as follows: In the next section we describe the overall architecture of VGE and the underlying service component and access models. We then describe application services, the associated QoS infrastructure and the negotiation of service-level agreements. This is followed by a description of data services with a special focus on data mediation as well as on distributed query processing. In addition, we provide an overview of the VGE client side programming environment. Finally, we report on the use of the presented technologies within the European projects GEMSS and @neurIST and discuss related work.

VGE Grid Infrastructure

The VGE Grid infrastructure comprises a generic service provision framework and a client-side Grid programming environment. The service provision environment enables service providers to expose compute intensive scientific applications available on HPC systems as well as distributed data sources as services that can be securely accessed on-demand by clients over the Internet. The client-side Grid programming framework offers a high-level application programming interface (API) that may be used to construct advanced Grid applications from application and data services.

Architecture

VGE adopts a service-oriented architecture and relies on standard Web Service technologies for virtualizing parallel applications and distributed heterogeneous data sources as services. VGE distinguishes two different types of services, application services and data services, which both are defined via WSDL and securely accessed using SOAP messages.

Application services virtualize compute intensive applications, usually parallel MPI codes available on clusters or other HPC systems. Using application services, clients may run applications on demand over the Internet and negotiate with service providers required QoS levels, for example, to ensure that an application job is completed before a certain deadline. VGE application services are comprised of generic service components providing interfaces for job execution, monitoring, data staging, error recovery and application-level quality of service support.

Data services virtualize data sources as Grid services, facilitating transparent access to and integration of heterogeneous data sources including relational data bases, XML data bases and flat files. Relying on advanced mediation mechanisms, data services provide transparent access to distributed data sources via a single integrated virtual schema. VGE data services are comprised of generic service components providing interfaces for query execution, data movement, and staging of result data.

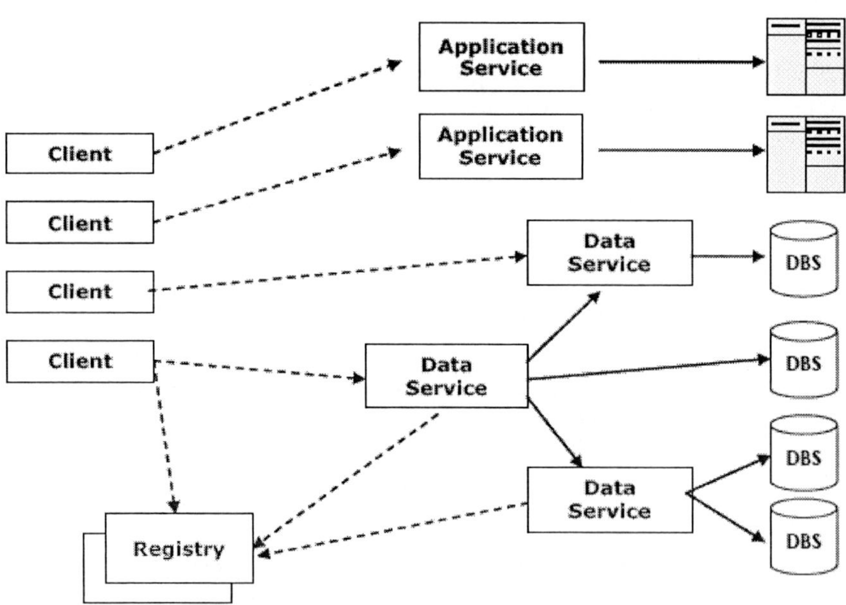

Figure 1. A VGE Grid comprising application and data services.

As shown in figure 1, a VGE Grid usually comprises multiple application and data services, multiple clients, one or more service registries for maintaining a list of service providers and the services they support, and a certificate authority for providing an operational PKI infrastructure and end-to-end security based on X.509 certificates.

The VGE environment provides mechanisms for service discovery based on services registries. Multiple service registries may be set up in order to enable service providers to publish their services, and clients to discover these services. VGE service registries are realized as Web Services. Service providers can describe their services using an arbitrary set of attributes (name/value pairs) which are published in the registry. These attributes are utilized during service selection to determine potential candidate services that might be able to fulfill a request.

VGE services, being Web Services, are hosted within a preconfigured service container, which comes with the VGE distribution. VGE client applications usually run on PCs or workstations connected to the Internet and make use of the VGE client API for interacting with services through the VGE middleware.

Figure 2 shows a layered abstraction of the VGE infrastructure. At the lower layer are a variety of data and information sources as well as simulation and modeling applications and the associated hardware infrastructure. These resources are virtualized through the Grid middleware layer and transparently accessed through the abstraction of application and data services by client applications. Application and data services are composed of generic service components providing basic capabilities for job and query execution, QoS, data mediation, data transfer, monitoring and error recovery.

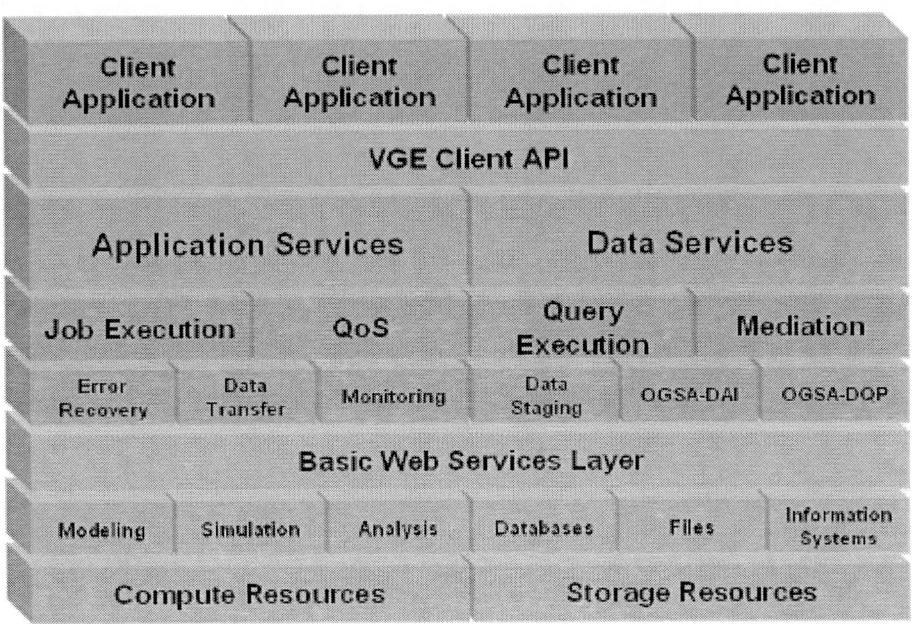

Figure 2. Layered abstraction of VGE.

The VGE middleware has been implemented mainly in Java and relies on standard Web Services technologies including SOAP, WSDL, WS-Addressing and WS-Security. VGE services comply with the Web Service Interoperability (WS-I) profile. The open-source frameworks Apache/Tomcat and Axis are utilized for service hosting and deployment. The VGE distribution is available for various Linux platforms as well as for Windows.

Service Component Model

VGE services adhere to a component model which supports the configuration and composition of application and data services based on a set of generic service components supporting common operations for data transfers between clients and services, job or query execution, data staging, monitoring, QoS management, and error recovery.

The service component model conceptually follows the WSRF model, which defines *a "generic and open framework for modeling and accessing stateful resources using Web services..."* [60].

The VGE service component model as depicted in figure 3 shows that a service is composed of a set of service components each accessing one ore more resources. A resource may be accessed by more than one service component in different ways, but each service component has only one associated resource. Every service component provides a separate WSDL document (interface) specifying the service component's operations. Consequently, the composite WSDL document, which is actually the union of all component WSDLs, provides the capabilities and operations of the overall service.

The communication among service components is handled by the VGE component framework, which provides mechanisms to directly access the operations of other service components via local method invocations as well as listeners and redirections of client SOAP calls using Apache Axis SOAP request and response handlers.

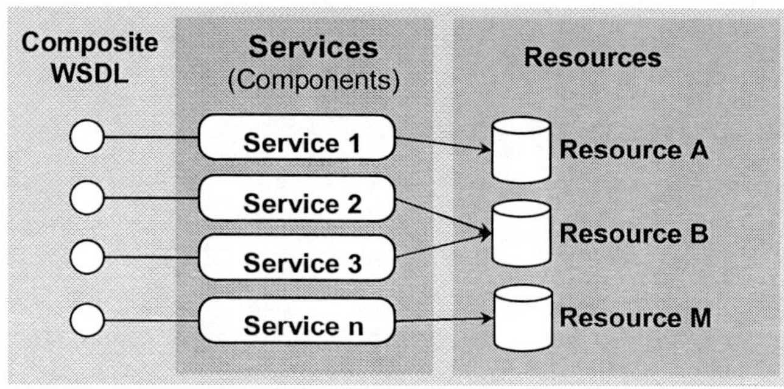

Figure 3. Service component model.

Service components provided by VGE include a *data transfer service component* for transferring files between a client and a service, an application execution service component for managing the execution of application jobs, a *QoS negotiation service component* for dynamic QoS negotiation, an *error recovery service component* for controlling check-pointing and restarting of an application, a *query execution service component* for managing queries to virtualized data sources, and a *data staging service component* for enabling direct data transfers between services.

The resources associated with these service components include directories within the service providers file system, applications and their associated files and scripts, the scheduler for managing the execution of jobs, and data bases and other information sources virtualized by means of data services.

Service Access Model

VGE relies on a purely client-driven approach for accessing application and data services via SOAP. All interactions of a client with services are initiated by the client and neither call-backs nor notification mechanisms are used. As a consequence, no site-firewall security compromises are necessary, since only one port for SOAP communication via HTTP (usually port 80) has to be open. For large file transfers SOAP attachments are utilized.

VGE services are inherently multi-threaded, i.e. if multiple clients access a service, a separate thread is generated for each client, and for each client a separate application job or data access is generated. This differs from other Grid service models which generate a separate service instance for each client using a factory pattern (cf. the now obsolete OGSI specification [11]). Session management and state handling is managed internally and transparently conceptually following the WSRF model but being implemented based on conversational identifiers and WS-Addressing mechanisms.

VGE application services may be configured on-demand to meet a client's QoS requirements based on a dynamic negotiation of service level agreements.

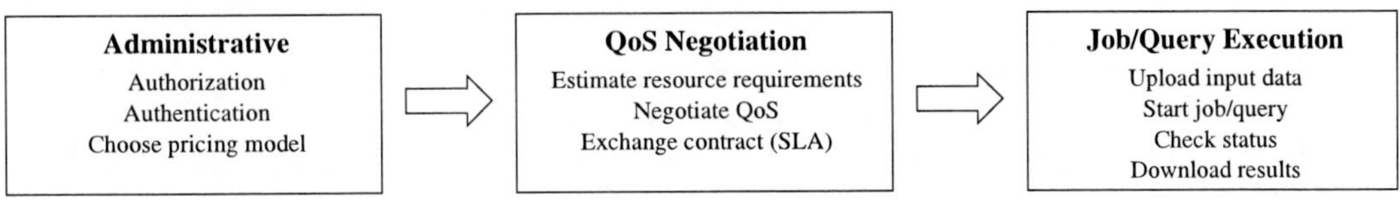

Figure 4. Service access workflow.

A basic VGE service access scenario is shown in figure 4. A client first performs administrative steps including authorization and authentication. The client then usually accesses a registry to find a set of candidate services and runs a QoS negotiation to determine a service that can fulfill its QoS constraints (e.g. execution time limits). Then the client uploads the input data, initiates job execution, and finally downloads the result.

Service Hosting

VGE provides a preconfigured service hosting environment based on the open-source frameworks Apache/Tomcat and Axis and a deployment tool for configuring and deploying services. The VGE deployment tool automates the provision of HPC applications and data sources as services. It offers an intuitive graphical user interface for enabling service providers to describe, configure, deploy and manage services without having to deal with the details of Grid and Web Service technologies.

The deployment tool enables to specify the service hosting environment, the security level, and to provide a service description. The description of an application service usually comprises the specification of input/output file names and of scripts for starting job execution and for gathering status information. Description of data services comprises the specification and configuration of the underlying data sources. VGE services support different security levels, including no security, basic security via SSL, and end-to-end security. Supported security technologies include PKI, HTTPS, WS Security and an end-to-end security protocol for separate encryption of sensitive portions of the transferred data. The information specified by the user with the deployment tool is stored internally in an XML service descriptor.Upon deployment of a service, a Web service with a corresponding WSDL interface is generated, deployed in the VGE hosting environment, and published in the VGE registry.

In a typical deployment scenario as shown in figure 5, services are deployed on a host in a de-militarized zone (DMZ). A second host with an installed Apache Web server is used for connecting VGE services with the Internet using a Tomcat connector (JK connector) between the Apache server and the Tomcat server. The resources accessed by service are usually located in the Intranet. In case of an application service, a resource will typically be an HPC application running on a cluster, which is accessed via an existing scheduling system. Data services usually access a DBMS hosted in the Intranet using a DB connector like JDBC.

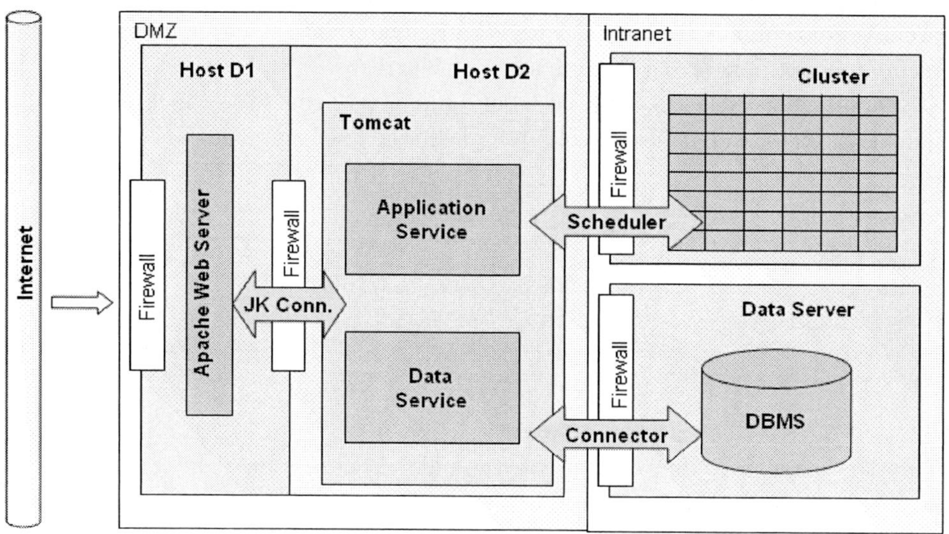

Figure 5. VGE service hosting scenario.

Virtualization of Scientific Applications

The VGE service provision framework enables service providers to virtualize HPC applications available on clusters or other parallel hardware as application services that can be accessed on-demand by clients over the Internet. Application services hide the details of their execution environment, providing abstract interfaces for managing job execution on remote computing resources.

The VGE application services infrastructure has been partially developed within the EU Project GEMSS [7], which devised a service-oriented Grid infrastructure that supports the Grid provision of advanced medical simulation. In order to enable the utilization of Grid-based simulation services during medical procedures, QoS support to ensure the timeliness of simulation results was a major requirement. Addressing these issues, VGE service may be configured with a flexible QoS negotiation service component, supporting dynamic negotiation of Service Level Agreements (SLAs) [3]. The QoS infrastructure enables clients to negotiate guarantees on service response times and price with service providers. The associated VGE QoS infrastructure supports the dynamic configuration of application services in order to meet the requirements of a client.

Application Services

Application services are constructed from a set of generic service components which provide common operations for controlling the execution of application jobs on a remote HPC system via a generic Web Services interface. In order to support time-critical usage scenarios VGE application services may be QoS-enabled relying on a client-driven negotiation of QoS guarantees based on Web Service Level Agreements [59]. The associated QoS infrastructure adopts a reservation based approach coupled with application specific performance models, advance reservation mechanisms, and client-driven negotiation of service level agreements.

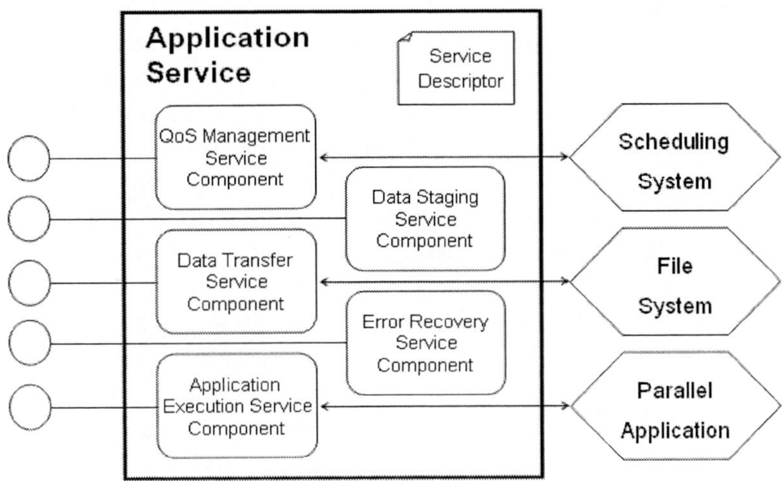

Figure 6. VGE application services.

As illustrated in figure 6, an application service may be composed of a *data transfer component*, which handles the transfer of input and output data between the client and the service, an *application execution service component*, which offers operations for executing application jobs on the associated compute resource(s) and for querying the status of an application job, and a *data staging service component* if input and/or output data should be transferred directly between services rather than between a client and a service. The QoS service component offers operations to support the dynamic negotiation of QoS guarantees as described in Section 3.2. Finally, an *error recovery service component* may be utilized in order to provide support for error recovery based on a checkpoint/restart mechanism.

The configuration and deployment of an application service for a specific application is usually accomplished using the VGE deployment tool as described later.

Data Transfer Service Component

The data transfer service component provides operations to transfer input and output data between a client and a service, usually in the form of ZIP files. A data transfer service component usually virtualizes a directory on the service provider's local file system where the corresponding application is installed. This directory is referred to as *working directory*.

The actual operations of this component are:

upload(input-filename, remote-input-filename)
download(remote-output-filename, output-filename)

Access to the working directory is under full control of the service provider and transparent to the client. For each client usually a separate sub-directory is generated within the working directory to store the input and output files. Clients do not have any means to access other parts of the service provider's file system.

Data Staging Service Component

The data staging service component provides operations to push and pull (stage) input and output data to and from one service to another service. Similar to the data transfer service component this component internally utilizes a directory on the local file system as resource. The actual operations of this component are defined as follows:

push(dest-service-URI, src-filename, dest-filename)
pull(src-service-URI, src-filename, dest-filename)

Application Execution Service Component

The application execution service component provides generic operations to control the execution of application jobs on a HPC system. It includes operations to start the execution of an application job, to monitor the status of a running job and eventually to kill a job if it does not terminate as expected. The application execution service component provides the following operations.

start()
getStatus()
kill()

When the operation start is invoked a corresponding start script is executed on the service side. The operation getStatus executes a pre-defined status-script and returns the output of this script to the client. The operation kill invokes a kill-script which terminates an application job.

Note that opposed to other Grid environments (like Globus or Unicore) clients have no way of sending executable scripts to a service nor do they have any means of controlling which scripts will be executed on the service providers machine. Only application providers can control which scripts are to be executed on their machines.

QoS Management Service Component

The QoS management service component provides a high level interface for QoS negotiation to clients, comprising the following operations:

requestQoSOffer(qos-request, request-descriptor)
cancelQoSOffer(qos-offer)
confirmQoSOffer(qos-offer)

QoS negotiation based on these operations is described in more detail later.

Error Recovery Service Component

Closely linked to the QoS management service component, this component provides support for check-pointing and re-starting of application jobs. To support error recovery, an application must already have checkpoint/restart functionality.

The actual operations of this component are:

checkpointUpload(filename)
checkpointDownload(filename)
restart()

Using these operations the client may download the latest checkpoint file of a failed application job and restart the application from this checkpoint using a different service provider.

QoS Management

QoS support for VGE application services is based on a flexible service-side QoS management component that can be configured by the service provider with an application specific performance model and a pricing model in order to determine the best possible QoS offer for a service request. To ensure the availability of computing resources for a service request, a resource manager, which provides an abstraction of a scheduling system with support for advance reservation, is utilized.

As can be seen from figure 7, the QoS manager interacts with the application performance model, the compute resource manager, and the pricing model using various XML-based descriptors as explained below. The QoS management component receives a QoS request and a request descriptor from a client, checks whether the client's QoS constraints can be met, and generates a corresponding QoS offer which is returned to the client. A *QoS request* document specifies the QoS constraints (e.g. earliest begin time of a job, latest finish time, price) of a client. A *request descriptor* contains input meta-data for a specific service request, typically describing the size of the application job to be run. A *QoS offer* specifies the QoS constraints offered by a service provider.

If the client decides to accept an offer, a corresponding QoS contract in the form of a service level agreement (SLA) is established and signed by both parties. The VGE QoS infrastructure utilizes a subset of the WSLA specification for representing QoS requests, QoS

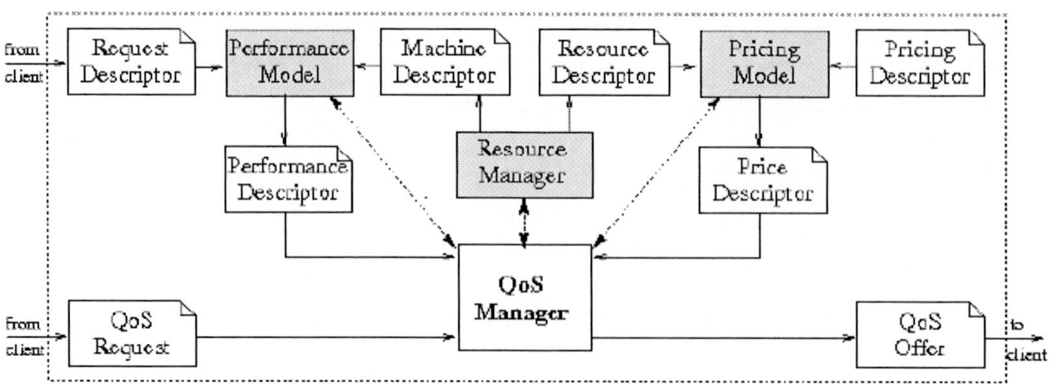

Figure 7. QoS management infrastructure.

offers and QoS contracts. Being machine readable, WSLA documents allow processing the QoS negotiation without the need of human intervention.

Performance Model and Pricing Model

The performance model is used during QoS negotiation to estimate the runtime and other performance relevant data for a service request. The performance model adheres to an abstract interface, taking as input a request descriptor and returning a performance descriptor. The request descriptor, supplied by the client, contains a set of request parameters, which characterize the input data of a specific service request. For example, in the case of an image reconstruction service, request parameters typically include image size and required accuracy. The returned performance descriptor usually contains estimates for the execution time, the required memory, and the required disk space. Assuming parallel MPI applications, the performance model is usually parameterized by the number of processors. It may be executed repeatedly with a varying number of processors until the time constraints set by the client are met, or the range of feasible processors as specified in the machine descriptor is exceeded.

The pricing model takes as input a resource descriptor and returns a price descriptor containing a concrete price for the resources specified in the resource descriptor. In the context of the GEMSS project, two alternative pricing models have been realized, a fixed price telephone pricing model where users are charged at a prearranged CPU hour rate, and a dynamic pricing model where the CPU hour rate is dependent on the current load levels the service provider is experiencing.

Performance models and price models only have to adhere to an abstract interface. The choice of model implementation, however, is left to the service provider, with each model implemented as a Java library that can be plugged in and selected dynamically. For example, a performance model could be implemented based on an analytical model, or where this is not feasible, a neural network or a database could be used to relate typical problem parameters to resource needs like main memory, disk space and execution time. The accuracy of performance models is critical to the ability of the QoS management system to select a large enough reservation to successfully run an application job.

Resource Manager

The resource manager provides an interface to the scheduler for obtaining information about the actual availability of computing resources. It is utilized internally by the QoS manager for creating temporary reservations during QoS negotiation. The compute resource manager takes as input the performance descriptor generated by the performance model, and generates a resource descriptor containing details about temporarily reserved resources. The resource descriptor is then used as input to the pricing model to determine the price for a service request. Different scheduling systems that provide support for advance reservation may be utilized with VGE, including the Maui scheduler [34] and the COSY scheduler [5].

Service Invocation

The invocation of application services usually comprises two phases, a QoS negotiation phase and the application job execution phase, as illustrated in the service invocation scenario shown in figure 8.

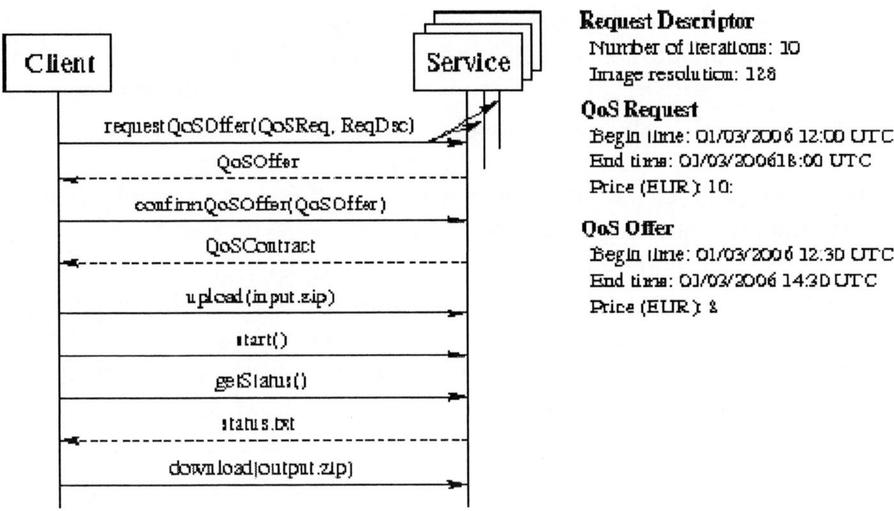

Figure 8. Basic service invocation scenario.

In an initial step, a client may access a registry to obtain a list of potential service providers for a specific application service. Then the client initiates a QoS negotiation to determine a service provider that can provide a service subject to the client's QoS constraints. The client invokes for each candidate service the operation `requestQosOffer`, passing along a request descriptor with input meta-data and a QoS request document with the required QoS constraints.

On the service side, the QoS manager executes the performance model and the price model to determine whether a client's time and price constrains can be met. If the required resources to run the client's job are available, a temporary resource reservation is made through the resource manager, and a corresponding QoS offer is returned to the client. On the client side, the QoS offers from different service providers are received and analyzed. The client confirms the best offer, or, if it is not satisfied with the offered QoS constraints, may set up a new QoS request with different constraints and start a new negotiation. If the client confirms an offer, the QoS manager confirms the temporary resource reservation made for the offer, signs the QoS contract and returns it to the client.

The client may then enter the actual service invocation workflow, which usually comprises the uploading of input data, starting of the remote application job, querying the status of the job, and finally, downloading the results.

Clients and service providers employ a relatively low level of trust in the negotiation. A service provider only makes a temporary reservation that will expire if the client takes too long to make a decision. Likewise service providers will be dropped from the negotiation if they fail to provide an offer in time.

Within the GEMSS project also more sophisticated negotiation strategies based on a closed-bid reverse English auction model have been realized by implementing different QoS negotiation service components [13].

Configuration and Deployment

An intuitive GUI-based deployment tool automates the task of exposing compute-intensive simulation applications available on clusters or other HPC hardware as Grid services that can be accessed on demand over the Internet. Usually no code changes are required, provided an application can already be executed in batch mode and that files in I/O operations are not accessed with absolute path names.

The deployment tool enables the service provider to specify the configuration details of an application (input/output file, job script, etc.) and the compute resources (memory, number of CPUs, etc.) that may be provided. Moreover, a set of QoS parameters which should be subject to QoS negotiation with clients (e.g., guaranteed begin time, guaranteed latest end time) may be specified. For each QoS parameter a corresponding QoS model (e.g. a machine specific performance model) has to be in place. The information specified by the user within the deployment tool is stored in an XML service descriptor and utilized by the VGE service provision environment to customize the generic service operations for the application at hand.

Upon deployment, the service provision environment automatically generates a Grid Service which encapsulates the application, publishes the corresponding WSDL document in a registry service, and deploys the service within the VGE hosting environment.

Figure 9 shows screenshots of the deployment tool during the configuration of a medical image reconstruction service. In a first step the service rpovider specifies the names of input and output files, the script to start an application, the working directory, which serves as root directory for all files transferred between clients and service, and a status script. In the second step, a set of request parameters (e.g., NumberOfIterations), a set of machine parameters (e.g. NumberOfProcessors) and a set of QoS parameters (e.g., BeginTime, EndTime) are specified.

Virtualization of Scientific Data Sources

The initial focus of Grid computing with an emphasis on compute-intensive tasks shifted in recent years towards more data intensive applications with significant processing of very

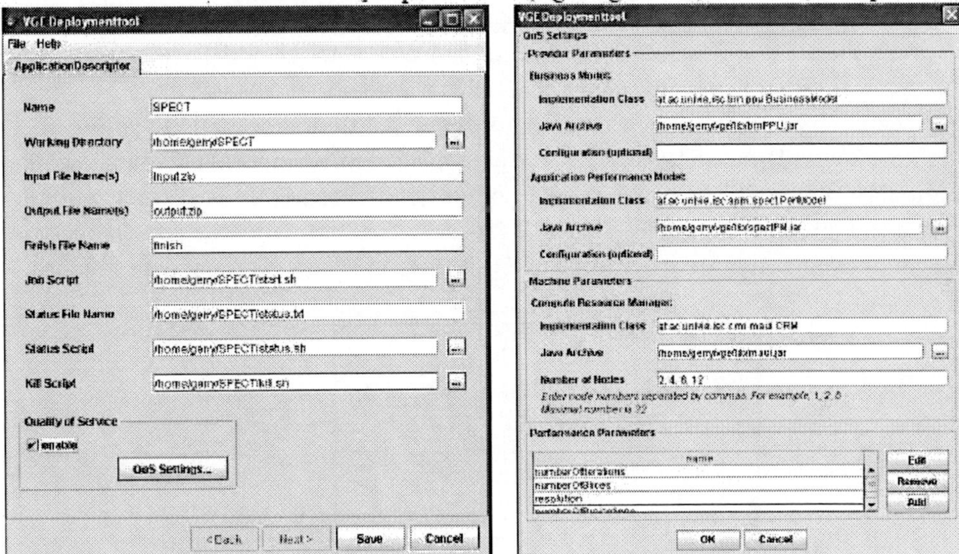

Figure 9. Configuration and Deployment of Application Services.

large amounts of data. Data management within Grids brings along additional complexities due the scale, dynamism, autonomy, and distribution of data sources [65]. In addition, more and more scientific disciplines are becoming data driven and scientists need to interact with large shared databases [63]. Grid-based scientific experiments produce more and more data that which increasingly is to be stored in structured form, often databases [29], as opposed to binary files in some proprietary format for non-shared single site usage. The need to overcome problems of accessing and integrating heterogeneous data, including issues pertaining to location, access rights and autonomy of data sources, and the requirement to discover new knowledge by combining existing data sources in new ways, becomes more and more important. All the complexities coming along with these issues, however, should be hidden from grid application developers via appropriate services.

Data Services

The VGE Grid middleware offers a generic service provisioning framework that supports the provision and deployment of data services. Data services virtualize heterogeneous scientific data bases and information sources as Web services, enabling transparent access to and integration of heterogeneous information sources including relational data bases, XML data bases and flat files. Data services resolve heterogeneities with respect to access language, data model and schema between different data sources, utilizing advanced data mediation and distributed query processing techniques based on OGSA-DAI [16], GDMS [61] and OGSA-DQP [17].

Data services support the same access patterns, client bindings, transfer protocols and security mechanisms as application services and are composed of basic service components as shown figure 10. The query execution service component provides common operations for managing remote query execution. A data transfer service component is utilized to transfer input data (usually specifying a query) and output data to and from a client. A data staging service component may be used if input and/or output data should be transferred directly between services rather than between a client and a service.

VGE data services may be set up in different configurations all providing the same interface to clients. Basic data services provide access to a single data source, while data mediation services establish a virtual data source for transparently accessing multiple data sources via a single virtual global schema. Data services usually provide an SQL-based query mechanism to clients and internally utilize OGSA-DAI to access physical data sources.

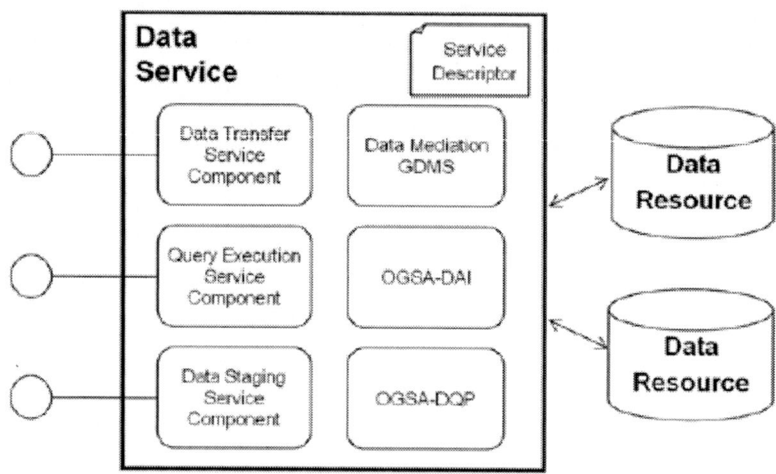

Figure 10. Data services.

Data Access Mechanisms

VGE data services have been developed on top of OGSA-DAI (Open Grid Services Architecture – Database Access and Integration), which is the de-facto standard Grid middleware assisting with access to and integration of data from separate data sources [16]. Clients access a VGE data service usually by uploading an OGSA-DAI perform document containing an SQL query, starting the query execution and downloading the results contained in an OGSA-DAI response document.

An OGSA-DAI *perform document* enables clients to describe the activities that they wish a data service to execute on data resources. The submitted requests can be based on multiple activities that can be combined and connected for querying, updating, transforming, delivering data, and schema extraction – similar to a simple workflow. Results and status information of the requested activities are delivered as OGSA-DAI *response documents*. Both perform and response documents are based on XML. OGSA-DAI offers a client framework for constructing perform documents and for processing response documents. This client API can be used for VGE data services as well.

An example of using OGSA-DAI is shown in figure 11, where the sqlQueryStatment activity executes a given query against the target relational data source and the following connected sqlResultsToXML activity defines the format in which the client expects the results, in our example the WebRowSet XML format defined by Sun.

```
Perform Document:
<perform xmlns="…" xmlns:xsi="…" xsi:schemaLocation="…">
   <sqlQueryStatement name="statement">
     <expression>
          select treatment from patient where id=10
     </expression>
     <resultStream name="statementOutputRS"/>
   </sqlQueryStatement>
   <sqlResultsToXML name="statementRSToXML">
    <resultSet from="statementOutputRS"/>
    <webRowSet name="statementOutput"/>
  </sqlResultsToXML>
</perform>

Response Document:
<response xmlns="…">
  <request status="COMPLETED"/>
  <result name="statement" status="COMPLETED"/>
  <result name="statementRSToXML" status="COMPLETED"/>
  <ns1:result name="statementOutput" status="COMPLETED">
        <![CDATA[<webRowSet>…</webRowSet>]]>
  </result>
</response>
```

Figure 11. Sample OGSA-DAI perform and response document.

The capabilities of OGSA-DAI can be extended and tailored by a user with additional functionality via extensibility points. Based on this mechanism, we have extended OGSA-DAI in order to provide support for data mediation (as described later) and for accessing other VGE data services as data resources.

In addition to the outlined OGSA-DAI access mechanism, VGE data services may be configured to support a *parameter-based* access mechanism. With the parameter-based mechanism the client sends a parameter file to the data service. On the service-side the parameter file is used to automatically fill a pre-defined query template, which is then used for accessing the data source. The result data of the query is then returned to the client in the form of a result data file. The structure and format of input parameter files and result data files are defined by the service provider during configuration of the data service depending on the requirements. The parameter-based access mechanism allows specialising a data service for a fixed query while the query-based mechanism allows more general queries.

Data Mediation

Clients often need to access related data from several distributed data sources. Access to such data sets should be possible in a way similar to accessing single data sources instead of forcing each application to interface directly with a set of individual databases and to resolve complex federation problems themselves. Therefore, VGE offers data mediation services to support transparent access to multiple heterogeneous mediated data sources via a single access point provided by a unified global data schema. The mediation process includes mechanisms for resolving heterogeneities with respect to access language, data model and schema, following a wrapper-mediator approach [21].

As Grids typically serve large scale applications spanning multiple administrative domains and producing an increasing amount of data, physical integration at one place is often not possible. Therefore we follow a virtual integration approach providing access to live data and functions, only temporarily materializing the result of queries at the time a query is posed. Our query reformulation approach follows the Global-as-View (GaV) [63] approach, where the global schema is described in terms of the local schemas. Since mappings between the local schemas and the global schema are fixed, changes in a local schema usually require revisions of the global schema. In the alternative Local-as-View [62] approach, local schemas are described as views of the global schema. This approach usually suffers from a more complex and costly query resolution process as compared to the simpler rule-unfolding of the GaV approach.

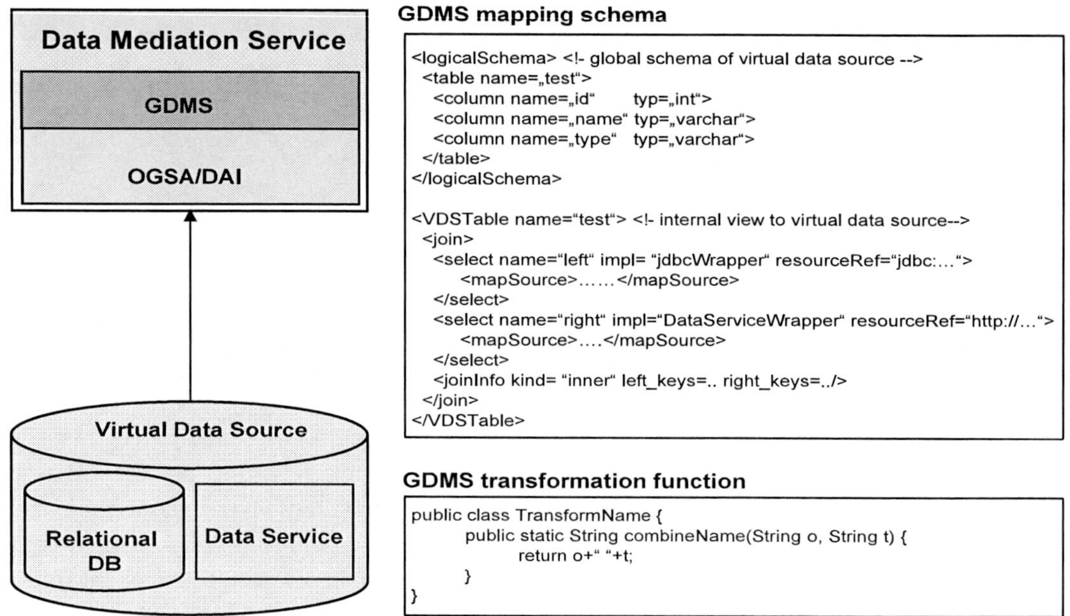

Figure 12. Query execution service component for a data mediation service.

VGE data mediation services utilize a Grid data mediation service (GDMS) component which is configured with a mapping schema (see figure 12). The mapping schema comprises the global logical data schema and mediation rules for decomposing a global query into queries against the target data sources and for combining the results. The mediation process can apply transformation functions on-the-fly written in Java for resolving several types of heterogeneities like different data representations, different data scales or different data structures. A simple example of such a transformation function is given in figure 12 illustrating how to combine parts of a name into a certain target format.

The mediated data sources are accessed internally using OGSA-DAI, which was extended by an additional data resource accessor for integrating other VGE data services as target sources permitting a hierarchical mediation structure. The fact that a data service integrates multiple data sources through mediation is transparent to clients. Clients can access the data mediation services as usual by uploading a perform document, starting query execution and downloading the results in an OGSA-DAI response document

Distributed Query Processing

In order to optimize access to multiple heterogeneous data sources, data mediation services support distributed query processing based on OGSA-DQP. Distributed query processing is transparent to clients and may be configured selectively by the service provider for certain data mediation services.

OGSA-DQP (Open Grid Services Architecture – Distributed Query Processing) is a Grid middleware based on OGSA-DAI supporting queries over OGSA-DAI data services and other services [17]. Data intensive requests to multiple data services can be parallelized and distributed by OGSA-DQP. OGSA-DQP contains a Grid Distributed Query Service, also called coordinator, which accepts client requests for queries to distributed data sources and generates a partitioned query plan. OGSA-DQP then uses multiple Query Evaluation Services (evaluators) each executing a partition of the generated query plan. Evaluators can execute the query plan partitions in parallel and then combine the results.

Note that opposed to VGE data mediation services, OGSA-DQP does not support a mediated global view over distributed data sources. As a consequence, using OGSA-DQP directly, would force clients to deal explicitly with the local data schemas and to specify how to integrate them within complex distributed queries. By combining the distributed query processing features of OGSA-DQP with the data mediation mechanisms of GDMS we seamlessly integrate the best of both systems.

Figure 13 illustrates VGE data services with distributed query processing support. In our approach GDMS performs the data mediation process including query reformulation from a query against the global schema into queries against the integrated data sources. The resulting query to the distributed data sources is then passed to OGSA-DQP for optimized distributed query processing. OGSA-DQP partitions the query plan using information about the individual data sources in order to optimize complex mediation operators like unions and joins. The OGSA-DQP coordinator controls the execution of the distributed query plan on a set of evaluation services. Each evaluation service processes a separate partition of the query plan. Finally, OGSA-DQP combines the results from the evaluation services and GDMS transforms the results according to the global schema.

The utilization of OGSA-DQP is transparent to clients and can be configured selectively by the service provider for certain data mediation services.

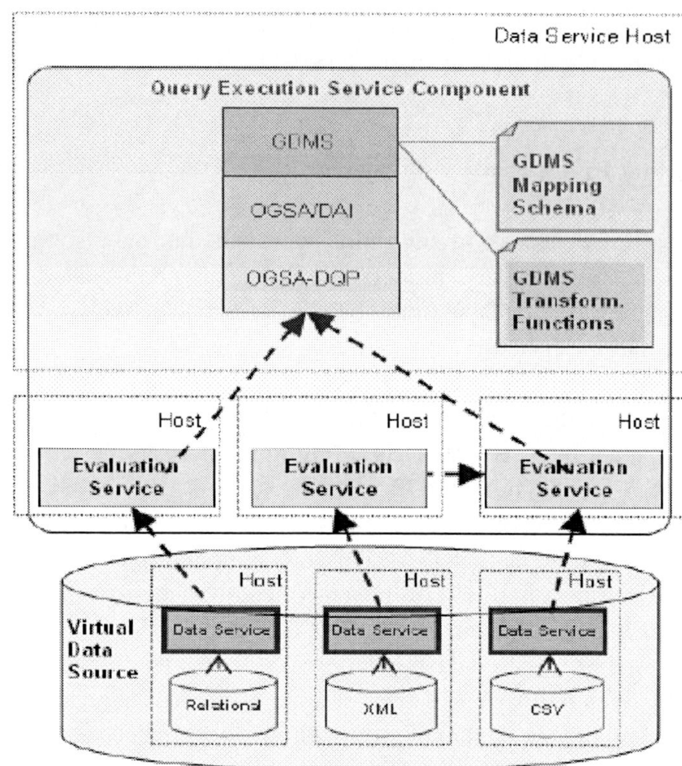

Figure 13. VGE data mediation service with distributed query processing.

Client Environment

Client applications that access VGE data services and application services could be built from scratch by only relying on the WSDL descriptions of services. Such a low-level approach would put the burden of interacting with VGE services on the client developer, resulting in complex and error-prone client applications. To overcome this obstacle, VGE provides a high-level client-side programming toolkit supporting languages, such as Java, C# and C++. Hence client-side applications that interact with VGE services can be built without having to deal with the complexities of Grid and Web Services technologies.

Beginning with an overview of the high-level client API, this section summarizes how the VGE client environment may be used to utilize application and data services. Moreover, we describe additional features that are provided to access services using a command-line interface or a Web-based interface.

High-Level Client APIs

In order to support the construction of client-side applications that interact with application services and data services, VGE provides a high-level client API with bindings for JAVA, C and C# (.NET). The client API hides most of the details of dealing with remote services from the client application developer.

The central abstraction provided by the client API is the ServiceProxy interface which specifies all methods required for transparently accessing application and data services from a client application. Moreover, the client API provides a set of classes for dealing with various aspects of the VGE environment at a high-level of abstraction hiding the details of the underlying interaction with VGE services.

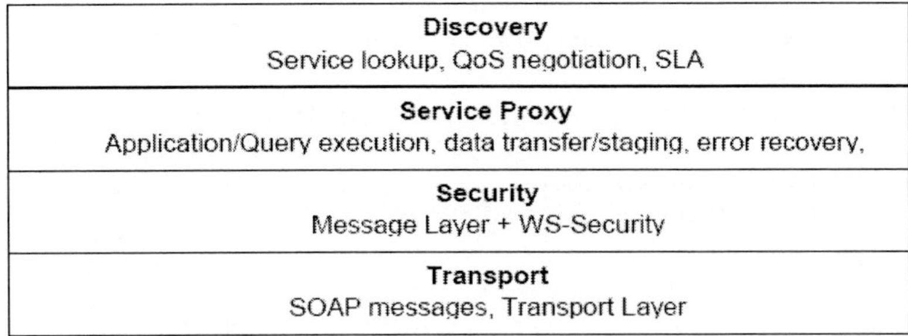

Figure 14. Client API abstraction layers.

As shown in figure 14, the programming interface is structured into several layers. The bottom layers implement data marshalling, message generation, signing, and encryption, while the top layers provide abstractions for service discovery and all service interactions provided by the ServiceProxy. Using these high-level abstractions, users can more easily develop client applications that interact with data and application service. The ServiceProxy interface comprises methods for handling transfers of input and output data, for job or query execution, for monitoring, as well as for error handling.

Client-Side Access to VGE Services

Different methods for discovering and selecting services are provided. As the simplest method, a service may be selected directly by providing a distinct service location, i.e. an end-point URI. A more sophisticated method of service selection requires specifying a set of service attributes (e.g. application category) which are used to locate services in VGE registries with matching attributes. Moreover, an application service may be selected from a set of potential services as the result of a QoS negotiation process subject to user specified QoS criteria. For this purpose the method getService may be invoked with a request descriptor and a QoS descriptor as input parameters. The details of accessing service registries and of QoS negotiation are handled transparently to the user by the API.

Figure 15 shows a simplified code excerpt for a typical client application accessing an application service that is dynamically discovered subject to certain QoS constraints specified within a QoS descriptor (qosDsc). The client specifies a QoS request containing QoS constraints (e.g. earliest begin time of a job, latest finish time, price) and a request descriptor containing input meta-data for a specific service request, typically describing the size of the application job to be run. Additionally the client has to specify a set of service attributes (e.g. application category) and an URL referring to a VGE registry for discovering appropriate

```
// initialize request descriptor
RequestDescriptor reqDesc = new RequestDescriptorImpl();
reqDesc.setPerformanceParameter(…)
…
// initialize QoS descriptor
QoSDescriptor qosDesc = new QoSDescriptorImpl();
qosDesc.setObjectiveBeginTime(...)
qosDesc.setObjectiveEndTime(…)
…
// initialize list of service registry and its attributes
String registryURI = " http://localhost/MyRegistry";
Attribute[] attributes = new Attribute[1];
attributes[0] = new Attribute("service.category", "MyCategory");

// initialize VGE proxy by QoS negoatiation
ServiceProxy serviceProxy = ServiceProxy.getService
(registryURI, attributes, qosDesc, reqDesc)

// upload local file (named upload.dat) with remote file-name "inputFile"
serviceProxy.upload(new File("upload.dat"), "inputFile");

// start remote service execution (start job)
serviceProxy.start();

String status = null;
// query the status of the service as long as the service runs
(is not finished) or an error occurs
while (((status = serviceProxy.getStatus()) != serviceProxy.FINISHED) and and
 (serviceProxy.getStatus() != serviceProxy.ERROR)) {
 // service runs ...
}
// finally download the results, save the remote file named
"outputFile" in the local file named
// "output.dat"
serviceProxy.download("output.dat", "outputFile");
```

Figure 15. Example of VGE client code interacting with an application service.

VGE services using the high-level client API. After a specific service has been selected, the client starts the application execution workflow including uploading of the input file, starting the application on the remote computing resource. The client queries the application status using the API and downloads the result files to the local machine.

A data service client looks similar to an application service client, except that the input and output files are OGSA-DAI perform and result documents. These XML documents are generated and manipulated using the OGSA-DAI API, which can be seamlessly integrated in the VGE client API.

Additional Features

Due to the miscellaneous use of the VGE system a number of additional features have been developed for certain environments. A command-line based interface has been realized to be used in scripts, where each operation can be executed separately using different command-line parameters. Opposed to this low level realization, a Web-based client is also available, which provides a simple graphical user interface in a web browser to be used in "drag'n'drop"-style to submit inputs, start jobs and download outputs.

Besides the already mention client APIs realized in Java, there are client bindings for .NET in C#, which enables even Office applications to use VGE Grid services via macros. C++ applications might also be extended to use VGE services utilizing a C++ binding based on gSOAP.

Additionally, VGE clients support several security mechanisms as provided by the VGE services and monitoring of job/query execution. The high-level client API is based on the service component model as outlined previously and therefore extensible with new service component functionalities.

In summary, the different VGE client interfaces enable client application programmers to utilize VGE service in a more convenient way.

APPLICATIONS

In this section we provide a brief overview of the EU project GEMSS [7], which developed a test-bed for Grid-enabled medical simulation services based on QoS-enabled application services as described earlier. Moreover, we outline the major objectives of the EU project @neurIST [1] which aims to develop an advanced Grid-based IT infrastructure in the biomedical domain.

Grid Enabled Medical Simulation Services

The GEMSS Project [7][2] was concerned with the development of a secure, service-oriented Grid infrastructure for the on-demand provision of advanced medical simulation and imaging applications as Grid services for improved pre-operative planning and near real-time surgical support. Key aspects of the GEMSS developments included negotiable QoS support for time-critical Grid services, flexible support for business models, and security at all levels in order to ensure privacy of patient data as well as compliance to EU law.

To enable the use of Grid services in a clinical environment, predictability of response times is of utmost importance. As a consequence, a flexible QoS support infrastructure for application services as described previously is required in order to provide support for explicit response time guarantees for simulation services which are executed remotely on a Grid host.

The GEMSS project focused on a business-oriented Grid model, where services are offered by service providers within an economic context. Unlike a traditional Grid model of free resource sharing, users do not provide resources to the community, but instead buy services from service providers. Service providers do not cooperate with each other nor disclose information about the availability of their resources, but wait until customers have a need to perform computations, and then propose a service level agreement.

The GEMSS project Grid-enabled six complex medical simulation applications including maxillo-facial surgery simulation, neuro-surgery support, radiosurgery planning, inhaled drug-delivery simulation, cardio-vascular simulation and advanced image reconstruction [10]. At the core of these bio-medical simulation applications are computationally demanding methods such as parallel Finite Element Modelling, parallel Computational Fluid Dynamics and parallel Monte Carlo simulation, which are realized as remote Grid services running on clusters or other parallel computing platforms. These services have been realized based on the VGE application service infrastructure.

The medical simulation jobs within GEMSS utilize patient data, which can only be legally released for processing by a hospital under very clear and restrictive conditions [57]. As a consequence, best practice security mechanisms are of paramount importance. In addition to the technical and legal requirements there are a number of practical considerations involved with medical simulation work. Some of the GEMSS applications require the patient to come in for some medical scans and stay at the hospital until the results of the medical simulations can be computed and analysed by medical physicists. Other applications require the availability of live simulation results during surgery. As a consequence, there is a need for high quality of service and a reservation capability to ensure sufficient compute resources are pre-booked in advance of patient consultations. These requirements could be successfully addressed with the flexible QoS infrastructure provided by VGE.

A major objective of GEMSS was also to address the legal requirements of dealing with patient specific data in a distributed environment represented by a Grid infrastructure. As our investigations showed, legal constraints (cf. EU Directive 95/46 which to wholly or partly automated processing of personal data [57]) have a major impact on how Grid technologies are applicable within a medical context. As a consequence, SLA negotiation is a major requirement as it allows the represenatation and exchange of an electronic contract, which can be used as proof that a service provider has accepted the legal responsibilities involved in processing of a job.

The GEMSS project, which successfully concluded in 2005, demonstrated that Grid technologies are applicable within a medical context, provided legal issues with respect to the processing of patient-specific data are taken into account carefully.

Integrated Biomedical Grid Infrastructure

The @neurIST project [1] aims to create an IT infrastructure for the management of all processes linked to research, diagnosis and treatment development for complex and multi-factorial diseases. Although the focus of @neurIST is on one such disease, cerebral aneurysm and subarachnoid haemorrhage, the core technologies are generic and transferable to other areas. The @neurIST infrastructure encompasses data repositories, computational analysis services and information systems handling multi-scale, multi-modal information at distributed sites. The @neurIST project bases its developments on a service-oriented Grid middleware, leveraging developments from GEMSS, VGE and InnerGrid [9], to facilitate access to computing and data resources by providing support for access control, advanced security, and quality of service guarantees.

Figure 16 shows a high-level overview of the @neurIST architecture. At the top layer, there are a set of integrative application suites, for integrative rupture and risk assessment, for linking genetics to diseases, for virtual endovascular treatment planning, and for multimodal data processing and image fusion. The middle layer constitutes the service-oriented Grid middleware which comprises a variety of advanced data and compute services.

On the one hand, @neuInfo comprises tools for constructing virtual data sources that enable transparent access to distributed heterogeneous biological and clinical data sources. Virtualization of data sources through Grid services is based on the VGE data services infrastructure and will be extended towards semantic, ontology-based systems, which are key to the integration of multi-level (from molecular to population) health related data. On the other hand, @neuCompute provides a compute grid infrastructure to accomplish computationally demanding modeling and simulation tasks utilizing VGE application services and InnerGrid services. At the lower layer are a large variety of data and information sources as well as simulation and modelling applications and the associated hardware infrastructure. These resources are virtualized through the Grid middleware layer and transparently accessed through the abstraction of services by the application suites.

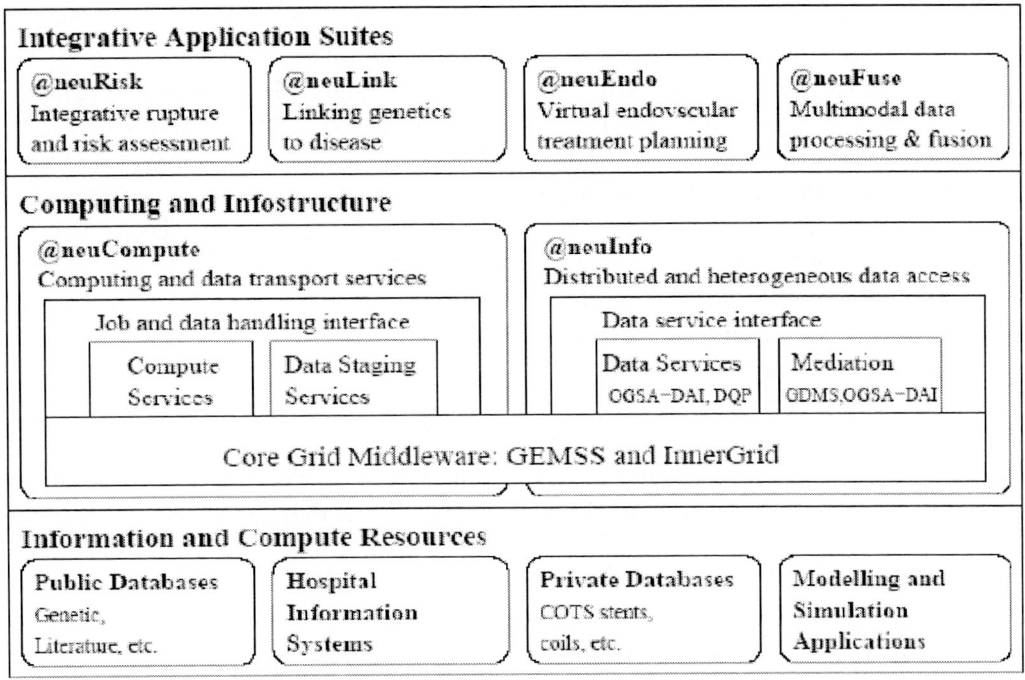

Figure 16. The @neurIST grid infrastructure.

Whereas the GEMSS project focused on the on-demand provision of compute-intensive simulation applications, the @neurIST middleware has the additional requirement of facilitating access to and integration of distributed heterogeneous data and information sources. For this purpose @neurIST utilizes VGE data services and relies on VGE's data mediation facilities for transparent access to distributed data sources via virtual global schemas. Going beyond the developments of GEMSS, @neurIST addresses the utilization of semantic technologies within a service-oriented Grid middleware. As a basic requirement, an

RDF/OWL-based domain ontology [4] has been developed to provide a common terminology capturing the semantics of all medical, bio-molecular, genetic and epidemiological entities related to aneurisms. Based on this ontology data and compute services will be semantically annotated, allowing to discover data services based on ontology concepts. Moreover, the domain ontology will serve as semantic glue for data integration. Annotating the schemas of data services with semantic information will facilitate the integration of semantically equivalent but structurally different data sources [13].

RELATED WORK

The OGSA Glossary of Terms defines virtualization as making a common set of abstract interfaces available for a set of similar resources, thereby hiding differences in their properties and operations, and allowing them to be viewed and/or manipulated in a common way. This has been done already for quite some time for physical resources, such as storage and servers, and has been extended in the last years by the Grid for workload and information virtualization [19].

There exist a number of grid infrastructures trying to ease the use of complex and heterogeneous gird environments. Some are based on WSRF like Globus [50] and UNICORE [53], others on plain WS technology like the EU project EGEE [51] middleware gLite [52] and the UK initiative OMII [49]. OMII Europe has the vision to harvest open-source, WS-based, Grid software from across Europe and to supply these key grid services in a form that will allow them to interoperate across the former mentioned heterogeneous infrastructures. Our VGE infrastructure uses plain Web Services technology and incorporates QoS components not available in the other infrastructures.

Moore and Baru [13] present an overview of ongoing research in the area of virtualization services for data intensive Grids. Initially focusing on files [18], there is now a trend to access and integrate more second-hand and public available data sets of multiple domains [8], often stored in pre-existing and independently operated databases [29]. OGSA-DAI (Open Grid Services Architecture – Database Access and Integration) is the de-facto standard Grid middleware assisting with access and integration of data from separate data sources [16]. OGSA-DQP (Open Grid Services Architecture – Distributed Query Processing) is a Grid middleware based on OGSA-DAI supporting queries over OGSA-DAI data services and other services [17]. For a good discussion on available grid data management systems and services please review [54].

Quality of service has been investigated in various contexts [40][44] while traditional QoS research usually focuses on network level QoS [46] with various techniques to guarantee service levels as an improvement to best effort protocols. The work in [40] focuses on QoS support for distributed query processing, providing significant improvements compared to best effort based systems. Our QoS infrastructure aims to guarantee service runtimes in advance via advance resource reservations. A good discussion regarding the demand for advance reservation in Grid applications can be found in [42] and [45]. In [39] the authors discuss how resource allocation in Grids affects global and local service level agreements based on enterprise applications, while our work focuses on medical applications. The work presented in [42] deals with a QoS-based Web Services architecture comprising QoS-aware components which can be invoked using a QoS negotiation protocol. Our work focuses on Grid provision of high performance computing application services as opposed to traditional Web Services.

In [39] a Grid QoS management framework is proposed that mainly focuses on service discovery based on QoS attributes. Conceptually, this work deals with both application-level and network-level QoS but does not specifically address response time and price guarantees for long running Grid services. In [43] a model for QoS-aware component architecture for Grid computing is proposed. The work in [41] proposes a SOAP message tracking model for supporting QoS end-to-end management in the context of WSBPEL [47] and SLAs. This work however does not address long running Grid services and SLA negotiation and the trade-off between response time and cost as supported by our QoS-aware services.

Data integration has been studied for quite a while [20]. For a recent report on state of the art middleware for data interoperability and management read [28]. A specialised platform for bioinformatics is SRS [48], allowing to integrate databanks and analysis tools.

In an environment like the Grid with its autonomous and heterogeneous sources with different capabilities, a wrapper-mediator approach [21] is often followed. OGSA-DAI can be seen as a wrapper while OGSA-DQP is the mediator (with a main concern on efficient query execution) combining the data. Optimization of this distributed/parallel query processing [22] is extremely relevant. Recently, adaptive query processing strategies [23] which can react to changes in the environment, e.g. data [25], network [27], computation [26], during execution gain more and more attention, also for the Grid [23]. Semantic technologies have already been used for quite some time to support semantic data integration [55]. More recently semantic registration and annotation [56] of data resources are used to support discovery and integration of relevant data sources.

Grid Technology is increasingly utilized in life sciences and medical applications are often mentioned as the "killer applications" for the Grid. Hence a number of projects deal with specific aspects in the field. In the bio-medical domain the EU BioGrid Project [30] aims to develop a knowledge grid infrastructure for the biotechnology industry. The main objective of the OpenMolGRID Project [31] is to develop a Grid-based environment for solving molecular design/engineering tasks relevant to chemistry, pharmacy and bioinformatics. The EU MammoGrid Project [33] builds a Grid-based federated database for breast cancer screening. The UK e-Science myGrid Project [32] develops a Grid environment for data intensive in-silico experiments in biology. While most of these projects focus on data management aspects, our efforts focus on the computational aspects of the Grid including Quality of Service (QoS) as well as virtualizing compute and data resources.

Other projects in the bio-medical field which also focus more on the computational aspect of the Grid include the Swiss BiOpera Project [35], the Japanese BioGrid Project [36], and the Singapore BioMed Grid [37]. The US Biomedical Informatics Research Network [38] initiative fosters distributed collaborations in biomedical science focusing on brain imaging of human neurological disorders and associated animal models.

The LEAD project (Linked Environments for Atmospheric Discovery [6]) addresses the integration of meteorological data, forecast models, and analysis and visualization tools in order to support the exploration of weather phenomena like severe storms. LEAD bases its developments on a service-oriented architecture that relies on a Web Services architecture similar to VGE.

CONCLUSION

In this chapter we presented the Vienna Grid Environment, a generic Grid infrastructure for virtualizing scientific applications and data sources as services that hide the details of the underlying software and hardware infrastructures. VGE is based on standard Web Services technologies and has been implemented based on corresponding open-source technologies. VGE application service virtualize scientific applications running on clusters or other HPC systems by providing common operations for transparently managing the execution of applications jobs on remote machines. The complex details of accessing HPC systems are hidden from users and client-side applications. VGE application services support an on-demand supercomputing model where the selection of service providers and the configuration of application jobs are performed automatically by the VGE middleware subject to user-specified QoS constraints. The associated QoS infrastructure is based on automated SLA negotiation, advance resource reservation, and application-specific performance models. VGE data services address the complex problems associated with access to and integration of distributed heterogeneous data sources. VGE data services virtualize distributed data sources by providing a global virtual schema to clients, hiding the details of the underlying physical data sources. VGE data services implement advanced data mediation techniques and provide support for distributed query processing. Application services and data services provide the same access patterns, client bindings, transfer protocols and security mechanisms as application services, simplifying their utilization within client-side Grid applications.

The VGE application service infrastructure has been successfully utilized within the European GEMSS project for the development of Grid-based medical simulation services. The VGE data service environment is currently being used in the European @neurIST project for the development of an advanced IT infrastructure that provides personalized risk assessment for multi-factorial diseases by integrating advanced analysis and simulation services with distributed heterogeneous data sources handling multi-modal, multi-scale information at distributed sites. Within the @neurIST project application services and data services are being further enhanced through the integration of semantic technologies, which characterizes a common future trend in Grid computing.

REFERENCES

[1] The @neurIST Project. Integrated Biomedical Informatics for the Management of Cerebral Aneurysms. EU Integrated Project, IST-2004-027703, http://www.aneurist.org/

[2] S. Benkner, G. Berti, G. Engelbrecht, J. Fingberg, G. Kohring, S. E. Middleton, R. Schmidt. GEMSS: Grid-infrastructure for Medical Service Provision, Methods of Information in Medicine, Vol. 44/4, 2005.

[3] S. Benkner and G. Engelbrecht. A Generic QoS Infrastructure for Grid Web Services. In *Proceedings of the International Conference on Internet and Web Applications and Services*, Guadeloupe, French Caribbean, February 2006. IEEE Computer Society Press.

[4] M. Boeker, H. Stenzhorn, K. Kumpf, P. Bijlenga, S. Schulz, S. Hanser. The @neurIST Ontology of Intracranial Aneurysms: Providing Terminological Services for an Integrated IT Infrastructure. AMIA 2007 Annual Symposium (AMIA 2007), Chicago, USA, November 2007.

[5] J. Cao, F. Zimmermann. "Queue Scheduling and Advance Reservations with COSY", *Proceedings of the International Parallel and Distributed Processing Symposium*, Santa Fe, New Mexico, 2004

[6] D. Gannon et al. Building Grid Portal Applications from a Web-Service Component Architecture, *Proceedings of the IEEE. Special Issue on Grid Technology*, 2005.

[7] The GEMSS Project: Grid-enabled medical simulation services. EU IST Project IST-2001-37153, 2002--2005. http://www.gemss.de/

[8] A. T. Tony Hey, "The data deluge: An e-science perspective", in *Grid Computing - making the global infrastructure a reality*. John Wiley and Sons, 2003, pp. 809–824.

[9] InnerGrid. Architecture and Services. http://www.gridsystems.com/images/pdf/IGV Architecture.pdf, 2006.

[10] D. M. Jones, J. W. Fenner, G. Berti, F. Kruggel, R. A. Mehrem, W. Backfrieder, R. Moore, A. Geltmeier. The GEMSS Grid: An evolving HPC Environment for Medical Applications, HealthGrid 2004, Clermont-Ferrand, France, 2004.

[11] I. Foster et al. The Open Grid Services Architecture, Version 1.5., Open Grid Forum, GFD-I.080, July 2006.

[12] I. Foster, C. Kesselman, J. Nick, S. Tuecke. "The Physiology of the Grid: An Open Grid Services Architecturefor Distributed Systems Integration", Open Grid Service Infrastructure WG, Global Grid Forum,2002.

[13] K. Kumpf, A. Woehrer, S. Benkner, G. Engelbrecht, J. Fingberg. A Semantic Mediation Architecture for a Clinical Data Grid. In *Grid Computing for Bioinformatics and Computational Biology*, Wiley Series in Bioinformatics, ISBN-13: 978-0-471-78409-8 - John Wiley and Sons, 2007.

[14] S. E. Middleton, M. Surridge, S. Benkner, G. Engelbrecht. Quality of service negotiation for commercial medical Grid services. *Journal of Grid Computing*, Springer Verlag, ISSN 1570-7873, 2007.

[15] R. W. Moore and C. Baru, "Virtualization services for data grids," in *Grid Computing - making the global infrastructure a reality*, John Wiley and Sons, 2003.

[16] M. Antonioletti et al., "The Design and Implementation of Grid Database Services in OGSA-DAI". *Concurrency and Computation: Practice and Experience*, Volume 17, Issue 2-4, Pages 357-376, February 2005

[17] M. N. Alpdemir et al., "OGSA-DQP: A Service for Distributed Querying on the Grid", *9th International Conference on Extending Database Technology*, LNCS, Volume 2992, pp858-861, 2004

[18] A. Rajasekar et. al, "Storage Resource Broker - Managing Distributed Data in a Grid", *Computer Society of India Journal, Special Issue on SAN*, Vol. 33, No. 4, pp. 42-54, Oct 2003

[19] M. Haynos, "IBM's take on Grid, virtualization and SOA", GridToday, 2006

[20] A.P. Sheth, J.A. Larson, "Federated database systems for managing distributed, heterogeneous, and autonomous databases",*ACM Comput. Surv.* 22 3 (1990) 183–236

[21] G. Wiederhold, "Mediators in the architecture of future information systems", 1992

[22] D. Kossmann, "The state of the art in distributed query processing", *ACM Comput. Surv.* (CSUR) 32 (4), 2000
[23] A. Gounaris, N. W. Paton, R. Sakellariou, A.A. Fernandes, "Adaptive Query Processing and the Grid: Opportunities and Challenges", 2004
[24] A. Deshpande, Z. Ives and V. Raman, "Adaptive Query Processing", *Foundations and Trends in Databases*, Vol. 1, No. 1, 1–140, 2007.
[25] P. Bizzaro, "Adaptive Query Processing: Dealing with incomplete and uncertain statistics", PhD Thesis, 2006.
[26] A. Gounaris, "Resource aware query prcoessing". PhD Thesis, 2005.
[27] H. Paques, L Liu, C. Pu, "Distributed Query Adaptation and its Trade-0ffs", 2003
[28] G. de la Calle, M. García, V. Maojo, Infobiomed Project, "State of the Art on Data Interoperability and Managmcnt", 2005.
[29] D. Pearson, "Data requirements for the grid", Global Grid Forum 5, 2002.
[30] Bala et. al, "BioGRID - An European grid for molecular Biology", *Proceedings of the 11th IEEE International Symposium on High Performance Distributed Computing*, 2002.
[31] F. Darvas, A. Papp, I. Bágyi, G. Ambrus-Aikelin, L. Urge L. OpenMolGRID, a GRID based system for solving large-scale drug design problems. Ed. Dikaiakos, M. D., In: Lecture Notes in Computers Sciences: Grid Computing, 2004.
[32] R. Stevens, A. Robinson, and C.A. Goble, "myGrid: Personalised Bioinformatics on the Information Grid", *Bioinformatics* Vol. 19, 2003.
[33] Amendolia et. al, "MammoGrid: A Service Oriented Architecture based Medical Grid Application", LNCS, 2004.
[34] Maui Cluster Scheduler. http://www.clusterresources.com/products/maui/
[35] BiOpera – Process Support for BioInformatics. ETH Zürich, Department of Computer Science. http://www.inf.ethz.ch/personal/bausch/biopera/
[36] Japanese BioGrid project, http://www.biogrid.jp/
[37] BiomedGrid Consortium, http://binfo.ym.edu.tw/grid/index.html
[38] Biomedical Informatics Research Network, http://www.nbirn.net/
[39] R. J. Al-AliI, A. Shaikhali, O. F. Rana, D. W. Walker, "Supporting QoS-based discovery in service-oriented Grids", In *Proceedings of the IEEE International Parallel and Distributed Processing Symposium*, 2003
[40] R. Braumandl, A. Kemper, D. Kossmann, "Quality of Serivce in an Information Economy", *ACM Transactions on Internet Technology*, Vol. 3, No. 4, Pages 291-333, 2003
[41] C. K. Fung et. Al, „A Study of Service Composition with QoS Management", In *Proceedings of the IEEE International Conference on Web Services*, 2005.
[42] D. A. Menasce and E. Casalicchio, "QoS-Aware Software Components", In *Internet Computing Online*, Vol. 8, No. 2, pp.91-93 , 2004
[43] S.B. Musunoori, F. Eliassen, R. Staehli, „QoS-aware component architecture support for grid". In *13th IEEE International Workshops on Enabling Technologies: Infrastructure for Collaborative Enterprises*, 2004.
[44] A. Oguz, A.T. Campell, M.E. Kounavis, R.F. Liao, "The Mobiware Toolkit: Programmable Support for Adaptive Mobile Networking". *IEEE Pesronal Communications Magazine, Special Issue on Adapting to Network and Client Variability*, 5(4), 1998.

[45] A. Sulistio, R. Buyya, "A Grid Simulation Infrastructure supporting Advance Reservation", *International Conference on Parallel and Distributed Computing Systems*, 2004
[46] Z. Wang, "Internet Quality of Service", Morgan Kaufmann, 2001
[47] OASIS Web Services Business Process Execution Language (WSBPEL) ,http://www.oasis-open.org/committees/wsbpel/
[48] T. Etzold, H. Harris S. and Beulah,"SRS: An integration platform for databanks and analysis tools in bioinformatics", *Bioinformatics Managing Scientific Data*, p 35–74, 2003
[49] Open Middleware Infrastructure Institute, http://www.omii.ac.uk/
[50] The Globus Toolkit, http://www.globus.org
[51] Enabling Grids for e-Science, http://www.eu-egee.org
[52] gLite, http://glite.web.cern.ch
[53] UNICORE, http://www.unicore.eu/
[54] VLDB 2003 Tutorial on Grid Data Management Systems and Services, Berlin, http://www.vldb.informatik.hu-berlin.de/ressources/vldb-2003-Tutorial-T4.pdf
[55] Wache et. al, "Ontology-Based Integration of Information - A Survey of Existing Approaches", 2001
[56] Shawn Bowers, Kai Lin, Bertram Ludäscher, "On Integrating Scientific Resources through Semantic Registration", 2004
[57] J.A.M. Herveg, Y. Poullet. 2003. Directive 95/46 and the use of GRID technologies in the healthcare sector: selected legal issues. In *Proceedings of the 1st European HealthGRID Conference*, Lyon, 229-236.
[58] Web Service Level Agreement (WSLA) Language Specification.
[59] http://www.research.ibm.com/wsla/WSLASpecV1-20030128.pdf, IBM 2003.
[60] Web Service Resource Framework. http://www.globus.org/wsrf/.
[61] Alexander Woehrer, Peter Brezany and A Min Tjoa. Novel mediator architectures for Grid information systems Journal for *Future Generation Computer Systems* - Grid Computing: Theory, Methods and Applications. January 2005
[62] A.Y. Levy, A. Rajaraman, and J.J. Ordille. Querying heterogeneous information sources using source descriptions. In *Proceedings of the Twenty-second International Conference on Very Large Data Bases* (VLDB'96), p 251 - 262, 1996.
[63] S. Chawathe, H. Garcia-Molina, J. Hammer, K. Ireland, Y. Papakonstantinou, J. D. Ullman, and J. Widom. The TSIMMIS project: Integration of heterogeneous information sources. In IPSJ Conference, Tokyo, 1994.
[64] J. Gray and A. S. Szalay. Where the Rubber Meets the Sky: Bridging the Gap between Databases and Science, *IEEE Data Engineering Bulletin*, Vol. 27. Iss. 4, 2004.
[65] V. Raman et. al. Data Access and Management Services on Grid, Database Access and Integration Services Working Group, Global Grid Forum (GGF) 5, 2002.

In: Cyberinfrastructure Technologies and Applications ISBN: 978-1-60692-063-3
Editor: Junwei Cao

Chapter 6

GRID RESOURCE BROKER FOR SCHEDULING COMPONENT-BASED APPLICATIONS ON DISTRIBUTED RESOURCES

Xingchen Chu[1], Srikumar Venugopal[2] and Rajkumar Buyya[3]

Grid Computing and Distributed Systems Laboratory
Department of Computer Science and Software Engineering
The University of Melbourne, Australia

ABSTRACT

This chapter presents the design and implementation of seamless integration of two complex systems component-based distributed application framework ProActive and Gridbus Resource Broker. The integration solution provides: (i) the potential ability for component-based distributed applications developed using ProActive framework, to leverage the economy-based and data-intensive scheduling algorithms provided by the Gridbus Resource Broker; (ii) the execution runtime environment from ProActive for the Gridbus Resource Broker over component-based distributed applications. It also presents the evaluation of the integration solution based on examples provided by the ProActive distribution and some future directions of the current system.

INTRODUCTION

Grids have progressed from research subjects to production infrastructure for real world applications. Grids such as TeraGrid [1], LCG [2] and EGEE [3] are being used by scientists

[1] E-mail address: xchu@csse.unimelb.edu.au
[2] E-mail address: srikumar@csse.unimelb.edu.au
[3] E-mail address: raj@csse.unimelb.edu.au

to run large-scale and data-intensive simulations thereby allowing them to explore larger parameter spaces than ever before. Current use of Grid technology mostly extends the batch-job paradigm wherein a job encapsulating the application requirements is submitted to a Grid and the results of the execution are then returned. However, the capabilities of Grids are better explored by active applications that can dynamically vary the parameter space based on different scenarios. This requires programming models that are able to go beyond the passive nature of batch-jobs and consider the Grid as a unified platform rather than a loose aggregation of dispersed heterogeneous resources. In this regard, component-based application development is a promising candidate as it allows Grid applications to be constructed out of loosely-coupled independent components that converse with each other through standard interfaces.

Many component-based softwarc frameworks for Grids have therefore been developed around the world. Examples of such frameworks are ProActive [4], XCAT [6] and SCIRun [7]. Such frameworks have to provide functions for handling the heterogeneity and dynamicity of Grid resources. These functions also have to be abstracted from the programming environment so that developers are able to concentrate on building applications. In this chapter, an integration of ProActive framework and the Gridbus Broker [8] is presented so that the former is able to take advantage of the latter's abilities in resource allocation, job scheduling and management, and support for different middleware. It discusses the challenges involved, and presents the integration process in detail. It also validates the integration by running the existing examples without any modifications. Finally, the chapter concludes and provides directions for future research.

BACKGROUND KNOWLEDGE

Before we present the integration solution proposed in this chapter, it would be better to mention about some technical details of the scheduling infrastructure for both systems in order to fully understand the reason why integration is necessary and important.

ProActive Grid Scheduler

ProActive [4] is a software framework for developing and deploying parallel and distributed applications. The programming model and APIs provided by ProActive greatly simplify the development and execution of those applications. Moreover, it provides a component framework as a reference implementation for GCM (Grid Component Model) for programming Grid applications as reusable components.

Among a lot of services provided by the ProActive framework such as execution environment, standardized deployment and component model, there is one very important service: Grid scheduler service which promises to schedule applications over heterogenous Grid resources. Let us take a look at how the Grid scheduler has been implemented within ProActive framework, as shown in figure 1. The upper layer user objects represent the normal terms such as jobs and resources. The Grid scheduler as well as the resource manager and job manager (known as the AbstractPolicy) are all in the scheduling layer. The actual scheduling algorithms are implemented as various policies which extends the AbstractPolicy class. As can be seen from the figure, four kinds of policies are provided by the ProActive framework.

Although the Grid scheduler implementation fulfils the basic requirement for simple scenarios, there is a lack of advanced scheduling algorithms that can be very helpful for scenarios requiring much more complicated resource allocation policies such as economic-based and data-intensive scheduling policies. This is where the Gridbus resource broker can help and the underlying rationale of this integration work.

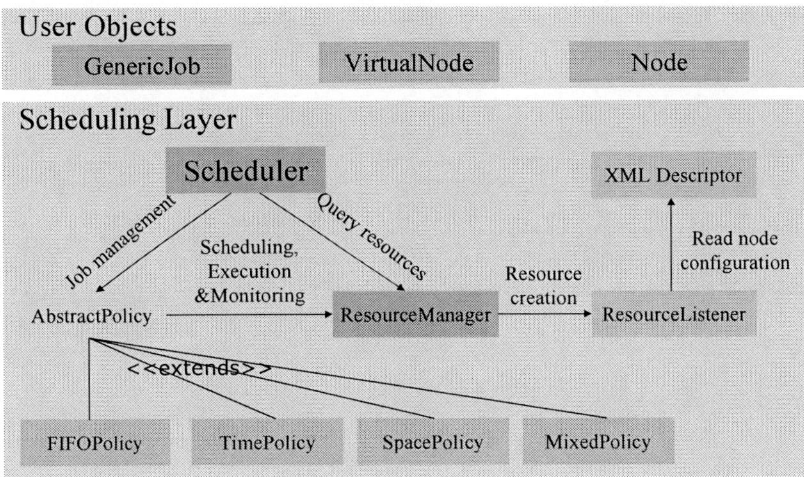

Figure 1. ProActive grid scheduler.

Gridbus Broker Scheduling Infrastructure

Gridbus Broker [8] is a user-level middleware that mediates access to distributed resources by discovering available computational resources, and scheduling jobs to best suitable resources. Users can choose various scheduling policies including simple round-robin and more advanced data-intensive or economy-based resource allocation algorithms [9].

As our integration concentrates on the scheduling infrastructure, it will skip other details related to the broker's application creation and job execution. The scheduling infrastructure provided by the Gridbus broker is composed of four main components including the Scheduler, Dispatcher, ServiceMonitor and JobMonitor, as shown in. Each object runs as a separate thread that exchanges information via the Broker's storage infrastructure. Various scheduling algorithms have been implemented by subclasess derived from the Scheduler class. The Dispatcher is responsible for dispatching jobs to heterogeneous resources. The SeviceMonitor and JobMonitor are responsible for monitoring resources and job status, respectively.

As the broker's point of view, the terms such as job and resources are described using Job and ComputeServer. The JobWrapper provides lifecycle methods that need to be invoked by the broker when the job status is changing. Subclasses of ComputeServer and JobWrapper always come out in pairs which provide a specific runtime environment for different grid middleware such as Globus Toolkit [10] and Alchemi [11]. As we will explain later in the chapter, we have implemented a specific ComputeServer and a JobWrapper for ProActive framework.

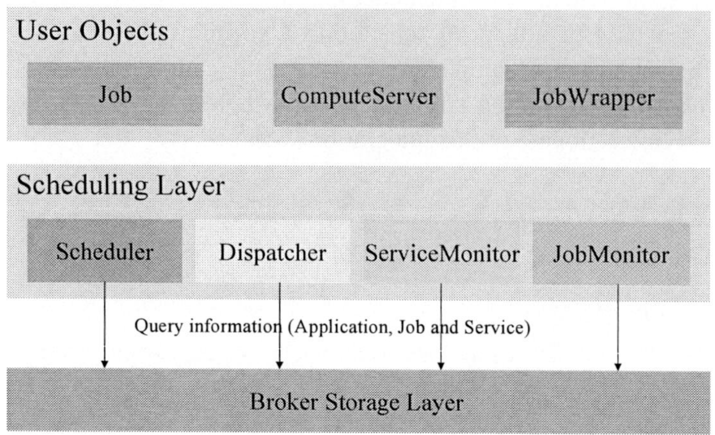

Figure 2. Overview of Gridbus Resource Broker scheduling infrastructure.

INTEGRATION CHALLENGES

As we mentioned, the objective of the integration is to make ProActive use the Gridbus Broker scheduling infrastructure. It in turns has positive effects on both sides. ProActive is able to leverage the economy-based and data-intensive scheduling algorithms provided by the Gridbus Broker. Gridbus Broker is also able to utilise the programming environment especially the component-based programming concept provided by ProActive. However, it is not an easy job as the two systems are both complex and have large codebase. There are several challenges we need to consider in designing a mature and realistic integration solution.

The first challenge we need to identify is on the potential impact of the integration on both systems. There should be no or minimum impact on both systems: (i) each complex system should have no or as little knowledge of each other as possible, and (ii) each complex system only need to concentrate on its own terms and conditions. It means that we should avoid direct dependencies between each other as much as possible, and the data or object representation of each system should remain the same.

The second challenge for the integration is the reuse of existing infrastructure and codebase provided by both systems. The goal is to maximise reusability and avoid modifying the existing source code. This can be divided into two aspects: (i) the scheduling infrastructure provided by Gridbus Broker should be reused without changing existing source codebase within the Broker, and (ii) the deployment and runtime execution infrastructure provided by ProActive should be reused to deploy and execute the applications.

In order to validate that the integration solution we have provided can meet all the challenges presented here, all the legacy applications that used to run with existing ProActive scheduling policy should work with the Broker's scheduling policy without recompiling and redeveloping the source code.

SYSTEM IMPLEMENTATION

As shown in figure 3, the integration solution we have proposed is purely based on basic object-oriented design principles such as inheritance and delegation. The top layer is the existing ProActive scheduling infrastructure which contains a scheduler, a job manager (the policy class) and a resource manager along with the resource listener that is used to acquire resources. The bottom layer is our proposed implementation for the integration. The ProxyPolicy, ProxyResourceManager and ProxyResourceListener extend the ProActive classes respectively. They are proxies and the links between ProActive and Gridbus Broker. Instead of using the ProActive objects, the proxy objects are responsible for initializing the relative components that exist in the Broker codebase. Various proxy policies can be implemented independently in order to enforce certain scheduling algorithms such as Round-Robin, and Cost/Time optimization that have already been implemented in the Broker.

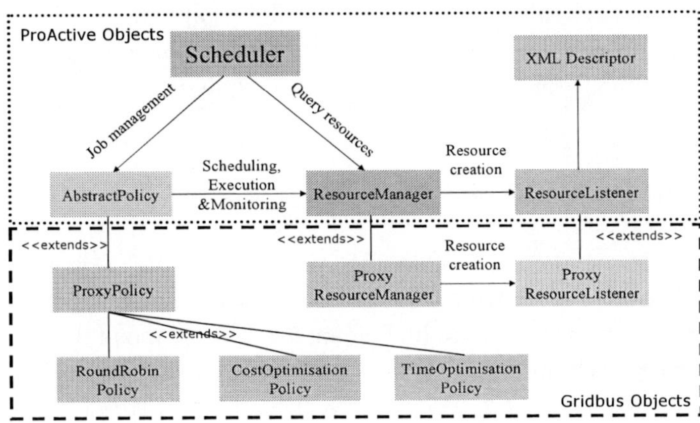

Figure 3. ProActive and Gridbus Broker integration design.

In order to seamlessly integrate the two systems together, we have to divide the implementation into two different aspects: (i) new classes that should belong to the ProActive codebase are responsible for replacing the existing scheduling infrastructure provided by ProActive dynamically, and (ii) new classes that should be part of the Broker codebase are responsible for wrapping the ProActive terms into the Broker's terms that can be scheduled and managed by the Broker's scheduling infrastructure.

Proxies for ProActive Scheduling Layer

Let us first examine the objects created belonging to the ProActive codebase.

- *Proxy Objects:* they bridge ProActive classes with Gridbus Broker classes, and internally delegate the responsibility to the corresponding behavior in the Gridbus Broker.
- *ProxyPolicy:* it is a subclass of ProActive AbstractPolicy class which is responsible for initialising the Broker runtime including JobMonitor, ServiceMonitor, Scheduler and Dispatcher. Besides, it also overrides the ProActive's job management functions

by delegating to the proper broker functionality. Its subclasses provide the information required by the Broker to match the scheduling algorithms and create an appropriate scheduler for a certain policy. For example, the RoundRobinPolicy is used to guide the Broker to utilise the basic round-robin scheduler that has already been implemented in the Broker scheduling infrastructure.

- *ProxyResourceManager*: this class extends the ProActive ResourceManager class. It overrides the ProActive's resource management functions using Broker's functionality.
- *ProxyResourceL*istener: as ProActive utilises the ResourceListener to acquire the resources from the XML deployment descriptor, the ProxyResourceListener simply overrides those implementation and add those resources into the Broker system by converting the terms from ProActive (such as virtual node and node) to the terms recognized by the Broker.

The proxies we have implemented that are fully compliant to ProActive scheduling infrastructure provide an alternative approach for the ProActive community that is promised to utilise Gridbus Resource Broker's advanced scheduling algorithms and infrastructure.

ProActive Wrappers for Gridbus Broker

We have already mentioned about the proxies for the ProActive's scheduling infrastructure that delegate ProActive's responsibilities to the Gridbus Broker. However, it is just one side of the coin, we also need to look at another side of the problem. As both systems use their own terms representing the same set of information such as resources and jobs. We have to also provide wrapper classes that translate different terms from one system to another system. As our goal of the integration is to utilise the Broker's advanced scheduling infrastructure for ProActive, it is very important to translate the terms from ProActive into the terms that the Broker can understand and manage.

As we have already seen the terms ProActive defined for the resources (virtual node, node) and jobs (GenericJob), an entity mapping has been implemented by wrapping proper objects into the Broker's terms as follows:

- *ProActiveJob*: this class extends the Broker's Job class and maintain a ProActive's GenericJob object internally. Whenever a GenericJob is created by the ProActive's job submission module, a ProActiveJob object will be created and queued for schedule by the Broker system.
- *ProActiveComputeServer*: it is the subclass of the ComputeServer class which is a representation of the compute resource for Broker. It internally keeps track of one virtual node and a list of nodes with that virtual node. Once the resource is created by ProActive deployment component, the ProActive compute server will be created by giving the virtual node and a list of nodes associated with the virtual node.
- *ProActiveJobWrapper*: the job wrapper is a unique class for the Broker that is used to execute a task using a given way. It provides lifecycle methods for a task that will be managed and executed by the Broker's scheduling infrastructure. This particular

implementation for ProActive is responsible for setting up the ProActive runtime environment and execute the job using the API provided by ProActive runtime.

Assembling the System

We have provided proxies for the ProActive system and also wrappers for the Broker system, the remaining work is to assemble the two independent parts into an integrated environment. Figure 4 demonstrates how the whole system works together.

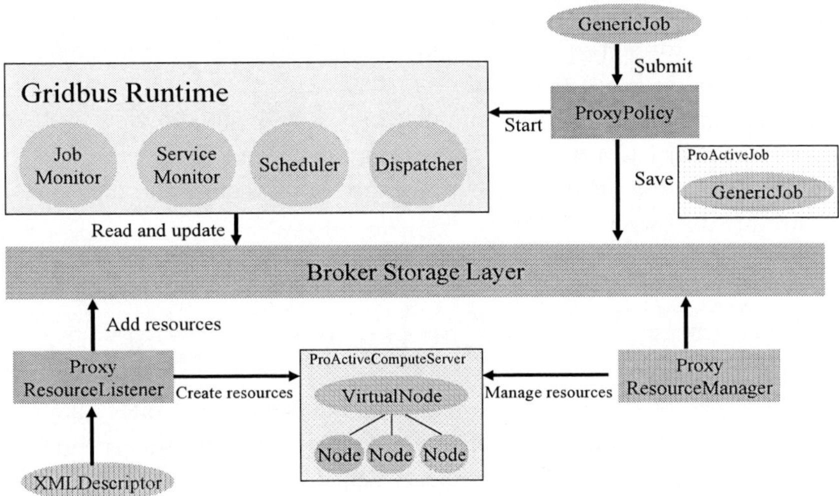

Figure 4. How the integration works.

As can be seen from the figure, the center part of the system is the Broker storage base, all the information related to the jobs and resources are pushed into the storage. It is the core part for decoupling ProActive and the Broker. The proxies at the ProActive side are totally unaware of the Broker runtime environment because they do not need the direct communication with those components such as scheduler, dispatcher and job/service monitor. The Broker on the other hand only need to retrieve information such as jobs and resources in the same manner without even knowing who has given those information.

The entire workflow can be described into three independent parts:

- *Resource Acquisition and Management*: the proxy resource manager is activated at the very beginning when the ProActive scheduler starts and initialises the proxy resource listener to acquire the resources from the XML deployment descriptor. The listener will create the ProActiveComputeServer objects based on the virtual node and nodes deployed by the XML deployment descriptor, and save those compute server objects into the Broker storage system. The proxy resource manager simply manages those resources via the Broker storage system.
- *Job Submission*: the ProActive user is able to submit their applications in terms of GenericJob objects to the ProActive scheduler just as before. Once the GenericJob objects have been submitted to the system, the proxy policy class will create the

ProActiveJob objects that wrap each GenericJob object. The ProActiveJob objects are saved into the Broker storage system for later scheduling.

- *Job Scheduling, Monitoring and Execution*: this part is fully controlled by the Broker scheduling infrastructure and totally transparent to the ProActive system. The proxy policy class will intialise the Broker runtime before any job submissions occur. The four Broker components including the Scheduler, Dispatcher, Job and Service Monitor will run as separate threads in the system. They periodically poll the storage system and retrieve or update required information. For example, the scheduler will try to find out all the newly submitted jobs and all the available resources, and schedule jobs over best resources based on the scheduling algorithms. If no job or no available resource has been found, it just waits for a certain time and polls it again in the next round. Similarly, the other three components work in the same manner.

The three independent parts work simultaneously without knowing each other. They obtain required data by querying the storage system, and update certain information accordingly. This separation enables decoupling for both systems, and they just concentrate on what they need to deal with.

VALIDATION

In this section, we look at how to validate our integration that meets the objective and solves the challenges, by running an application developed using the ProActive framework and being scheduled via the Gridbus broker's scheduling infrastructure. The integration solution is valid unless the following conditions are satisfied:

- Legacy applications developed using ProActive should work without changing and recompiling the source code. The only acceptable modification will be the configuration file that needs to be used to connect to the Grid scheduler.
- The dependencies between each system should be minimized. Users from ProActive should be able to dynamically choose which scheduler to use, either the existing one or the Gridbus broker's scheduler. And the Gridbus broker should not be aware of any ProActive runtime information.
- Reuse the existing infrastructure at both sides without adding new features. The job scheduling should be delegated to the Gridbus broker, and the ProActive's application deployment and job execution need to be reused.

C3D Application

The application we are using is the C3D which is a Java benchmark application measuring the performance of a 3D raytracer renderer distributed over several Java virtual machines. Image rendering is the process of generating an image from a model by means of computer programs. The model is described as a three dimensional object that can be understood by computer programs. Information such as geometry, texture, lighting, viewpoint and shading may be carried by the objects. Ray tracing is just one of the algorithms that can be used to render the image, which is a brute-force method calculating the value of each pixel [12]. In general, the ray tracing rendering approach is too slow to consider for the realtime image rendering. The C3D application makes use of the distributed rendering engines working in parallel which is able to accelerate the speed of the normal ray tracing rendering approach. Users can interact through messaging and voting facilities so that they can choose a scene that is rendered using a set of distributed rendering engines.

Launch the Application

This application has been provided by the ProActive distribution. We have configured a scheduler node with 5 nodes (pure JVM) which uses the default round-robin scheduler and asks the C3D application to connect to the scheduler node. We also monitor this application via the IC2D monitor GUI, as shown in figure 5. From the user's point of view, there is nearly no difference between this run and the one without the Gridbus broker except that the RoundRobinPolicy and ProxyResourceManager have been presented in the scheduler virtual node show on the IC2D panel.

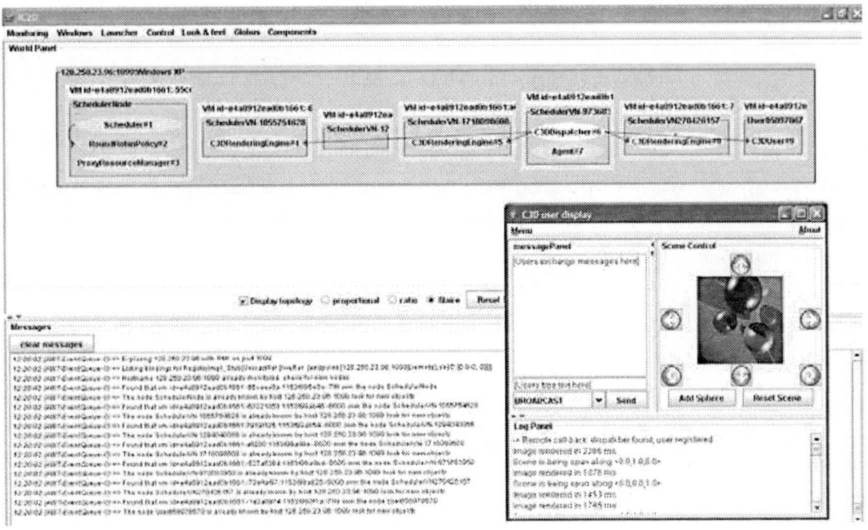

Figure 5. C3D application using Gridbus Broker in IC2D monitor.

As the application can be run without modifying any source code, the only thing we need is to connect the application to the scheduler node via an XML configuration file provided by the ProActive's deployment infrastructure. It satisfies the first condition we have set for the validation.

The dependencies between each system have been minimized as well. The following code shows how to start the grid scheduler with the broker's proxy classes.

The ProActive still uses it grid scheduler and it does not know about the existence of the broker. The implementation classes related to the broker have been dynamically loaded via Java reflection by giving the name of those classes. Different policies can be easily applied by giving a different class name that delegates to the appropriate policy implementation. So we could say that the integration also fulfils the second condition.

```
public class StartScheduler{
   public static void main(String [] args){
     String policyName =
           "org.objectweb.proactive.scheduler.gridbus.policy.RoundRobinPolicy";
     String resourceManager =
           "org.objectweb.proactive.scheduler.gridbus.ProxyResourceManager";
     Scheduler.start(policyName,resourceManager);
   }
}
```

The last condition has also been satisfied by our implementation. All the implementations we have done for the integration are simply providing proxies and wrappers for both system. No new features are introduced into either of them.

Since all the conditions we have set for the validation have been achieved, we could conclude that our proposed integration not only works with the two complex systems, but is also a valid solution that does not comprise the user's prospective.

Remarks of the Integration

To evaluate a successful integration solution, only considering whether it is working or not is not sufficient. It is very important and necessary to look at what impacts the integration has made to each of the independent systems. In this section of the chapter, we will examine the impacts that our work brings to each system.

Impacts on ProActive

At the ProActive side, the integration solution provided in this chapter has the least impacts on the original ProActive system both at the system level and the application level. First of all, it is necessary to check the impacts on the ProActive's system level functions. In terms of the changes of the source code for the ProActive existing codebase, it only requires tiny changes in the scheduler class that has been modified to add a new overload start method supporting dynamic configuration of proxy resource manager and proxy policy. Other than that, the most relevant and important infrastructures such as the deployment service, the runtime execution environment and the scheduler services have been reused. In other words, the reusability of the system has been greatly achieved with this integration solution which has minimised the impacts on the system level.

Another aspect to look at is the application level, or the necessary changes that might be required by the client applications developed using the ProActive framework. As the previous section has shown the C3D application is totally are unaware of the existence of the Gridbus broker at all. This is achieved by decoupling the process of creating the resource manager and policy from the Scheduler class via the new start method. It is very important to understand that the implementation of the broker's scheduling service including the proxy resource manager and proxy policy are plugged at runtime via the modified scheduler class not at compile time which eliminates the recompilation process of existing legacy applications requiring the original scheduling service. This decoupling helps to reduce the unnecessary dependencies between the client and the scheduling service, which means that the client applications can safely ignore the existence of the underlining scheduling service.

The last notable aspect is to check how many extra dependencies (extra jar files) that are required by the ProActive system to perform the scheduling via the Gridbus broker. this is so important to mention is because the integration approach adopted can largely affect the size of the jar files required by certain configuration. With inappropriate approach, the runtime system will require a huge amount of jar files on the classpath and it will cause more problems such as class conflicts if the two systems use the same open source products with different versions. The integration solution in this chapter considers this problem. Only one Java jar file which is the broker runtime library is needed for ProActive to work with the Gridbus broker.

Impacts on Gridbus Broker

The impact of this integration on the broker is even lesser. The architecture of the Gridbus broker is very flexible in which the scheduler plays a core role, and various types of runtime Grid middleware such as Globus, Alchemi, SGE, Condor and PBS can be plugged into its runtime environment at runtime via the configuration. The broker itself acts as a middle-man who is responsible for matching jobs to heterogeneous resources. In order to implement a middleware that can be supported by the Gridbus broker, only few classes are required to be extended including the ComputeServer class representing the grid resource and the JobWrapper class which contains the job execution logic for that particular middleware.

The integration solution extends this idea of the broker and makes ProActive as another type of Grid middleware. The benefit of this decision is that all the classes related to ProActive can be an independent project that does not need to change the original broker's source. At runtime, the broker is totally unaware of the existence of ProActive and just treats it as a type of normal Grid resource. The broker schedules and dispatches jobs as before without worrying about the terms defined in ProActive such as GenericJob, Virtual Node and Node.

Conclusion and Future Work

This chapter presented the design, implementation and evaluation of an integration solution for enabling schedule component-based applications on global Grids by utilising the Gridbus resource broker. The integration solution provided in this chapter strictly follows the object-oriented design principles, which seamlessly makes the two complex systems collaborate without knowing each other. Any legacy applications running under the original runtime can still work without modifying and recompiling the source code. The glue between the two systems has been provided via the configuration file and the execution environment loads it dynamically at runtime through the existing deployment service. To validate whether the integration meets the objective, a legacy 3D raytracer renderer application along with the ProActive monitor has been tested via the integration.

From the validation, this chapter concludes that the integration solution has minimum impacts on both systems. They all can work independently and the community users of the ProActive can freely choose either to use the Gridbus broker scheduling service or not to worry about it at all. Although this integration solution has demonstrated the feasibility and potential benefits of scheduling component-based applications via Gridbus broker, a much closer integration is required in the future to facilitate the economy-based scheduling services. The job model of the ProActive has to be extended to support QoS parameters such as budget and deadline via a configurable way. Moreover, the provided implementation only focuses on the RMI runtime provided by ProActive, a much more comprehensive testing needs to be done with other types of runtime environments provided by ProActive.

ACKNOWLEDGEMENT

We would like to thank the anonymous reviewers for their valuable comments. This work is fully supported by the international linkage research grant from the Australian Department of Education, Science and Training (DEST).

REFERENCES

[1] R. Pennington, Terascale Clusters and the TeraGrid , Invited talk, Proceedings for HPC Asia, Dec 16-19, 2002, pp. 407-413.

[2] I. Bird, L. Robertson and J. Shiers, Local to Global Data Interoperability - Challenges and Technologies, 20-24 June 2005, CERN, Geneva, Switzerland.

[3] EGEE, Enabling Grids for EScience, http://www..eu-egee.org.

[4] R. Bramley, K. Chiu, S. Diwan, D. Gannon, M. Govindaraju, N. Mukhi, B. Temko, and M. Yechuri, A Component Based Services Architecture for Building Distributed Applications. In Proceedings of the 9th IEEE international Symposium on High Performance Distributed Computing, Pittsburgh, PA, USA, August 2000, IEEE Computer Society, Washington, DC, 2000.

[5] F. Baude, L. Baduel, D. Caromel, A. Contes, F. Huet, M. Morel and R. Quilici, Programming, Composing, Deploying for the Grid, GRID COMPUTING: Software Environments and Tools, Jose C. Cunha and Omer F. Rana (Eds), Springer Verlag, January 2006.

[6] D. Gannon, et al., Programming the Grid: Distributed Software Components, P2P and Grid Web Services for Scientific Applications. Cluster Computing 5(3):325-336, Springer-Verlag, Berlin, Germany, 2002.

[7] S. G. Parker, and C. R. Johnson, SCIRun: a scientific programming environment for computational steering. Proceedings of the 1995 ACM/IEEE Conference on Supercomputing (SC '95), San Diego, California, United States, December 04 - 08, 1995. ACM Press, New York, NY, 1995.

[8] S. Venugopal, R. Buyya and L. Winton, A Grid Service Broker for Scheduling e-Science Applications on Global Data Grids, Concurrency and Computation: Practice and Experience, 18(6): 685-699, Wiley Press, New York, USA, May 2006.

[9] S. Venugopal and R. Buyya, A Deadline and Budget Constrained Scheduling Algorithm for eScience Applications on Data Grids, 6th International Conference on Algorithms and Architectures for Parallel Processing, Oct. 2-5, 2005, Melbourne, Australia.

[10] I. Foster and C. Kesselman, The Globus Project: A Status Report, Proceedings of IPPS/SPDP'98 Heterogeneous Computing Workshop, 1998, pp. 4-18.

[11] A. Luther, R. Buyya, R. Ranjan, S. Venugopal, Alchemi: A .NET-Based Enterprise Grid Computing System, Proceedings of the 6th International Conference on Internet Computing (ICOMP'05), June 27-30, 2005, Las Vegas, USA.

[12] G. H. Spencer and M. V. R.K. Murty (1962). General Ray-Tracing Procedure . *J. Opt. Soc. Am.* 52 (6): 672–678.

In: Cyberinfrastructure Technologies and Applications ISBN: 978-1-60692-063-3
Editor: Junwei Cao

Chapter 7

CROWN: A SERVICE GRID MIDDLEWARE FOR E-SCIENCE

Jinpeng Huai[1] ***and Chunming Hu***[2]
Beihang University, Beijing 100083, P. R. China

ABSTRACT

In the past few years, the Grid computing paradigm has emerged as an instance of cyber infrastructure, promising to enable resource sharing and collaborating across multiple domains. In the research community there has been an intense interest in designing and studying of such system.

CROWN (China R and D Environment Over Wide-area Network) project is an e-Science project funded by China Natural Science Foundation Committee, and China 863 High-tech Program. The main goal of CROWN project is to empower in-depth integration of resources and cooperation of researchers nationwide and worldwide. CROWN was started in late 2003. The main goal of CROWN project is to build the middleware infrastructure and wide area testbed to support computation intensive, data centric e-Science applications.

Recently, with the evolution of Web services, the service-oriented architecture has become a significant trend for grid computing, with OGSA/ WSRF as the de facto standards. CROWN has adopted the service-oriented architecture, connecting large amount of services deployed in universities and institutes. Up till mid 2007, lots of applications in different domains have been deployed into CROWN grid, such as gene comparison in bioinformatics, climates pattern prediction in environment monitoring, etc. The long-range goal for CROWN is to integrate home user resources in a fully decentralized way with a robust, scalable grid middleware infrastructure.

In this chapter, based on a proposed Web service-based grid architecture, a service grid middleware system called CROWN is introduced. As the two kernel points of the middleware, the overlay-based distributed grid resource management mechanism is proposed, and the policy-based distributed access control mechanism with the capability of automatic negotiation of the access control policy and trust management and

---

[1] E-mail address: huaijp@buaa.edu.cn
[2] E-mail address: hucm@buaa.edu.cn

negotiation is also discussed in this chapter. Experience of CROWN testbed deployment and application development shows that the service-oriented middleware can support the typical scenarios such as computing-intensive applications, data-intensive applications and mass information processing applications.

BACKGROUND

In the year 2004, with the development of Grid technologies in both the academic and industry, the Natural Science Foundation Committee of China (NSFC), which is one of the main funding parties in China, announced its e-Science Program named *Network-based e-Science Environment.* As one of the Grid related research program, this program is also referred as the *NSFCGrid.* The program started in 2004 and ended at the end of 2007. The main goal of this program is to build up a virtual science and experiment environment to enable the wide-area research corporation such as large-scale computing and distributed data processing. The research projects are organized into three layers: basic theory and principles, general test bed and pilot applications. Different with CNGrid and ChinaGrid, NSFGrid pays more attention on the fundamental research of Grid related technologies.

CROWN (http://www.crown.org.cn) [1] is the brief name for ***C**hina **R**esearch and Development environment **O**ver **W**ide-area **N**etwork*, one of the main projects in NSFCGrid program, which is led by Beihang University (BUAA), with several partners in China, such as National University of Defense Technology of China (NUDT), Peking University, Computer Network Information Center of China Academy of Science (CNIC, CAS) and Tsinghua University. The output of CROWN project may fails into three parts: the middleware set to build a service-oriented Grid system, the testbed to enable the evaluation and verification of Grid related technologies, and the applications.

CROWN service Grid middleware is the kernel to build an application service Grid. Basic features of CROWN are listed as follows. First, it adopts an OGSA/WSRF compatible architecture [2]; second, considering the application requirements and the limitation of security architecture of OGSA/WSRF, more focus is put on the Grid resource management and dynamic management mechanism in the design stage, and a new security architecture with distributed access control mechanism and trust management mechanism which are proposed to support the sharing and collaborating of resources in a loosely coupled environment is proposed in CROWN.

Under the framework of CROWN project, we also created the CROWN testbed, integrating 41 high performance servers or clusters distributed among 11 institutes in 5 cities (by April 2007). They are logically arranged in 16 domains of 5 regions by using the CROWN middleware. The testing environment is growing continuously and it becomes much similar to the real production environment. The CROWN testbed will eventually evolve into a wide-area Grid environment both for research and production.

CROWN is now becoming one of the important e-science infrastructure in China. We have developed and deployed a series of applications from different disciplines, which include Advanced Regional Eta-coordinate numerical prediction Model (AREM), Massive Multimedia Data Processing Platform (MDP), gViz [3] for visualizing the temperature field of blood flow, Scientific Data Grid (SDG) and Digital Sky Survey Retrieval (DSSR) for virtual observatory. These applications are used as test cases to verify the technologies in CROWN.

CROWN Middleware

Design Motivation

Since many researchers are focusing on technologies and applications of Grid, the interoperability and loosely coupled integration problem between different Grid systems are now becoming a hot topic. At the same time, the application and standardization of Web services technology are developed rapidly, and service-oriented architecture (SOA) becomes an important trend in building a distributed computing environment for wide area network, which helps the merging of Grid and Web services. Recently, Open Grid Service Architecture [2] and Web Service Resource Framework (www.globus.org/wsrf/) were proposed and have become one of the fundamental technologies in Grid competing. SOA and related standardization work provide an important methodology to the research and application of Grid technology. First, the resources are encapsulated into services with standardized interfaces, supporting the unified service management protocol, which helps to solve the problem caused by the heterogeneity of resources; second, the resources are utilized through a service discovery and dynamic binding procedure, which helps to set up a loosely coupled computing environment. But the current resource management mechanism is not enough for all the Grid application scenarios because of the distributed and autonomic resource environment, and the existing security mechanism cannot pro-vide features like privacy protection and dynamic trust relationship establishment, which embarrass the further application of Grid technology.

Actually, not only the Grid computing, but also the Peer-to-Peer computing and the ubiquitous computing try to explore the Internet-oriented distributed computing paradigm. The common issue of these computing paradigms is how to use the capability of resources efficiently in a trustworthy and coordinated way in an open and dynamic network environment. As we know, Internet (especially the wireless mobile network) is growing rapidly, while it is deficient in the effective and secure mechanism to manage resources, especially when the resource environment and relationship between different autonomic systems are changing constantly. At this point, three basic problems such as cooperability, manageability and trustworthiness are proposed. The cooperability problem is how to make the resources in different domains work in a coordinated way to solve one big user's problem. The manageability problem is how to manage heterogamous resources and integrate the resources on demands in a huge network environment, which is a basic condition of building an Internet-oriented distributed computing environment. The trust-worthiness problem is how to set up the reliable trust relationship between cross domain resources when they are sharing and collaborating.

From year 2002, based on our previous work on Web service supporting environment, and OGSA/OGSI compatible service Grid middleware WebSASE4G, a WSRF compatible CROWN Grid Middleware is proposed. According to the three basic problems, several key issues have been explored, such as resource management of service Grid, distributed access control, cross-domain trust management, Grid service workflow and service orchestration based software development.

Architecture Design

Grid computing started from the meta-computing in the 1990s. In recent years, Grid was changed from Metacomputing to computing Grid and service Grid, but the basic architecture of such a computing paradigm was not changed much. In 2001, the five-layer sand-glass architecture was proposed and accepted generally. With the application and standardization of Grid, the architecture and its supporting technology be-come an important research issue. OGSA is a service-oriented architecture, which adopts the service as the unified resource encapsulation format to provide better extensibility and interoperability between Grid resources. WSRF refines the service interface and interoperating protocols of OGSA, and makes the OGSA a Web service-compatible implementation framework, which helps the merging of Grid and Web service technology more smoothly.

Actually, the five-layer sand-glass architecture just proposed an abstract functionality structure for service Grid. OGSA/WSRF provided an implementation framework based on service concept, and a set of Grid service interfaces, neither of which discussed the design principles, middleware component definitions, and the detailed solutions for access control and trust management in service Grid. In this chapter, we analyze the requirements from typical Grid applications, and provide a detailed introduction of CROWN middleware and its architecture, design principles and kernel technologies based on OGSA/WSRF service Grid architecture.

Generally, there are three kinds of services in an OGSA/WSRF service Grid: general services, resource encapsulating services and application-specific services. General services, such as Grid job broker service, meta-scheduling service, and Grid information services (GISs) are an important part of a service Grid middleware. In a typical application Grid scenario (see figure 1), a user first submits job to the meta-scheduling service, then gets the job status and result from the job broker service; meta-scheduling service retrieves the resource requirements from the job description language, queries GIS to discovery the necessary service, submits the job to the service and traces the status of the job execution.

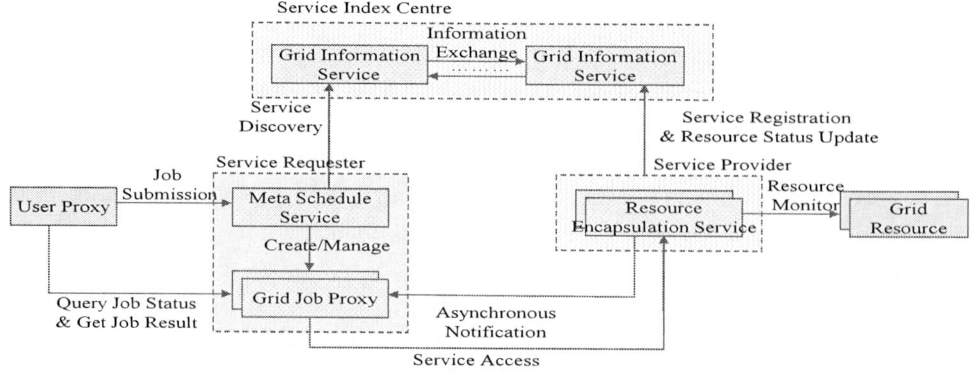

Figure 1. CROWN-based service Grid application pattern.

Based on the above analysis, we provide a layered architecture for CROWN and compare it with the five-layer sand-glass architecture (see figure 2). In the figure 2, service Grid middleware covers resource layer, collective layer and application layer. In our sys-tem design, services in resource layer (e.g., resource encapsulation service), collective layer (e.g., Grid information service, meta-scheduling service and job broker service) and part of services in application layer are OGSA compatible Grid services, using information service to obtain the capability of registration and dynamic discovery. Grid application support tools are used to enable the interoperation between the collective layer and application layer (for example, the application-specific developing framework, routines, and Web-based Grid application portals, etc.).

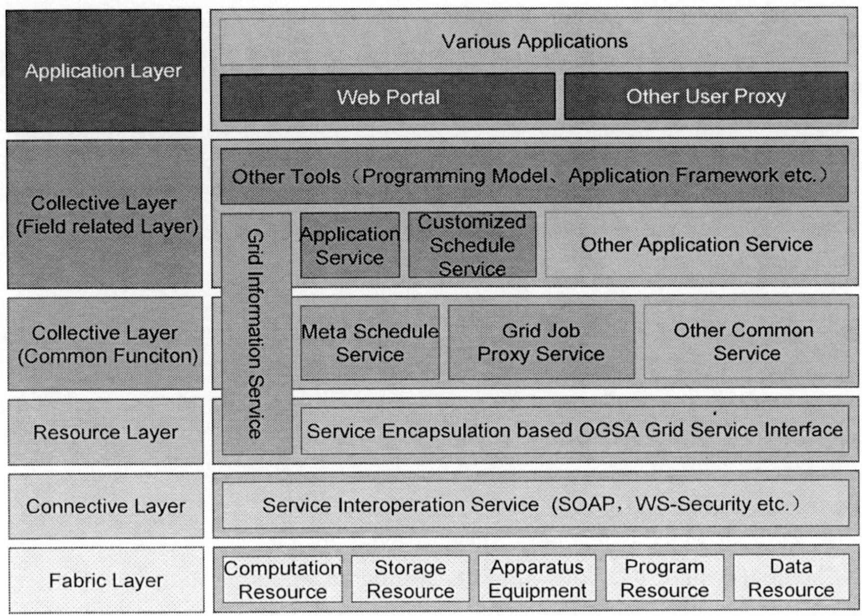

Figure 2. CROWN-based service Grid application.

Components in CROWN

There are 11 components in total in CROWN service Grid middleware analyzed as follows in detail.

Node Server

Node Server [4] provides a basic runtime environment for Grid service. Using Node Server, the underlying resources can be encapsulated into Grid services. Node Server provides all the generic functionalities when running a service instance, such as SOAP message processing, service instance management; instance/invoke lifecycle management and notification mechanism. Based on Globus Toolkit 4.0, Node Server adds lots of features such as remote and hot service deploying, resource status monitoring and re-porting, logging and remote control and management. By adding security modules, Node Server can provide features like PKI/Kerberos based authentication, fine-gained authorization, trust management and automatic trust negotiation (ATN), which could guarantee the security and privacy effectively when resources are used by remote users or cross-domain resources.

Resource Locating and Description Service (RLDS)

CROWN Resource Locating and Description Service (RLDS) is a distributed information service system for service registration and discovery. Multiple RLDS instances use information exchange and topology maintenance protocol to build the hierarchical architecture and the overlay network at runtime as needed to get better performance of re-source management and service discovery.

CROWN Scheduler

CROWN Scheduler is a meta-scheduling service in CROWN, queues and schedules user's jobs according to a set of predefined strategies, interoperates with RLDS to get current service deployment information and job status, uses predefined scheduling policy (random policy, load balancing policy, etc.) to do the matchmaking, and performs the service invocation. CROWN scheduler supports two types of job, POSIX application invocation and Grid service invocation. Job submission description language (JSDL) is used to describe the QoS requirements and security demands of the jobs.

CROWN CommSec

CROWN CommSec is a plug-in for Node Server and a generic Web service to provide the basic security communication feature such as building and verifying of certificate chains. Administrators can edit the predefined policy file according to complex security requirements to provide independent, extensible and feasible security solutions.

CROWN Authz Service

CROWN Authz Service is a generic service using an XACML based authorization policy description language and provides the capability of authorization decision and policy management. It supports the multi-granularity access control policy and domain access control policy.

CROWN CredMan

CROWN CredMan is used to manage user credentials. Through the agent certificate issue, the identified subjects can be managed especially when the job is submitted from the Web portal or in the mobile networks.

CROWN CredFed

CROWN CredFed contains a plug-in for Node Server and a credential mapping service. It can be used to map credentials from different security infrastructures (such as PKI and Kerberos) to enable the identity mapping between two security domains. Administrators can modify the mapping policy to control the behavior of CredFed.

CROWN ATN

CROWN ATN contains a plug-in for Node Server and a set of generic services. The Automatic Trust Negotiation (ATN) is to establish the trust relationship between strangers in Internet, protecting the privacy (for example, the information on attributes-based certificates and the negotiation policies) of both sides. It provides a security decision and trust management mechanism for open network environment.

CROWN Portal and Rich Client Framework

These two tools, CROWN Web Portal and Rich Client Framework, provide a unified user interface to service Grid to support the job parameter configuration, job submitting, JSDL generating and result demonstration. CROWN Portal provides a Web-based interaction model for the applications, and Rich Client Framework provides a Java-based application framework for applications, which has extensive visualization and interaction demands (such as complex visualization). The framework can be customized according to the application scenario to speed up the application development.

CROWN Designer

CROWN Designer is a Grid service developing and deploying tool based on Eclipse platform. A set of wizard and dialogs provided make the development and deployment of Grid service more easily. By using the remote and hot deploy feature of Node Server, the designer provides drag and drop features to deploy the GAR file. In the near future, more service orchestration tools will be integrated into the CROWN Designer.

CROWN Monitor

CROWN Monitor is an Eclipse RCP based client tools written in Java. It is used to retrieve, store and analyze events/information from different Grid entities, and to show current runtime information using map and chart. We can also adjust parameters of the tool to change the monitor behavior to the target service Grid systems.

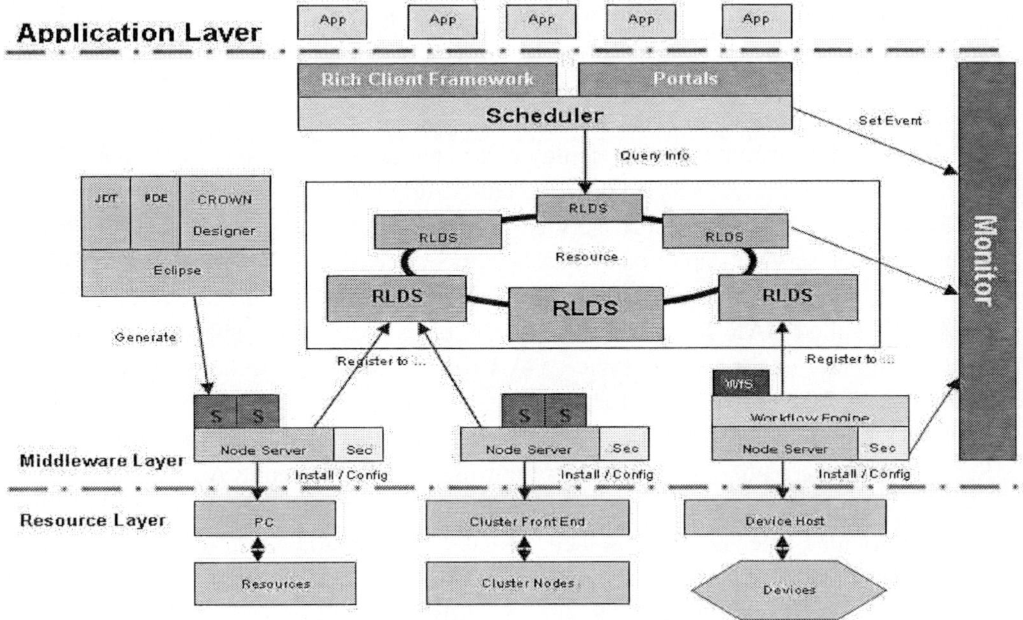

Figure 3. Service Grid Design Principle based on CROWN middleware.

Based on these middleware modules, CROWN is designed under the guide of figure 3. There are three layers in a service Grid. CROWN middleware connects resources in a resource layer. Application Grid uses Web portal or other customized client interfaces to submit jobs and solve the user's problem. First, Node server should be deployed on each resource to support service deploy and runtime management; second, all the Grid re-sources are divided into multiple domains, at least one RLDS instance should be deployed into each domain, and all the RLDS instances have to be configured into a predefined heretical architecture to form a distributed information systems; third, CROWN scheduler will be deployed into the Grid, to get the job request from the user and to find the proper services for each job; finally monitoring and developing tools simplify the building procedure of a Service Grid and its applications.

Features Overview

CROWN adopts an OGSA/WSRF compatible architecture, with the following features.

Overlay-Based Distributed Resource Management

Overlay technique is an effective way to support new applications as well as protocols without any changes in the underlying network layer. The basic idea of the overlay net-work is to build a new network over the existing physical network nodes according to some selected logical rules. In CROWN, resources are managed by an information service overlay consisted by a set of RLDS services [5]. These RLDS instances are linked with each other according to a tree-based topology with carefully selected short-cuts, exchanging resource and request information with their logical neighbors. Such a fully decentralized structure can provide better performance with the avoidance of single point of failure at information systems.

Remote and Hot Deploy with Trust (ROST)

We developed the remote and hot deploy technique with trust support, called ROST [6].

Traditionally, the remote service deployment is supported in a cold fashion, which means to deploy a new service, and the service runtime environment needs to be restarted. Therefore, the hot service deployment has become increasingly important, which does not need to restart the runtime environment while deploying services. To achieve this feature, an archive format called GAR file (Grid Archive) is proposed to encapsulate all the necessary files and configurations for a Grid service. GAR file can be moved to target service container through SOAP/HTTP protocols. Target service container receives the GAR file and uncompresses it to update the container information without stop the container.

Security issues are guaranteed through the trust negotiation using ATN (autonomic trust negotiation) technique. ATN is a new approach to access control in an open environment, which, in particular, successfully protects sensitive information while negotiating a trust relationship. With ATN, any individual can be fully autonomous. Two individuals, which are not in different security domains, try to set up a trust relationship by exchanging credentials according to respective policies.

With the availability of remote and hot service deployment, many applications will benefit, such as load balancing, job migration and so on.

JSDL-Based Job Submission and BES-based Job Scheduling

JSDL (Job Submission Description Language) and BES (Basic Execution Service) are adopted in CROWN scheduler, with extension to Web service-based Job submission. Job can be submitted to CROWN scheduler via any JSDL compatible clients, such as GridSAM, and gLite using a BES interface. Interoperability demonstrations are proposed in AHM 2006 and SC 2006, organized by HPCP (High Performance Computing Profile) working group in OGF.

Security Architecture Supporting Domain Interoperability

CROWN uses a federate construction to form the virtual organization. We use the term region to denote the area with homogenous security infrastructure such as PKI or Kerberos, and term domain to denote the area of autonomous organization. When Grid services are deployed in different domains, each domain may have their own security concerns about the services. CROWN provides a fine-grained and extensible architecture that maximizing the separation of service administrators and service developers.

Beside this, CROWN enables that the same implemented service can be deployed into those PKI domains as well as Kerberos domains without having to modify the source code of the service. Furthermore, CROWN-ST also supports users from domains with heterogeneous security infrastructures to access the resources from other domains.

Resource Management in CROWN

Overview

CROWN employs a three-layered structure of resource organization and management [5], as illustrated in figure 4, based on the characteristic of e-Science applications and the resource subordination relationship. The three layers are Node Server, RLDS (Resource Locating and Description Service), S-Club and RCT.

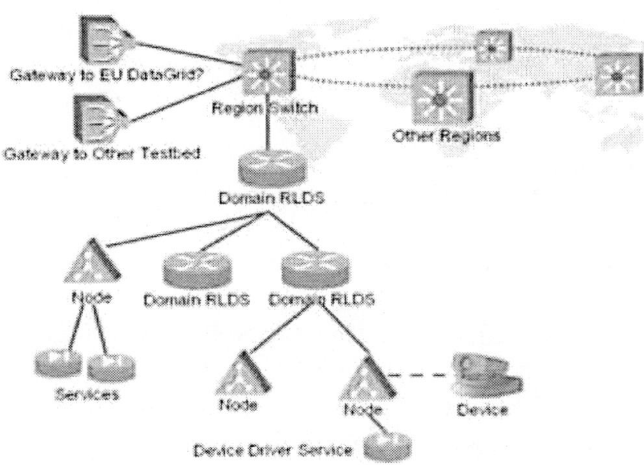

Figure 4. CROWN resource organization and management.

In CROWN, before a computer becomes a Node Server (NS), it must be installed with CROWN middleware. The service container is the core component in CROWN middleware, which provides a runtime environment for various services. Each NS usually belongs to a security domain. Every domain has at least one *RLDS* to provide information services, and *RLDS* maintains the dynamic information of available services. S-Club and RCT are used for more efficient resource organization and service discovery.

Resource Sharing Using Node Server

All kinds of heterogeneous resources are encapsulated into CROWN Nodes, and services are deployed on these nodes to provide a homogeneous view for upper middleware to access the resources.

CROWN Node Server is implemented on the basis of GT4 Java WSRF core, and figure 5 shows its system architecture. GT4 provides a stable implementation of WSRF specification family and a light-weight embedded runtime. However, these basic functions are not enough to satisfy the requirements for service container in real grid environments.

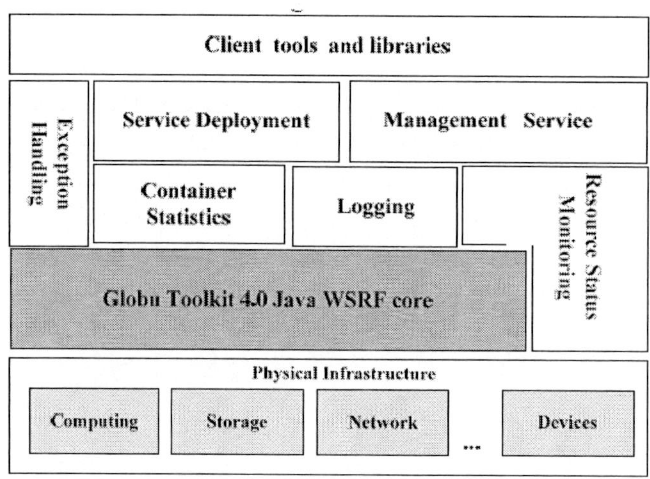

Figure 5. System Architecture of Node Server.

Resource Management Using RLDS

CROWN defines domain, region, and gateway to management heterogynous resources on the wide area network.

Nodes are organized into different domains according to their subordination or resource catalogs. A Resource Locating and Description Service (*RLDS*) is deployed in each domain as the center of grid information management. Domains may contain several sub-domains such that all RLDS come up with a tree like topology.

For flexibility, the concept *region* is introduced to coordinate different domain trees. A region switch is deployed in each region such that flexible topologies and information sharing mechanisms can be applied among regions.

To connect with other grid systems, we deploy several gateways on the edge of the CROWN system.

S-Club and RCT

CROWN employs a service club mechanism, called S-Club, for efficient resource organization and service discovery.

S-Club is used to build an efficient overlay over the existing GIS (Grid Information Service) mesh network [8, 9, 7]. In such an overlay, GISs providing the same type of services organized into a service club. An example of such a club overlay is shown in figure 6, where nodes C, D, E, G form a club. A search request could be forwarded to the corresponding club first such that search response time and overhead can also be reduced if the desired result is available in the club.

Intuitively, to set up a club requires information exchange, and clubs need to be maintained dynamically because new GISs may join and some existing GISs may leave. Also, it is possible that some types of services become less popular after the club is built. Therefore, the system has to be careful on the trade-off between the potential benefit and the cost incurred. In general, the usage of services is not uniformly distributed. Some types of services can be very popular and others may not. When/how clubs are constructed/destroyed will be key issues in S-club scheme.

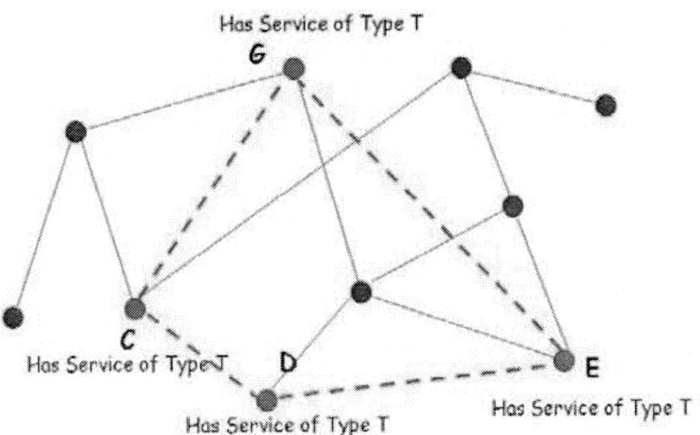

Figure 6. An example of Service Club.

Assuming any search request is firstly sent to a GIS close to the user. On receiving a search request for a specific service type, the GIS checks locally whether there has been a club for this type. If yes, the GIS forwards the request to the club, which will be flooded within the club only. If there is no club for this type, however, the GIS floods the request throughout the mesh network.

When a new GIS joins the GIS network, it has no idea what clubs are there. But since it has at least one neighbor in the underlying mesh network, it can ask one of its neighbors for the information of existing clubs. Namely, it simply copies the information of clubs from its neighbor. For more detailed discussion, please refer to [7].

Besides S-Club, there is a RCT (Resource Category Tree) for the third layer's resource management. Computational resources are usually described by a set of attribute-value pairs. Among all attributes of a computational resource, one or several attributes is choose to characterize the resource capacity of meeting application resource requirements as primary attributes (PA). An overlay called RCT (Resource Category Tree) is used to organize computational resources based on PAs.

Grid applications can be characterized by their requirements for computational resources, e.g., computing intensive and data intensive applications. In turn, categorizing computational resources based on certain resource characteristics that can meet application resource requirements. By doing so, resource discovery is performed on specific resource categories efficiently. For example, resources with huge storage can better serve a data intensive application, thus they can be organized together based on an overlay structure.

Furthermore, according to the observation, the values of most resource attributes are numerical, e.g., values of disk size. And attributes whose values are not numerical can be converted to be numerical through certain mathematical methods. Based on this consideration, RCT adopts an AVL tree (or balanced binary search tree) overlay structure to organize resources with similar characteristics. The attribute that can best describe the characteristic of resources organized by an RCT is named a primary attribute or PA. Figure 7 is an example of RCT. The chosen PA is available memory size, and the value domain of available memory ranges from 0MB to 1000MB.

Compared with traditional AVL, each node of RCT manages a range of values, instead of a single value. Each node only needs to maintain its connection with direct child nodes and parent, and operations like registration, updating and query can start from any node. Unlike in traditional AVL structure, higher-level nodes of RCT are not required to maintain more information or bear more load than those in lower levels, which provide the basis for RCT to scale easily.

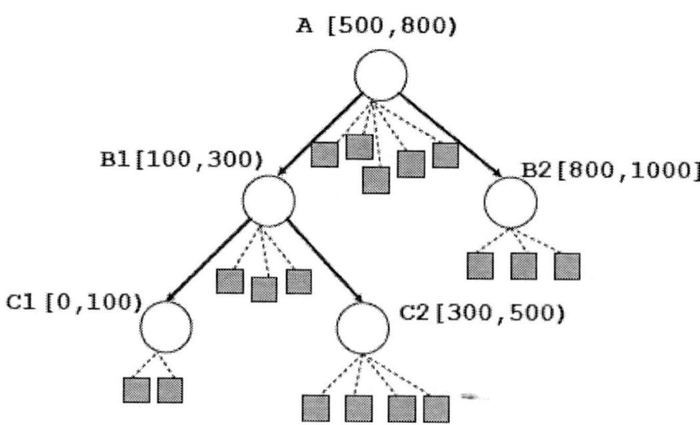

Figure 7. An example of RCT.

Suppose D is the value domain of the PA of an RCT. Each node n of an RCT is responsible for a sub range of D, or Dn. All resources with PA values belonging to Dn register themselves to node n. We name each RCT node an HR (Head of a sub Range). And terms of “HR n” and “node n” will be used interchangeably in the rest of this chapter. In figure 1, the circles denote HRs, while the squares below an HR denote computational resources registered with an HR.

Suppose N is the total number of HRs in an RCT, lc(n) and rc(n) are the left and right child nodes of HR n respectively. Since an RCT is a binary search tree, following observations can be found.

$$D_i \cap D_j = \varphi \;,\; \forall_{i,j} \in [1,N] \qquad (2)$$

$$D_{lc(i)} < D_i < D_{rc(i)} \;,\; \forall_i \in [1,N] \qquad (3)$$

Di < Dj if the upper bound of Di is less than the lower bound of Dj, e.g., [1, 2] < [3, 4].

If the ranges of node i and node j, i.e. Di and Dj, are adjacent, node i is referred to as a neighbor of node j, and vice versa. If node i is a neighbor of node j and Di< Dj, node i is called left neighbor of node j (denoted by L-neighbor(j)); node j is called right neighbor of node i (denoted by R-neighbor(i)). Note there are two exceptions: the leftmost HR and the rightmost HR, the former has no left neighbor and the latter has no right neighbor.

As shown in figure7, C2 and B2 are neighbors of A, while C2 is L-neighbor of A and B2 is R-neighbor of A. Note that C1 does not have left neighbor and B2 has no right neighbor.

As resources are owned and managed by different resource providers, providers may define different Pas for their resources, which results in constructing multiple RCTs. In figure 8, we present a 2-layer architecture for organizing resources across resource providers by using RCT. In the lower layer, each resource provider defines a set of PAs that can best describe their resources. Based on PAs, resources are organized through a certain number of RCTs. To enable wide area resource discovery across different providers, an RCT index service (RIS) is deployed by each service provider in the upper layer. RIS is a basic service that stores information about PAs of a provider and entry points of RCTs. RISs can be implemented, e.g., as web services or grid services, and find each other using services like UDDI.

In practice, a resource may have many attributes, but only a few of them are chosen as primary attributes. So there will not be too many RCTs. When a query request cannot be satisfied by a resource provider, the RIS will contact other RISs to recommend another resource provider for further discovery operations.

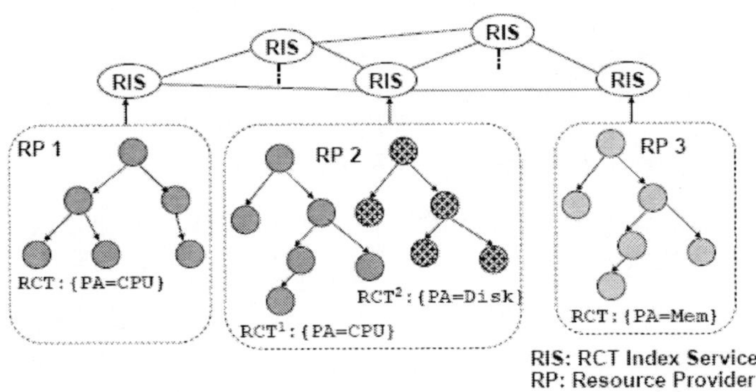

Figure 8. Resource organizing with RCT.

SECURITY ARCHITECTURE IN CROWN

Overview

CROWN provides a hierarchical security solution to secure virtual organizations established via CROWN middleware system. There are three levels of security mechanisms in CROWN Security Architecture, including node level, domain level, and region level security mechanisms. CROWN uses a federated construction to form the virtual organization, and the architecture of Security is designed accordingly as shown in figure 9. Term region is used to denote the area with homogenous security infrastructure such as PKI or Kerberos, and term domain to denote the area of autonomous organization.

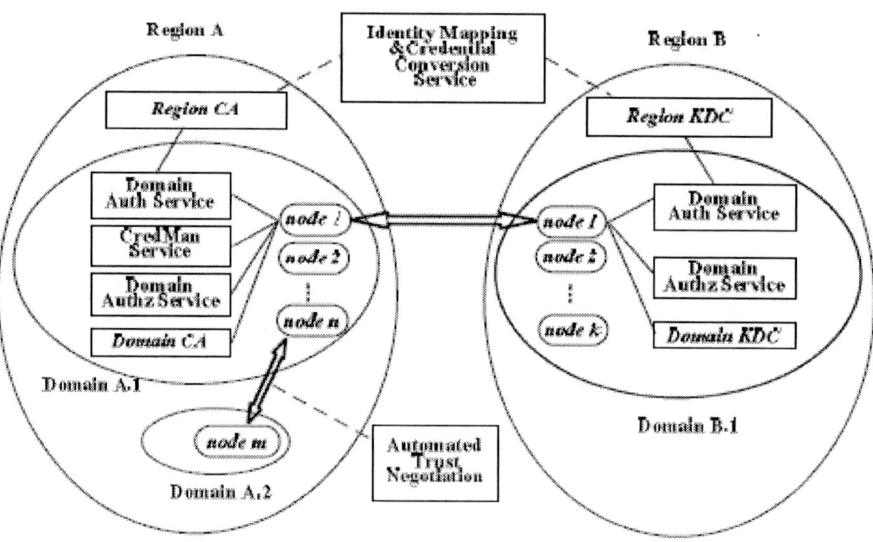

Figure 9. Architecture of CROWN security.

In order to wrap, share and protect the raw resources in autonomous domains, CROWN node should be deployed in the domain. It's the responsibility of the CROWN node to accept or intercept resource requests from grid users and do the security control. The raw resources to be protected are located in what we called protected area which may be a physical area or a conceptual area. The deployment of CROWN Security should insist that every access to the resources in protected area is mediated by a CROWN node.

Node Level Security

In CROWN node, CROWN Security implements communication security, fine-grained access control, basic message level security, such as encryption/decryption, signing/verification, and authentication, authorization mechanisms is provided. Moreover, other new functionalities can easily be extended in this architecture due to its flexibility. CROWN Security is highly flexible through configuration, which facilitates administrators to specify fine-gained security policies for each service. For example, it is feasible to apply various security processing modes, such as a signature-only mode, or an encryption-with-signature mode, etc., to different services, even different methods or method parameter values in a service, in the same node.

Currently, CROWN Security supports two kinds of security infrastructure, Kerberos and PKI. Therefore X.509 certificate [14] and Kerberos ticket are both supported in authentication module. Both Kerberos and PKI authentication are implemented as a WSSecureConversation [15] service which conforms to GSS-API standard [16]. For instance, a service deployed in Kerberos region can use CROWN Security to authenticate and authorize users according to their Kerberos credential, and in the mean time the same service can also be deployed in PKI region with only slight configuration adjustments made by administrator. This feature is the essential infrastructure to support further credential federation among regions.

In particular, during the dynamic trust establishment between two unknown nodes located in different security domains, the sensitive credentials or access control policies may be disclosed. In CROWN Security, a dedicated ATNService, namely Automated Trust Negotiation Service, which comply with WS-Trust standard, is provided to preserve privacy for the nodes. If the service requester has a trust ticket issued by target service, then the trust can be established without negotiation. Otherwise, trust negotiation will be triggered, where the negotiation strategy enforcer in the ATNService will determine where and which credentials should be disclosed. Specially, an advanced trust chain construction component, which holds by trust management with various delegation credentials, is supported in ATNService.

Domain Level Security

Although some security functions such as authentication and authorization are implemented as a node level security mechanism in CROWN Security. Sometimes, it is a huge burden for administrators to maintain authentication and authorization policies on enormous number of CROWN nodes in each domain. Therefore, several fundamental security services are provided by CROWN Security with the intention to ease the security administration and reduce the administration burden, including the authentication service (AuthService) and the authorization service (AuthzService). For example, a centralized authorization service can be deployed in a domain, and this authorization service will serve for all grid services reside in the CROWN nodes in this domain to make authorization decision.

Furthermore, CROWN Security provides a credential management service (CredManService) as a MyProxy [11] replacement in CROWN middleware. CredManService allows users to access their credentials anywhere, anytime, even when they are on a system without Grid infrastructure or without secure access to their long-term credentials as MyProxy does. However, CredManService is implemented as a grid service, and it is decoupled with underlying security mechanisms. This actually benefits to the administrators with immeasurable flexibility to tailor different security configurations for different service deployment. On the other hand, MyProxy is heavily coupled with SSL as a session security mechanism and a built-in access control model which is hard coded and inflexible to extend.

The domain administrator can deploy these services selectively. These services are all implemented as an extension to the WS-Trust standard [17], which has a policy-based design, therefore they are highly adaptable and easy to configure.

Region Level Security

Region level security mechanism in CROWN Security is realized by credential federation service (CredFedService). In a multi-party collaboration, users in one region may have fundamental problem accessing services provided by other region because they have different authentication method as well as different format for user credential, such as X.509 certificate and KerberosV5 [18] ticket. A credential conversion mechanism is an essential enabling mechanism to establish profound collaboration among multi-parties. For example, CredFedService can be employed as a bridge between PKI region and Kerberos region. Therefore users from one region can access the resources across different security infrastructure via the policy-based identity mapping and credential conversion feature provided by CredFedService. CredFedService is also implemented as a grid service, which is decoupled with underlying security mechanisms. Administrators can adapt different security configuration as well as identity mapping policy to their own requirements.

Design and Implementation

As discussed above, CROWN Security presents an extensible framework and implements basic communication security component inside CROWN Node. CROWN Security also provided four other components based on the framework, including Credential Management, Policy Based Authorization, Trust Management and Negotiation, and Credential Federation. Implementation of CROWN Security is tightly integrated with the CROWN NodeServer, which is the core component of CROWN Middleware system. Basic function of CROWN Security comes together with CROWN NodeServer and several fundamental security services are available as grid service archives, which can be remotely deployed into a CROWN NodeServer through ROST service [10].

Before diving into design and implementation details, security structure of CROWN node will be discussed in following subsection, which is essential design to realize flexible and adaptable features of CROWN Security.

Security Structure for CROWN Node

Figure 10 depicts the internal security structure for CROWN node. The design of this structure is much inspired by Axis although the purpose of CROWN Security is message level security processing rather than SOAP message processing. The security processing depends on a configurable security chains, and generally two chains, which deal with the request and response messages of service respectively, are configured for every grid service. For the sake of simplicity, there is only a processing chain shown in figure 10. Each handler in the processing chain is in charge of some specific security functions. The grid services developers, deployers and administrators can customize grid services protection by merely configuring these chains. The configuration is stored in security descriptors.

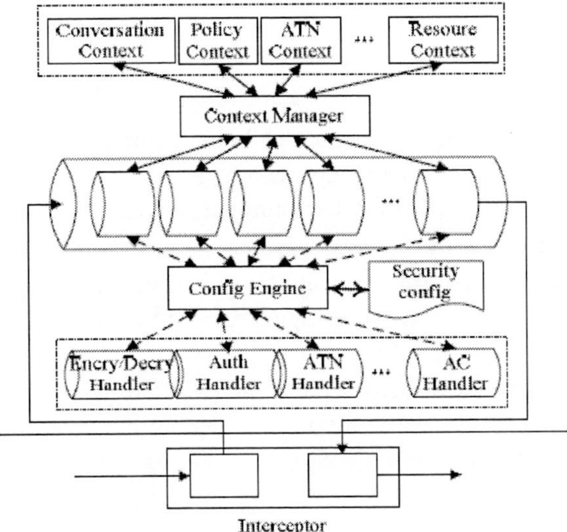

Figure 10. Security architecture of CROWN node.

When the message interceptor embedded in CROWN node intercepts request or response message for a grid service, it will call the engine with information related to this message. Then the engine will generate an appropriate chain by means of a configuration engine according to the security descriptor, and invoke the chain to process the message, i.e., invoke each handler in the chain in a sequential order. After invoking a handler, the engine will choose continuing the process or terminating it according to the current result. The configuration engine can also be used to cache the instantiated processing chains and handlers in order to achieve better performance.

Besides the information related to the request or response message, some other information such as session keys, states of automated trust negotiation, properties of resources are also needed by the handlers to finish their processing. In CROWN Security, the information is called security contexts and is classified and managed by a context manager.

It should be noted that most handlers currently implemented in CROWN Security follow the policy-based design. For some handlers, such as authentication handler and authorization handler, there are two editions available with different modes, namely call-out mode and stand-alone mode. For instance, a standalone authentication handler will follow the authentication policy specified in a node-local security descriptor with all the policy decisions made locally, while a call-out authentication handler will merely read the locations of access point of a centralized authentication service and consult the service for policy decisions.

As discussed above, to wrap, share and protect the raw resources in autonomous domains, CROWN nodes should be deployed into the hosts of the domain. The software to be installed is called CROWN NodeServer, which is the core component of CROWN Middleware system. NodeServer is implemented based on GT WS-Core container with various new features and extensions, such as remote and hot service deployment, monitoring and management service, etc.

The security structure for CROWN Node is tightly integrated with the CROWN NodeServer. Some function of CROWN-ST comes together with CROWN NodeServer as security handlers, which can be configured and customized by administrators in security processing chains.

Communication Security

Communication security module consists of both security handlers and security services which can be used to secure corresponding messages between nodes, including encryption, decryption, signature, authentication handlers, authentication service, and secure-conversation service. All handlers provided by CROWN Security conform to WS-Security standard in terms of SOAP message encryption and signature. Moreover, WS-Policy [12] language is used to express different policies for message processing which makes CROWN Security highly flexible.

CROWN Security currently supports three modes of message level security, which are username token mode, secure-message mode, and secure-conversation mode. The first two modes are similar with those implemented in GT4, which complies with WSSecurity. Furthermore, our secure-conversation mode supports using both X.509 certificate and KerberosV5 ticket as user credential for authentication and encryption, which conforms to WSSecureConversation [15], WS-Trust [17], and IETF GSS-API standards [16].

Policy Based Authorization

Policy-based authorization module in CROWN Security implements policy decision point in both handler and service. We adopt XACML [13] (eXtensible Access Control Markup Language) to express fine-grained access control policy in AuthzService. By using SAML assertions, the AuthzService can make authorization decision based on user attributes rather than identity.

In figure 11, authorization module intercepts each request sent to target service, then collects attribute certificates signed by attribute authority for both user and service to form a request context, which is conducted by policy decision point to make a authorization decision for the request. As mentioned previously, authorization policy is coded with XACML language and managed by domain administrator.

Credential Management

Grid Portals are increasingly used to provide user interfaces for grid. Through these interfaces, users can access a grid conveniently. When security is taken into account, it is required that the user can access his credentials in a secure and convenient way anytime, anywhere. Credential management module, which consists of a CredManService and

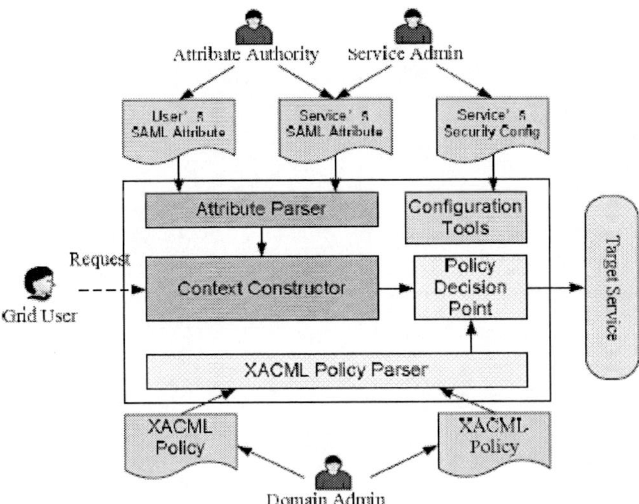

Figure 11. Structure of the Authorization module.

corresponding client tools, in CROWN Security, known as CredMan, is designed to meet this requirement. At first, a user would use a CredMan client command, named *credman-init*, to visit the CredManService, and delegate a set of proxy credentials which are signed by the user's permanent credential to the service repository. At a later time when the user's credential is needed, the user, or service acting on the behalf of the user to get a proxy credential delegated from the proxy credential stored in the repository.

In CROWN, client tools can be integrated with the portal, that is, a user can access his credential through portal. By using the tools, a user can easily delegate to and retrieve credential from the repository. Moreover, some client tools are provided for the user to manage the credential stored in the repository. In order to protect the credentials in the service repository, CredManService provides a protected mechanism in which user can specify authentication information and retrieval restrictions to protected his credentials in the repository.

Trust Management and Negotiation

A dedicated ATNService can be deployed with the target service to support the trust negotiation with the service requester.

As illustrated in figure 12, a series of procedures are involved in the trust negotiation. When the client requests the target service which protected by trust negotiation service, it will firstly initialize an *ATNEngine* through local *RedirectHandler*. Upon receiving the negotiation requests from client, the service provider will create an *ATNEngine*, too. The state of negotiation will be stored in *ATNContext*. Then, the two participants may disclose their credentials according to the provider's policy; or policies for sensitive credential. This process will be interacted until a final decision ('success' or 'failure') is reached. If the negotiation succeeds, ATNService will return a success status, and the context will be updated accordingly. The requester can insert the session id into SOAP header and sign it before sending to the target service. The target service will verify the authenticity of session id through its AuthzHandler, and allow the access if the verification succeeds.

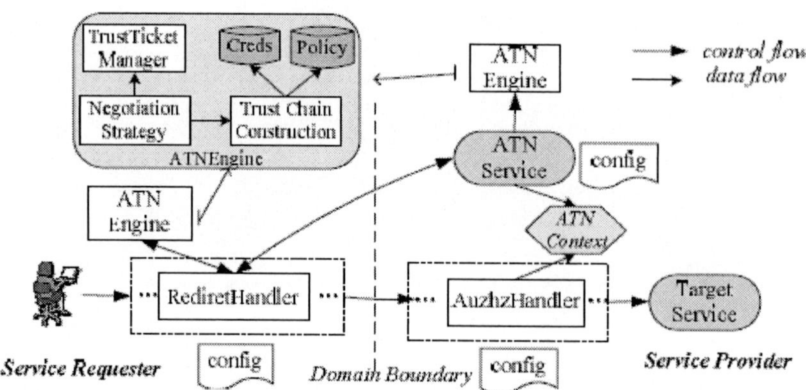

Figure 12. ATNService and ATNEngine.

Credential Federation

Credential Federation component is provided as a grid service called CredFedService. The function of CredFedService is to convert X.509 certificate to KerberosV5 ticket according to specified identity mapping policy, and vice versa. Figure shows relationships and data flow among the modules inside CredFedService implementation.

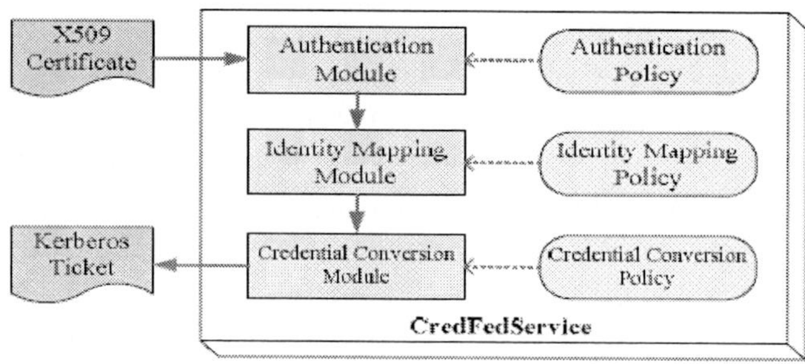

Figure 13. CredFedService.

The input of CredFedService is user's credential, and the output is a new credential in a format different with the input credential. Figure 13 demonstrated a procedure of mapping a X.509 credential to a Kerberos credential. First, the input credential is processed by the authentication module, which is realized by secureconversation mode offered by underlying communication security component of CROWN-ST, to verify whether the user is the real owner of this credential. If so, the credential is then forwarded to identity mapping module, which will map the identity of user to another domain based on mapping policy. Then the new identity will be processed by credential conversion module to generate a new credential for the user. Finally this credential is returned to the user by CredFedService.

As shown in figure 13, each module has its corresponding policy that can be customized by administrator of CredFedService.

Summary

CROWN Security provides a finegrained and extensible framework enabling trust federation and trust negotiation for resource sharing and collaboration in an open grid environment. We also demonstrate the performance of our implementation through comprehensive experimental studies. CROWN Security aims to satisfy some security requirements of dynamic distributed resources sharing and integration, but much works remains to be done.

TESTBED AND APPLICATIONS OF CROWN

CROWN is now becoming one of the important e-science infrastructure in China.

We have developed and deployed a series of applications from different disciplines, which include Advanced Regional Eta-coordinate numerical prediction Model (AREM), Massive Multimedia Data Processing Platform (MDP), gViz for visualizing the temperature field of blood flow, Scientific Data Grid (SDG) and Digital Sky Survey Retrieval (DSSR) for virtual observatory. These applications are used as test cases to verify the technologies in CROWN.

AREM uses Grid as a tool to study and refine the numerical prediction model of weather and climate. Several numerical models are worked out by meteorologists during their research and prediction work. Typically these models use the raw weather data from national meteorology authority as input, and simulate the weather transformation according to the laws of atmospheric physics and fluid dynamics. The output can be used as a prediction result of the future weather. The simulations are all based on complex numerical calculations and need large quantity of computing power and storage capacities. By using the resource organization, job scheduling technologies provided by CROWN, we successfully developed AREM research system. We encapsulated the Fortran complier, visualization tools (GrADS) and the simulation framework of AREM as services, and a unified raw weather data center is also deployed. Meteorologists can submit simulation jobs to the system and refine the numerical models according to the results. Since the jobs are executed by using the resources provided by CROWN testbed, the execution procedure can be parallel, the execution time can be much reduced and the efficiency of weather system research and prediction model refinement is improved.

Large amount of storage capability and computing power is needed when performing multimedia data processing, such as content recognition of voice or video. Traditionally a central processing model is applied and pieces of data are collected and processed in a single point. When the input data increase, this method provides little scalability especially for the real-time applications. We combine the service Grid technologies with the massive data processing and implement the MDP platform for multimedia data processing. MDP has been deployed into CROWN and provides service since 2005. We encapsulate the related algorithms into services and deploy then on many Grid nodes. Users can provide many ways of multimedia data and submit jobs to the Grid scheduling system. After analyzing the work load of Grid nodes available resources can be found automatically and data can be processed by invoking corresponding services. Since the platform is deployed in a wide-area environment, we also introduced the trust management and negotiation mechanisms. These technologies protect the user data and make the processing trustworthy. By using MDP, the resources that can be used to process multimedia data increase apparently, and the throughput and dependability of processing can be much improved.

CROWN interoperates with other Grid middleware through specifications. The testbed also links to some famous Grid testbeds. For example gViz application is deployed both in CROWN and White Rose Grid (WRG) which is a part of UK National Grid Service (NGS). We demonstrated the application on UK e-Science All Hands Meeting 2005 to show the interoperability of heterogeneous and autonomic Grid systems.

The experience of system development and deployment mentioned above shows that CROWN provides the capability of resource management, distributed access control and trust management and negotiation. It can bc used to support applications that are computation intensive and/or data intensive. 11 applications are deployed into CROWN by April 2006 and more than 25000 requests are processed.

ACKNOWLEDGEMENT

Part of this work is supported by grants from the China Natural Science Foundation (No. 90412011 and 60703056), China 863 High-tech Programme (Porject No. 2006AA01A106), China 973 Fundamental R and D Program (No. 2005CB321803) and National Natural Science Funds for Distinguished Young Scholar (No. 60525209). We would also like to thank Jianxin Li, Tianyu Wo and Hailong Sun and other members in CROWN team at Beihang University for contributions to design and implementations of CROWN middleware, testbed, applications and related resource management technologies mentioned in this chapter.

REFERENCES

[1] "CROWN, http://www.crown.org.cn"

[2] I. Foster, C. Kesselman, J. Nick, and S. Tuecke, "The physiology of the Grid: An Open Grid Services Architecture for distributed systems integration", http://www.globus.org/research/papers/ogsa.pdf

[3] Ken Brodlie, David Duce, Julian Gallop, Musbah Sagar, Jeremy Walton, Jason Wood. "Visualization in Grid Computing Environments". *Proceedings of IEEE Visualization 2004*, pp155-162.2004

[4] Hailong Sun, Wantao Liu, Tianyu Wo, Chunming Hu, "CROWN Node Server: An Enhanced Grid Service Container Based on GT4 WSRF Core", *Fifth International Conference on Grid and Cooperative Computing Workshops*, pp. 510-517, 2006

[5] Jinpeng Huai, Tianyu Wo, and Yunhao Liu, "Resource Management and Organization in CROWN Grid," in *Proceedings of the 1st international conference on Scalable information systems*, 2006.

[6] Jinpeng Huai, Hailong Sun, Chunming Hu, Yanmin Zhu, Yunhao Liu, Jianxin Li," ROST: Remote and hot service deployment with trustworthiness in CROWN Grid", *Future Generation Computer Systems* Volume 23, Issue 6 (July 2007)

[7] J. Frey and T. Tannenbaum, "Condor-G: A computation Management Agent for multi-Institutional Grids," *Journal of Cluster Computing*, vol. 5, pp. 237, 2002.

[8] W. Hong, M. Lim, E. Kim, J. Lee, and H. Park, "GAIS: Grid Advanced Information Service based on P2P Mechanism," in *proceedings of the 13th IEEE International Symposium on High Performance Distributed Computing (HPDC-13)*, 2004, pp. 276-277.

[9] A. Iamnitchi, I. Foster, and D. C. Nurmi, "A Peer-to-Peer Approach to Resource Location in Grid Environments," in *proceedings of the 11th IEEE International Symposium on High Performance Distributed Computing (HPDC-11)*, 2002.

[10] H. Sun, Y. Zhu, C. Hu, J. Huai, Y. Liu, and J. Li, "Early Experience of Remote and Hot Service Deployment with Trustworthiness in CROWN Grid," presented at *6th International Workshop on Advanced Parallel Processing Technologies (APPT 2005)* 2005.

[11] J. Basney, M. Humphrey, and V. Welch, "The MyProxy Online Credential Repository," *Software: Practice and Experience*, vol. 35, pp. 801-816, 2005.

[12] S. Bajaj, D. Box, and D. Chappell, "Web Services Policy Framework," 2005.

[13] T. M. Simon Godik, "OASIS eXtensible Access Control Markup Language (XACML)," 2003.

[14] R. Housley, W. Ford, T. Polk, and D. Solo, "Internet X.509 Public Key Infrastructure Certificate and CRL Profile," 1999.

[15] S. Anderson, J. Bohren, and T. Boubez, "Web Services Secure Conversation Language," 2005.

[16] J. Linn, "Generic Security Service Application Program Interface, Version 2," 1997.

[17] S. Anderson, J. Bohren, and T. Boubez, "Web Services Trust Language," 2005.

[18] C. Neuman, T. Yu, S. Hartman, and K. Raeburn, "The Kerberos Network Authentication Service (V5)," 2005.

In: Cyberinfrastructure Technologies and Applications ISBN: 978-1-60692-063-3
Editor: Junwei Cao

Chapter 8

SEMANTICS-ENABLED SERVICE DISCOVERY FRAMEWORK IN A PAN-EUROPEAN PHARMACEUTICAL GRID*

***Changtao Qu*[1], *Falk Zimmermann*[2], *Kai Kumpf*[3], *Richard Kamuzinzi*[4], *Valérie Ledent*[5] and *Robert Herzog*[6]**

[1,2] IT Research Division, NEC Laboratories Europe, NEC Europe Ltd.
Rathausallee 10, D-53757 Sankt Augustin, Germany
[3] Fraunhofer Institute for Algorithms and Scientific Computing
Schloss Birlinghoven, D-53754 Sankt Augustin, Germany
[4,5,6] Department of Molecular Biology, Universite libre de Bruxelles
Rue des Professeurs Jeener et Brachet 12, B-6041 Gosselies, Belgium

ABSTRACT

We present the design and implementation of a semantics-enabled service discovery framework in a pan-European pharmaceutical Grid: SIMDAT, an industry-oriented Grid environment for integrating thousands of Grid-enabled biological data services and analysis services. The framework consists of three major components: the OWL-DL-based biological domain ontology, OWL-S-based service annotation, and semantic matchmaker based on the ontology reasoning. Built upon the framework, workflow technologies are extensively exploited in the SIMDAT to assist biologists in (semi)automatically performing *in silico* experiments. We present a typical usage scenario through the case study of a biological workflow: *IXodus*.

---

* This chapter is an extended version of "C. Qu, F. Zimmermann, K. Kumpf, R. Kamuzinzi, V. Ledent, and R. Herzog, Semantics-Enabled Service Discovery Framework in the SIMDAT Pharma Grid, IEEE Trans. on Information Technology in Biomedicine, vol. 12, no. 2, March 2008, pp.182-190".

1 E-mail address: qu@it.neclab.eu

2 E-mail address: zimmermann@it.neclab.eu

3 E-mail address: kumpf@scai.fraunhofer.de

4 E-mail address: rkamuzinzi@ulb.ac.be

5 E-mail address: vledent@ulb.ac.be

6 E-mail address: rherzog@ulb.ac.be

INTRODUCTION

EU FP6 Grid Research Projects: Overview

In Sept. 2004, the EC (European Commission) launched 12 research projects in the area of Grid technologies under FP6 (the sixth Framework Programme), which received 52M Euro of EU (European Union) funding. These projects are intended to investigate the key approach to Grid research, which, as envisioned by the EC, combines "technology push", i.e., developing underlying technologies and interoperability standards, with "application pull", i.e., developing the enabling technologies needed for real world applications such as modeling, simulation, data mining and collaborative working tools. The bulk of the EU funding went to four flagship projects, respectively, the SIMDAT (Data Grids for Process and Product Development using Numerical Simulation and Knowledge Discovery, http://www.simdat.org), the NextGRID (Architecture for Next Generation Grids, http://www.nextgrid.org), the Akogrimo (Access to Knowledge through the Grid in a Mobile World, http://www.mobilegrids.org/), and the CoreGRID (The European Research Network on Foundations, Software Infrastructures and Applications for large scale distributed, GRID and Peer-to-Peer Technologies, http://www.coregrid.net), which each received an EU contribution of around 9M Euro. Together with other eight smaller projects launched in Sept. 2004 as well as 20 additional Grid projects launched in Sept. 2006 with 70M Euro of EU funding, the EU FP6 Grid research projects bring together dozens of universities, research institutes, large and small companies from across Europe to muster the "critical mass" of expertise and resources necessary to trigger change. As illustrated in figure 1, these projects cover almost all technologies that are critical to the next generation Grid, and the dissemination of the research results is also intended to go beyond Europe, driven by several continual projects also under FP6, such as the EC-Gin (Europe-China grid InterNetworking, http://www.ec-gin.eu/), the EchoGrid (EC-China strategic GRID Roadmap, http://echogrid.ercim.org/), the Grid@Asia (Advanced Grid Research Workshops through European and Asian Cooperation, http://www.gridatasia.net/), which were launched in the beginning of 2007 with the focus on cooperation with China on Grid Technologies[7].

SIMDAT and the SIMDAT Pharma Grid

As one of the four flagship projects of the EU FP6 program, the SIMDAT is basically an "application pull" project, which aims at developing generic Grid technologies for the solution of complex application problems in four industrial application sectors: Automotive, Pharma, Aerospace and Meteorology [1]. In overall, seven key technology layers are identified as important to achieving SIMDAT's objectives, respectively [1]:

- An integrated Grid infrastructure offering basic services to applications and higher level layers;
- Transparent access to data repositories on remote Grid sites;

[7] For detailed information about the EU FP6 Grid research projects, we refer to the EU IST (Information Society Technologies) website under http://cordis.europa.eu/ist/grids/projects.htm.

- Management of virtual organizations;
- Scientific workflow;
- Ontologies;
- Integration of analysis services; and
- Knowledge services.

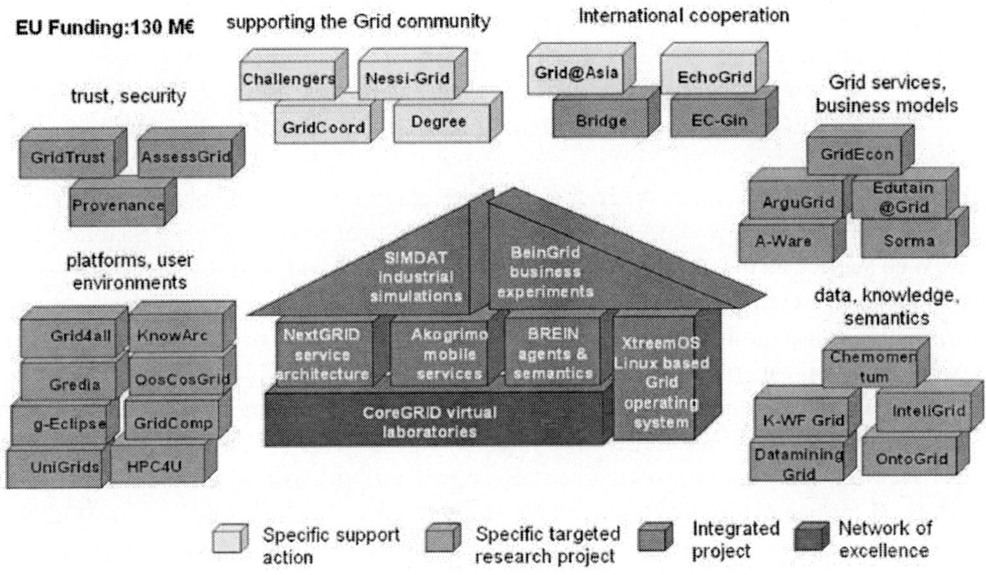

Figure 1. Overview of the EU FP6 Grid research projects.

These technologies are developed within the SIMDAT, and, according to different application requirements, are further adapted to each of four application domains.

In the SIMDAT Pharma sector, the focus is on the provision of a global knowledge Grid for life sciences, which is able to integrate thousands of Grid-enabled biological data services and analysis services to support collaboration between research institutes, e.g., within the EMBnet (European Molecular Biology Network, http://www.embnet.org), and, in particular, between businesses, e.g., between biotechnology companies such as Inpharmatica Ltd. (http://www.inpharmatica.co.uk/) and pharmaceutical companies such as GlaxoSmithKline (http://www.gsk.com/). Until the end of 2007, two testbeds have been set up in the SIMDAT Pharma Grid, respectively:

- *Distributed in silico analysis and annotation of biological sequences*: This testbed is focused on the analysis and annotation of a *cDNA* databank using a large number of traditional bioinformatics tools, including the EMBOSS toolkit and access to BLAST sequence comparison services. The working scenario is typical of the work conducted by biotechnology and pharmaceutical research organizations in early stages of drug design, and specifically in understanding the biological targets of the drugs. The tools used are made available by many organizations, and each tool is encapsulated as a Grid service. There may be multiple service providers for each service; thus, the focus of the testbed is to enable dynamic, rather than static, binding of different Grid service instances.

- *Collaborative B2B (Business to Business) Pharma application*: This testbed is focused on the interaction between a large pharmaceutical company, namely, GlaxoSmithKline, and a biotechnology company, namely, Inpharmatica, which is capable of providing specialized screening services. The working scenario is typical of the future B2B collaborations for drug discovery. Grid technologies are used to coordinate the data flow and invocation of remote services between both organizations within a secure environment providing resource management and service accounting facilities. Furthermore, different from the first testbed, in the B2B testbed, remote Grid services are transparent to end users, which implies that the services can be checked and audited by service requestors.

Within the SIMDAT Pharma Grid, one of the essential requirements is on the discovery of biological services. Nowadays more and more biological resources, e.g., Soaplab [2], have chosen Web service as the standard interface. However, as biological resources are essentially distributed and heterogeneous [3], the discovery of biological resources/services in bioinformatics is becoming increasingly challenging in the face of flooded, interlinked, and intertwined biological resources. Specific to the SIMDAT, biological services in the SIMDAT Pharma Grid are typically provided by several famous bioinformatics tools including the SRS (Sequence Retrieval System) [4], the MRS (Maarten's Retrieval System) [5], the BLAST (Basic Local Alignment Search Tool) [6], the InterProScan [7], and the EMBOSS (European Molecular Biology Open Software Suite) [8], etc. While these biological services are basically, or can be mapped to, standard Web services, in the SIMDAT, they are further wrapped through an industry-strength Grid middleware, namely, GRIA (Grid Resources for Industrial Applications) [9], thus possessing enhanced capability in security, QoS (Quality of Service), and business management. The service discovery in the SIMDAT has to deal with not only a large number of services from different categories and administrative domains, but also lots of service features including both bioinformatics and Grid ones. Furthermore, with the introduction of some advanced experimental methods in the SIMDAT, in particular, biological workflow [10][11], the service discovery is becoming increasingly crucial, as advanced service discovery, i.e., (semi)automatic service discovery, selection, composition, and invocation, as defined in [12], is deemed the key enabling technology of workflow [10].

Due to a strong capability for knowledge modeling, integration, and reasoning, semantic technologies, in particular, ontologies, are increasingly recognized as a key technology in bioinformatics to represent sophisticated relationships between biological resources, and further enable some more advanced resource discovery patterns than simple keyword-based ones [13][14][15]. However, restricted by the technology development, most of previous semantic service discovery frameworks in bioinformatics such as Semantic MOBY [13], Feta [14], and Grimoires [15] are rather lightweight in the sense that they have not yet made full use of semantic technologies to support advanced discovery functionalities such as matchmaking (c.f. "Related Work"). With the special purpose to support (semi)automatic biological workflow in the SIMDAT, we design and implement a service discovery framework, which, in comparison to previous projects, is more focused on semantics-enabled service matchmaking. Taking advantage of the state-of-the-art semantic technologies, the framework highlights an OWL-DL (Web Ontology Language - Description Logic) [16] based bioinformatics domain ontology, OWL-S [12] based service annotation, and semantic matchmaker based on the ontology reasoning. Within Grid computing environments, the framework can easily be integrated with most of the mainstream Grid middleware through its Web service, WSRF (Web Service Resource Framework)/GT4 (Globus Toolkit) [17][18], and the GRIA [9] interface.

DESIGN OF THE SEMANTICS-ENABLED SERVICE DISCOVERY FRAMEWORK IN THE SIMDAT PHARMA GRID

As *in silico* experiments conducted in large knowledge discovery processes are both cumbersome and error-prone if run manually, workflow technologies are extensively exploited in the SIMDAT with the purpose to enable *in silico* protocol designers to devote more to the core scientific problem than to dealing with the complex deployment of biological services. By means of the SIMDAT workflow toolkit, biologists are expected to be effectively shielded from the technical middleware. They can compose the experiment procedure using “abstract” task components, describe their requirements on each component, and the workflow enactor is then responsible for (semi)automatically discovering/selecting service instances according to the user requirements, and further managing the execution of the workflow. As *in silico* experiments are basically designed at a high level of abstraction following such a workflow approach, the workflow components and commonly used workflow patterns can easily be reused. This can, on the one hand, enable rapid development of new *in silico* experiments, but, on the other hand, it may also reduce the requirements on users’ domain knowledge for designing complex *in silico* experiments.

For the (semi)automatic workflow execution in the SIMDAT as described above, the service discovery is of vital importance. As all “abstract” workflow components have to be dynamically bound to “concrete” service installations at run-time, the workflow enactor depends on the service discovery module to discover actual service instances and further select the most suitable one. Functionally, the service discovery framework has to provide a mechanism to annotate and advertise service instances, represent user requirements/queries, and discover service instances in terms of user requirements and service annotations. Going a step further, it is also desired to support some advanced functionalities such as the service matchmaking and advanced result ranking, which are deemed indispensable for the workflow enactor to (semi)automatically select optimal services [10].

In order to implement such an advanced service discovery framework in the SIMDAT, we exploit a semantic approach, in which ontologies are adopted as a key technology to represent biological domain knowledge, describe different service features, define query protocols, and further serve as the basis for advanced service discovery and matchmaking. This design is essentially different from some popular Web service or Grid service registries such as the UDDI (Universal Description, Discovery, and Integration) [19] and the ebXML [20], which are generally not able to handle semantic service annotation/discovery in terms of domain ontologies. It is also distinct from some semantic extensions to above registries such as [15] and [21], whose discovery capability is basically restricted by the rigid data model of either the UDDI or the ebXML. In our framework, we adopt the OWL-S as a more generic, extendable, and semantic-rich data model to describe and discover services, and also define an OWL-DL-based domain ontology to provide formal semantics for representing biological domain knowledge. As the service description data model (i.e. OWL-S), service annotations, and domain ontologies are uniformly in the OWL-DL format, the semantic service discovery and matchmaking can be realized by means of the OWL-DL reasoning.

In figure 2, we illustrate the design of the semantics-enabled service discovery framework in the SIMDAT Pharma Grid.

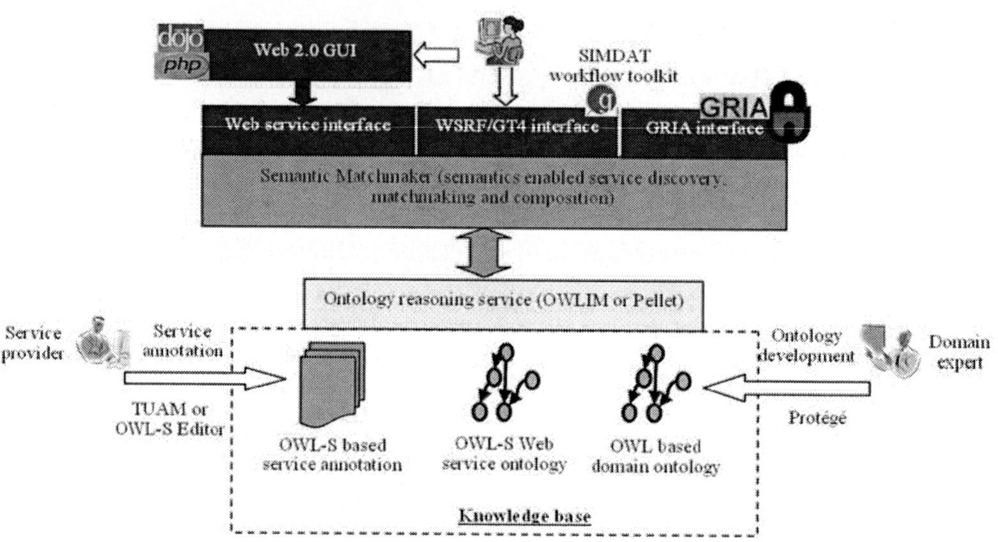

Figure 2. Design of the semantics-enabled service discovery framework in the SIMDAT Pharma Grid.

OWL-DL-Based Bioinformatics Domain Ontology

The ontology in (bio)informatics is, according to Schulze-Kremer, “a concise and unambiguous description of principle relevant entities with their potential, valid relations to each other” [22]. In the biological domain in general, ontologies have traditionally served data integration purposes [23][24]. For this purpose, a plethora of ontologies from diverse subfields have been developed mostly independently of each other, which implies that there is no common upper ontology. Most ontologies try to describe well-defined sub-domains of biology from the level of taxonomic genera down to molecular biology [25]. Most of these ontologies deal with entities from the world of living beings instead of the tools required to analyze them.

Unlike some popular biological/biomedical domain ontologies such as the BioPax (Biological PAthway eXchange) [26] and the FMA (Foundational Model of Anatomy) [27], etc., the SIMDAT domain ontology is not intended to model quasi-complete knowledge of a specific domain such as biological pathway in the BioPax, but to provide inter-domain knowledge for annotating and discovering service features of interest, e.g., data processed and produced through service I/O (Input/Output). This position determines that the domain ontology may not necessarily be large and complex, but it has to fit the biologists’ lines of thought when they look for services for specific biological tasks in *in silico* experiments. Just as described in [3], such an ontology should be able to "initiate a knowledge-driven line of questioning spanning the entirety of the expensive biological data space". Here the "entirety" also implies that the domain ontology should be scalable, extendable, interoperable, and (easily) maintainable. In the SIMDAT, we design such a domain ontology with joint efforts from biologists and informaticians.

From the biologists' perspective, three types of biological ontologies are identified as necessary to address the biologists' lines of thought for discovering biological services:

- *Bio Data Ontology*: Are we dealing with sequences, structures, amino acids, proteins, profiles, mutations, etc.? This ontology draws some concepts from the SO (Sequence Ontology) [28], and is clearly needed for annotating service features such as I/O. The most significant classification here is made between primary molecular data types (sequences of nucleic or amino acids) and higher order structures or derived data (e.g., alignment data). For each data type, a range of possible data formats or representations can be assigned. Note that we do not exploit the complete SO because it also represents biological sequence features that are attributes of such sequences such as whether a fragment of *DNA* plays a role as an *exon* or *intron*. Usually, these are derived data that have to be computed on demand or are stored as optional annotations to the primary data in databases, which implies that these do not play a role for the classification of services which concentrate on the primary data.
- *Bio Tool Ontology*: Are we looking for databank queries or alignment, phylogenetic tree tools, etc.? This ontology borrows some relationships from the myGrid ontology [14], but takes its major part from the EMBOSS tools classification hierarchy (http://poblano.health.unm.edu/Software/EMBOSS/Apps) [7] for categorizing biological analysis services. For data services (abstracted databases), it also references the SRS classification ontology [4]. This classification is not a simple inheritance tree, but rather contains some multiple inheritance. Thus, alignment tools like BLASTx must be listed below both protein and nucleic_acid tools. This "feature" cannot be avoided when one tries to accurately model various aspects of certain software tools that play a role depending on the task at hand.
- *Bio Taxonomy Ontology*: Which organisms are we dealing with? This ontology may refer to other related classification systems such as the one provided by the NCBI (National Center for Biotechnology Information) [29], and is clearly relevant for representing biological domain knowledge. For example, a database query could focus on specific *DNA* sequences from primates and mouse for later alignment.

As the SIMDAT domain ontology is directly used by biologists to annotate and discover services, it is thoroughly engineered during the design phase with the goal to facilitate the end user's usage. First of all, we try to make it straightforwardly interpretable for biologists through keeping it as sparse as possible. As illustrated in figure 3, at the top level of the Bio Tool Ontology, only two principal types of services are distinguished, i.e., computation service and database service. Two relationships are then modeled to link the Bio Tool Ontology to the Bio Data Ontology: one describes the input and the other the output data type, thereby providing basic workflow enabling information. Taxonomic information is only entered at the level of database instances as a NCBI taxonomy ID string. The rationale behind this is that, in general, only databases will sometimes contain taxon-specific information.

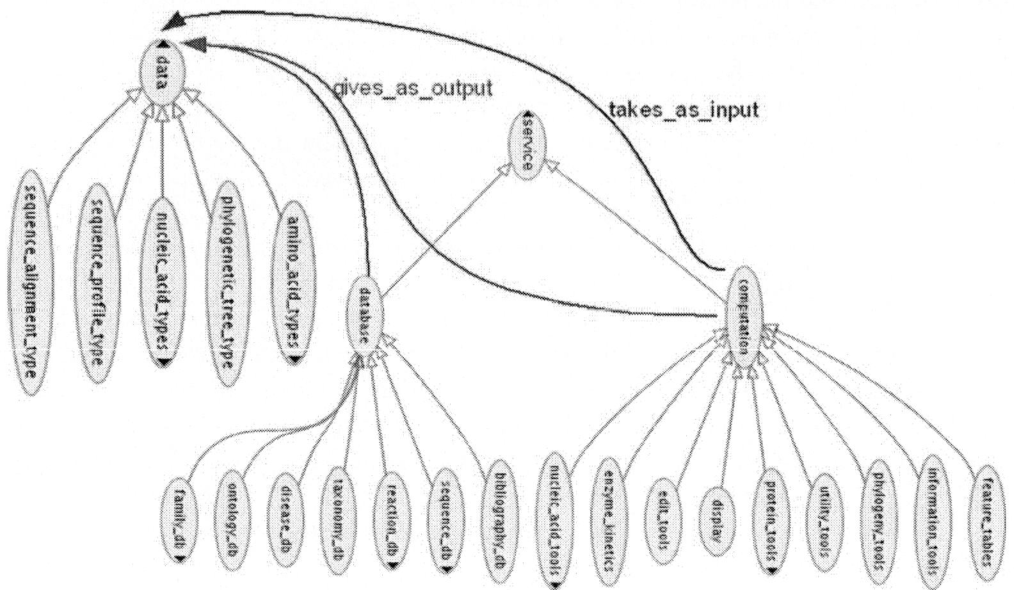

Figure 3. Top level concepts and relationships modeled in the SIMDAT domain ontology.

Second, different from the common practice of the ontology development as in the BioPax [26], all real service types in the SIMDAT domain ontology are also represented as class instances so that a service annotation only has to establish an "is_a" mapping between the service and the according instances. The class and consequently predefined instances are designed to accurately reflect the static or unvarying qualities of the service. Having predefined instances does not preclude the possibility of having an annotation strategy that lets the annotator create new service class instances on the fly. This is however not intended to be manually done by users, rather by a dynamic annotation tool, the Dynamo (Dynamic Annotator Module). As detailed in the following, the Dynamo can create new service instances including both predefined static service features and dynamically changing ones.

In correspondence to different service types modeled in the domain ontology, several service query patterns are provided by the SIMDAT workflow toolkit through a GUI (Graphical User Interface) to assist biologists in constructing (semantic) queries based on the controlled vocabularies defined in the domain ontology. Depending on user's domain knowledge and query interest, a query may only be pointed to semantic service features, e.g., *find services that belong to a class <A> of tools that allow data types <B><C> as input and give at least data type <D> as output*. Such a semantic query can further be extended to point to nonsemantic, bioinformatics service features, e.g., *services using <E> databank released after "2008-01-01"*, or nonsemantic, Grid-related service features, e.g., *services provided by a site with more than 4G memory*, etc. Note that depending on users' domain knowledge, the "semantic" vocabularies used in queries may be rather "shallow" or rather "deep" in the taxonomy tree. Correspondingly, the matching degrees of the same service instance may vary a lot, representing different semantic similarities between the user requirements and service annotations.

From the informatician's perspective, a scalable, extendable, interoperable, and (easily) maintainable domain ontology is designed using the W3C (the World Wide Web Consortium) standard ontology language OWL [16], and more precisely, the OWL-DLP (OWL Description Logic Programs) [30], which is the strict subset of the OWL-DL. Although the OWL-DLP is usually considered not to be suited for modeling complex biological/biomedical domain knowledge [31], we find it possible to restrict the SIMDAT domain ontology within the OWL-DLP, as at the current project stage we do not see the need for modeling complex relationships in the domain ontology for the service discovery purpose. However, with increasing requirements on biological service discovery, in particular, with the need for importing complex biological domain ontologies such as [26], [27], and [28], the sole OWL-DLP reasoning is deemed not to be sufficient. Therefore, in parallel to the OWL-DLP reasoner OWLIM [32], a full-fledged OWL-DL reasoner Pellet [33] can also be used in the framework to support full OWL-DL reasoning, though at the cost of system performance. As the semantic matchmaker is independent of underlying ontology reasoners thanks to a predefined uniform interface, it can easily switch between OWLIM and Pellet in order to leverage different levels of reasoning capability.

OWL-S-Based Service Annotation and the Annotation Toolkits

Among the WSMO (Web Service Modeling Ontology) [34] and the WSDL-S (Web service Description Language Semantics)/SAWSDL (Semantic Annotation for WSDL and XML Schema) [35][36], the OWL-S is one of the major initiatives in the SWS (Semantic Web Service) community, which is purposed to enable (semi)automatic discovery, selection, composition, and invocation of Web services [12]. As comparatively the WSMO is not yet mature enough and the WSDL-S/SAWSDL is conceptually much weaker (c.f. "Related Work"), we choose the OWL-S as the standard data model to describe different service features based on the SIMDAT domain ontology.

OWL-S and the OWL-S-Based Service Annotation

The OWL-S is an OWL-based Web service ontology which provides a core set of markup language constructs for describing the properties and capabilities of Web services in unambiguous, computer-interpretable form. As illustrated in figure 4, OWL-S itself consists of three subsets of ontologies [12]:

- The service profile for advertising and discovering services ("what the service does");
- The service model for describing detailed service operations ("how the service works");
- The service grounding for specifying the service invocation details such as the communication protocol, message formats, etc. ("how to access the service").

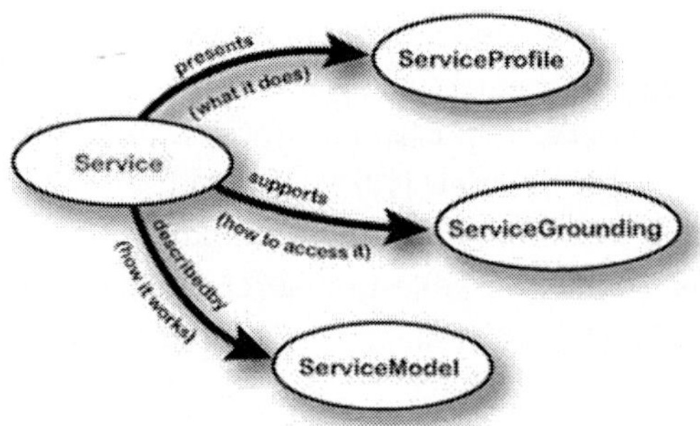

Figure 4. Top level structure of the OWL-S.

As the typical usage of the service discovery framework is to support (semi)automatic service discovery and selection through the SIMDAT workflow toolkit, we leave most of the complex OWL-S service model annotations such as "pre-condition", "result", as well as complicated control flow and data flow unaddressed, but specifically focus on the OWL-S service profile and service grounding annotation. According to [37], due to considerable complexity of *in silico* experiments, the automatic service composition in bioinformatics is still of less requirement in comparison to the semiautomatic one. However, the cost of annotating the OWL-S service model for the automatic service composition/orchestration would be rather expensive.

As illustrated in figure 5, the OWL-S service profile consists of several properties which can be used to annotate various service features. According to the current service annotation/discovery requirements in the SIMDAT, the following OWL-S service profile properties are used for annotating services:

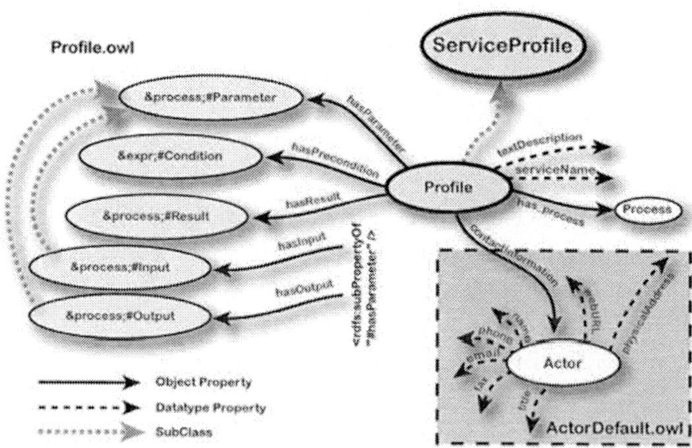

Figure 5. Structure of the OWL-S service profile.

- *serviceName, textDescription*: Used to annotate general service information.
- *contactInformation*: Used to annotate service provider information based on the OWL-S ActorDefault ontology (http://www.daml.org/services/owl-s/1.2/Actor Default.owl).
- *serviceClassification*: Used to annotate the service classification referring to the Bio Tool Ontology.

- *serviceCategory*: Used to annotate the service category referring to the Bio Taxonomy Ontology.
- *hasInput, hasOutput, and hasParameter*: Used to annotate service I/O referring to the Bio Data Ontology.
- *serviceParameter*: As the major extension point of the OWL-S service profile, *serviceParameter* is heavily used to annotate service features like QoS, security, and associated databanks, etc.

For the service grounding annotation, we mainly annotate services' WSDL grounding, and, respectively, using *grounding:WsdlAtomicProcessGrounding/grounding:wsdlService* to annotate the service invocation point, and *grounding:WsdlAtomicProcessGrounding/ grounding:wsdlDocument* to annotate the URL (Universal Resource Locator) of the service WSDL document. Such a grounding annotation is sufficient for the SIMDAT workflow toolkit to automatically invoke services through the GRIA middleware.

OWL-S-Based Service Annotation Toolkits: the TUAM and the Dynamo

As the OWL-S-based service annotation involves a sophisticated service upper ontology, namely, the OWL-S, an important issue to address in the service discovery framework is how to simplify the service annotation process with efficient tooling support. While Protégé's OWL-S Editor [38] is rather handy for informatician/bioinformaticians to annotate biological services, it requires considerable knowledge and familiarity with the OWL-S and the OWL-S-based service annotation model. In the SIMDAT, a bioinformatics-specific service annotation toolkit, namely, the TUAM (Tool for Universal Annotation and Mapping) [39], is designed and developed, which can better be used by biologists to create the OWL-S-based service annotations by means of an intuitive ontology browsing/navigating interface and an annotation GUI that can hide the OWL-S-based service annotation model from service annotators.

Unlike usual semantic Web annotation toolkits such as the Ontomat-Annotizer [40] and the Annotea (http://www.w3.org/2001/Annotea/), etc., which are mainly focused on annotating Web documents, the TUAM is designed to annotate different types of bioinformatics resources including biological services, database schema tables, ASCII-files, and spreadsheets, etc. It allows establishing and persisting arbitrary relationships between any number and kind of data sources in a *n:m* fashion, as well as effectively building a semantic network that can comprise unstructured or pre-structured data and terminologies with the expressiveness power from thesauri to OWL ontologies. Although the TUAM has previously been used in similar environments for the deep annotation of biological data services [39], it has not yet been applied to deal with the OWL-S service annotation model and Grid service descriptions represented in the extended WSDL format as in the GRIA. Thus, in the SIMDAT, the TUAM is mainly extended to accommodate the OWL-S model, as well as represent annotated service features through a GUI. In figure 6, we show a screenshot of annotating a BLASTp service in the TUAM based on the SIMDAT domain ontology. As we can see, by means of the TUAM, the complexities of the OWL-S are completely hidden from the annotator who only needs to concentrate on matching services with corresponding concepts from the ontology. The navigation/browsing of the domain ontology is also greatly simplified through a GUI component as shown in the top right part of the screenshot.

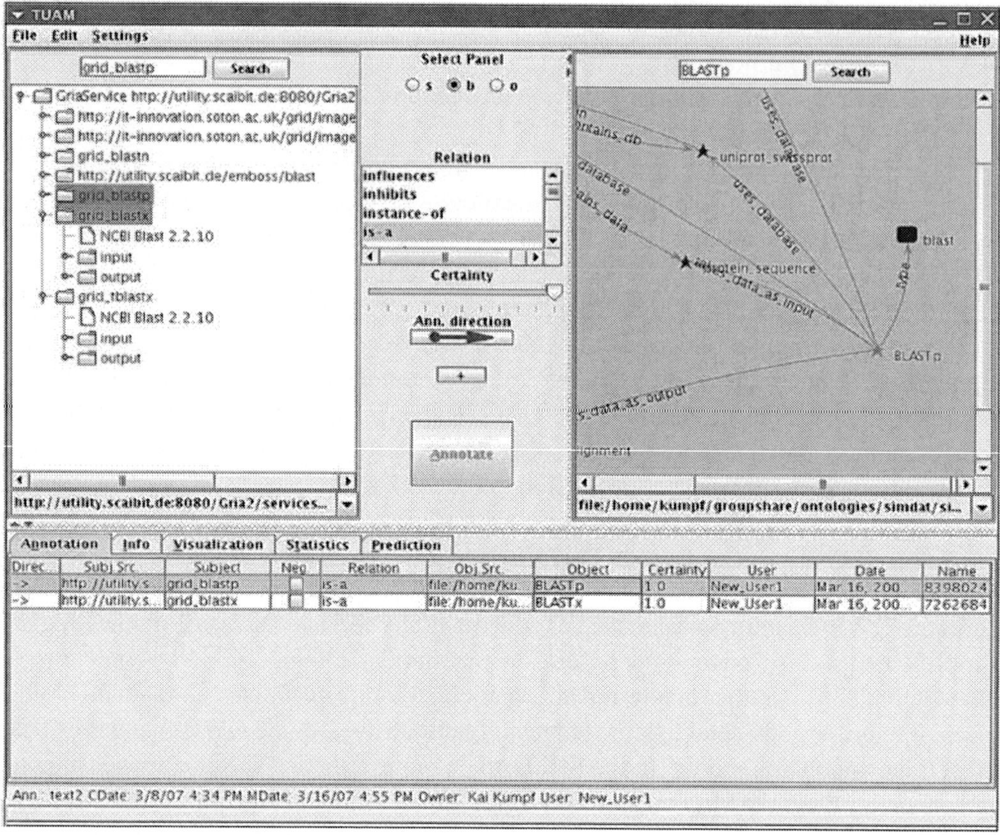

Figure 6. Semantic annotation of a BLASTp service in the TUAM based on the SIMDAT domain ontology.

Unlike usual service features such as service I/O, which are relatively stable, biological services usually have some rapidly changing service features such as databank entries, databank release date, etc. As these dynamic service features are sometimes of great interest of biologists while discovering biological services, they are critical to the service discovery and matchmaking, at least as equally important as static service features such as the service classification. In order to keep these dynamic service features short-periodically updated in the knowledge base, in the SIMDAT, besides the TUAM, a complementary annotation tool called the Dynamo (Dynamic Annotation Module) is developed to automatically update dynamic service features.

In contrast to the TUAM, Dynamo itself is not semantics-aware. It is merely a software agent that can be triggered by databank updating events and sequentially updates service annotations by inserting the dynamic data items and then republishing the annotations. At present, the Dynamo can work with two popular sequence retrieval systems, i.e., the SRS [4] and the MRS [5]. In figure 7, we illustrate the typical deployment of the Dynamo in the service discovery framework.

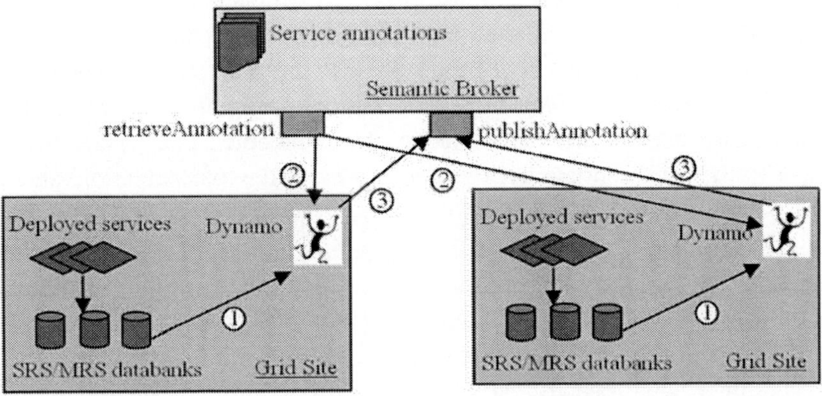

Figure 7. Deployment of the Dynamo in the service discovery framework.

Semantic Matchmaker Based on the Ontology Reasoning

The semantic matchmaker, called the SB (Semantic Broker) in the SIMDAT Pharma Grid, is the kernel component of the service discovery framework, which is basically a service registry built upon ontology reasoning services. Unlike usual nonsemantic or semantic service registries such as the UDDI [19] and the Grimoires [15], the SB is focused on the service matchmaking. The users represent their requirements on services in terms of the OWL-S service profile, the SB can then compute the matching degree of each discovery result in term of the semantic similarity between the user requirements and service annotations. At this point, the SB is also a bit different from some OWL-S-based generic semantic matchmakers such as ones mentioned in "Related Work". Specifically designed for biological service discovery, the SB matchmaking algorithm is more extensive. It can include any interested service features into the matchmaking process.

The SB matchmaking algorithm is basically a modification/extension to the semantic matchmaking algorithm proposed in [41], which identifies four types of semantic "degree of match" between the "requirement concept" *Creq* and "advertisement concept" *Cadv* in terms of the minimal distance between concepts in the taxonomy tree:

- *Exact*: If *Creq* = *Cadv*, then *Creq* and *Cadv* are equivalent, the matching degree between *Creq* and *Cadv* is *exact*. If *Creq subClassOf Cadv*, the matching degree is still *exact* under the assumption that by advertising *Cadv* the service provider commits to provide features consistent with every immediate subtype of *Cadv*.
- *Plug-in*: If *Cadv* subsumes *Creq*, the matching degree between *Creq* and *Cadv* is *plug-in*, which denotes a weaker relation between *Creq* and *Cadv* than the *exact* relation.
- *Subsume*: If *Creq* subsumes *Cadv*, the matching degree between *Creq* and *Cadv* is *subsume*, which denotes a weaker relation between *Creq* and *Cadv* than the *plug-in* relation
- *Fail*: If no subsumption relation exists between *Creq* and *Cadv*, the matching degree is *fail*.

In our algorithm, the *plug-in* relationship is first dropped taking into account specific features of the biological service annotation in the SIMDAT. Typically, as biological domain experts can clearly differentiate between services at the leaf concept level of the domain ontology, the occurrence of the *plug-in* relationship can efficiently be avoided by means of the service annotation using either leaf or second-leaf concepts. As a replacement of the *plug-in* relation, we further differentiate between two cases of the *exact* relation, and denote *Creq subClassOf Cadv* as a new type of matching degree *exact-*, which is weaker than the *exact* relation but stronger than *subsume*. Such a differentiation makes sense for the biological service discovery, as generally it is not a common practice in biological domain ontology design to enumerate all subclasses of a concept. In this case, as a service annotated using second-leaf concepts does not necessarily hold the assumption proposed in the original algorithm, it is rather beneficial to degrade the *Creq subClassOf Cadv* relationship to make the semantic matchmaking more "accurate".

Besides semantic service features, nonsemantic service features are also included into the SB matchmaking algorithm. These features typically include Grid ones such as the memory size of a site, and bioinformatics ones such as databank entries, databank release date, etc. Generally, nonsemantic service features are not modeled in the domain ontology, instead they are restricted by the XML Schema simple data types [42]. Thus, the matchmaking of nonsemantic service features is either "equal" or "greater-than/after" or "less-than/before", which implies that if nonsemantic service features are present in queries, they are always considered as "hard" requirements. In the SB matchmaking algorithm, we always handle nonsemantic service features prior to semantic ones in order to possibly avoid costly computation for the ontology reasoning.

The result ranking in the SB matchmaking algorithm is principally based on the average matching degree. However, if the matching degrees of several discovery results are equivalent, the secondary ranking will be conducted, which typically uses one of the user-preferred, nonsemantic service features as the secondary ranking criterion, e.g., the memory size of a site, databank release date, etc.

Framework Usage

The first SIMDAT Pharma Grid testbed, i.e., "distributed *in silico* analysis and annotation of biological sequences" testbed, has been deployed since Sept. 2005 in five Grid sites across three European countries. As two Grid sites in Belgium are affiliated with the EMBNet, they directly extend the reach of the SIMDAT Pharma Grid to the whole Europe. In the following we describe the usage of the service discovery framework in the SIMDAT Pharma Grid based on the case study of a pilot biological workflow designed in the SIMDAT: *IXodus*.

IXodus Biological Workflow

The Lyme disease has been identified in the 1990s as a significant source of human and animal pathology in temperate areas of the world (North America, Central and Western Europe). It is caused by the bite of a tick of genus *Ixodes*, infected by the pathogen bacterium

Borrelia burgdorferi. The study of the man-parasite interactions is an active research area, as about 20% of the ticks have been found infected by this bacterium. To better understand the infectious process frequently associated with this organism, a biological workflow called *IXodus* is designed in the SIMDAT to deal with the characterization of genes expressed in the salivary gland of the tick *Ixodes ricinus* at various stages of the host-parasite interaction.

The *IXodus* workflow is fed with a batch of nucleic sequences coming from the systematic sequencing of thousands of salivary gland *cDNAs*. As detailed in figure 8 through the UML (Unified Modeling Language) 2.0 diagram, the workflow itself consists of two distinct phases:

1) During the first phase, each sequence of the batch is first analyzed for possible redundancy with sequences already residing in the project database. This is performed by using the BLASTN algorithm that compares the new sequences with a databank composed from all previously entered sequences to identify matches of 100% between one sequence and a portion of the other. At the end of the first phase, all project sequences and their existing relationships are stored in a relational database management system.
2) Once a sequence from the current batch has been considered as relevant for further analysis, it is submitted to the main analysis part of the workflow, which basically consists of three steps:
   2a) A comparison of the nucleic sequence with the UNIPROT public databank of protein sequences by using the BLASTX algorithm. Again, a decision is taken about a possible relevant hit, by using the BLAST E-value and percentage of identity reported by the program. The analysis outputs of the analyzed sequences are reported accordingly in the databank. If no hit is found, the analysis continues with the next step.
   2b) The sequence is submitted to the EMBOSS GETORF program in order to extract the longest ORF (Open Reading Frame). This ORF is then processed by the InterProScan package. A total of 13 different analysis methods (Generalised profiles, hidden markov models, etc.) are applied to identify functional elements (motifs). If no hit is found, the sequence is submitted to the next step.
   2c) A sequence comparison with the current release of the EMBL public databank is performed thanks to the BLASTN algorithm. At this step of the analysis, all information relative to matches between the new sequence and EMBL sequences is stored in the project database.

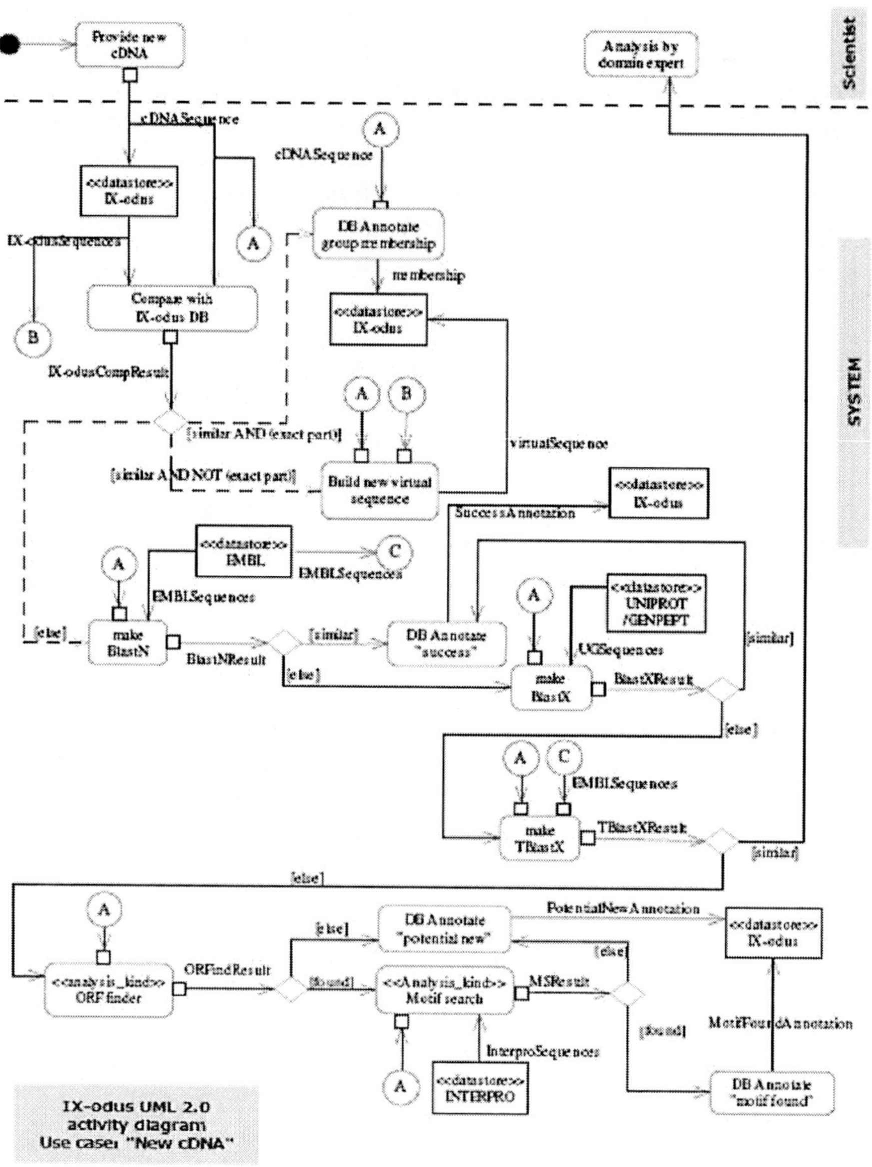

Figure 8. *IXodus* biological workflow represented through the UML 2.0 diagram.

Deployment of the Service Discovery Framework in the SIMDAT Pharma Grid

In figure 9, we illustrate the deployment of the service discovery framework in the SIMDAT Pharma Grid.

On each Grid site, either the SRS or the MRS is installed and configured to host different biological data services and analysis services. Each service is wrapped through a GRIA wrapper program, and further interacts with other Grid components/services through the GRIA middleware. All communications between Grid components/services are secured through the E2E (End-to-End) security module, as detailed in [43]. In addition, the access to each Grid component/service is also secured through the PBAC (Process Role-based Access Control) module of the GRIA middleware [9].

The SB itself is also deployed as a GRIA service, which exposes four operations to end users: *publishAnnotation*, *removeAnnotation*, *retrieveAnnotation*, and *getServiceMatchings*. Although we have developed a Web 2.0 interface to enable end users to directly interact with the SB through a GUI, this interface only serves for educational purpose. Basically, the SB is a "non-face" component, which is not intended to be directly exposed to end users. Just as illustrated in figure 9, in the SIMDAT, there are only three components that directly interact with the SB. While the TUAM and the Dynamo contact the SB for publishing and updating service annotations, the SIMDAT workflow toolkit interacts with the SB through the *getServiceMatchings* operation to send queries and sequentially get matchmaking results.

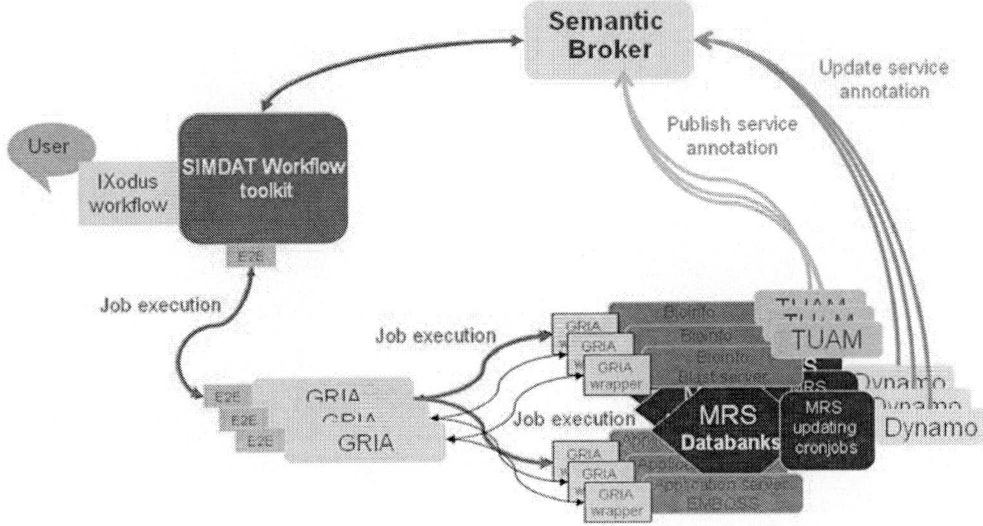

Figure 9. Deployment of the service discovery framework in the SIMDAT Pharma Grid.

It is worth noting that the access to different SB service operations is controlled through the PBAC module. In terms of the PBAC, each interaction with the SB is based on a specific user X.509 certificate. For users allowed to access the SB, they are dynamically assigned a process role, either as *SB_User* or as *SB_Annotator*. Whereas the *SB_User* role only allows to access the *getServiceMatchings* operation, the *SB_Annotator* role allows to access all SB operations. This implies, for example, if a user would like to publish a service annotation through the TUAM, he must first ask the SB administrator to acquire the *SB_Annotator* role for invoking the *publishAnnotation* operation.

Framework Usage: the *IXodus* Case Study

The *IXodus* biological workflow is executed within the SIMDAT Pharma Grid testbed, in which all analysis methods, public sequence databanks, and private project sequences used in *IXodus* are distributed over the Grid as the GRIA services. For the design and execution of

IXodus, users have to follow the procedure as illustrated in figure 10, which demonstrates the typical usage of the semantics-enabled service discovery framework through the SIMDAT workflow toolkit.

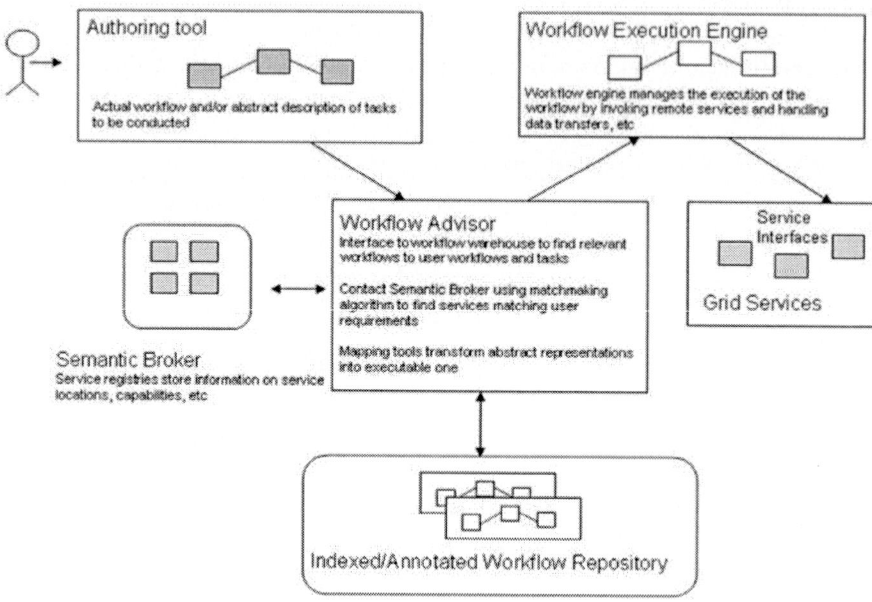

Figure 10. Usage of the semantics-enabled service discovery framework through the SIMDAT workflow toolkit.

For the design of the *IXodus* workflow, the biologists do not need to have any knowledge of deployed service instances. Through operating on sets of "abstract" task components in the workflow authoring tool, the users can directly construct the workflow at a high level of abstraction, just like illustrated in figure 11.

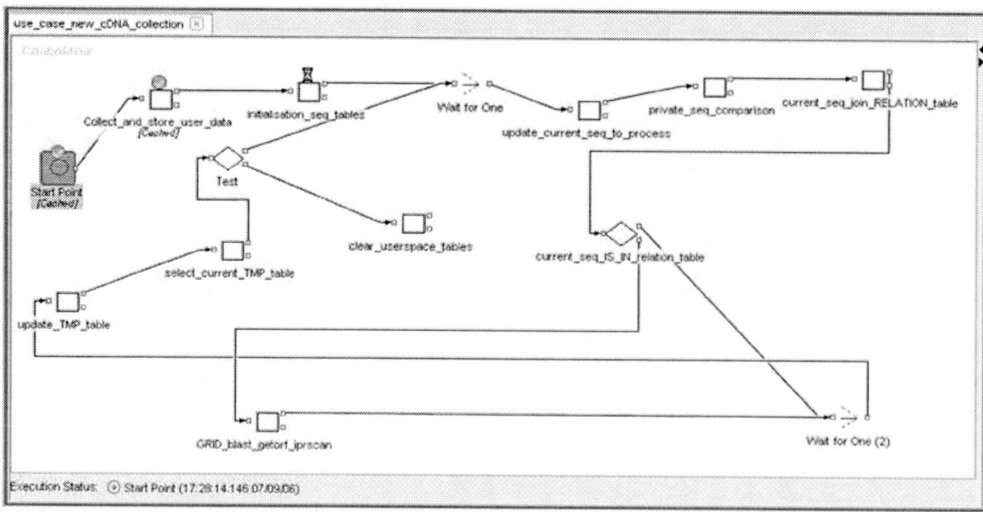

Figure 11. An example "abstract" workflow produced through the SIMDAT workflow authoring tool.

On each "abstract" component, the users can specify their requirements in terms of predefined semantic query patterns. Taking *IXodus* task 2a) as an example, the requirement might be to *request a service instance that belongs to "similarity_search" class of tool, uses "embl" databank released after "2008-10-01", allows "nucleic_acid_sequence" as input, and gives at least "sequence_pairwise_alignments" as output.* Note that we can also add additional service QoS requirements here. This can easily be done through a generic service QoS query pattern defined in the workflow authoring tool [10].

In order to bind each "abstract" task component to a "concrete" service instance, the workflow adviser will contact the SB, which may return a set of candidate services in a ranking list, e.g., sets of BLASTx, tBLASTx instances from different providers. This is a dynamic discovery process, which implies that service instances are discovered in term of not only the semantic similarity, but also actual service states including "heartbeat detection". The service selection is then either optionally semiautomatic, i.e., the users choose their preferable service instance according to actual service information, or fully automatic, i.e., the workflow adviser automatically chooses the first service instance in the ranking list. After all "abstract" components are instantiated, the "concrete" workflow will be submitted to the workflow execution engine, which can parse and validate the workflow, perform meta-scheduling operations between different services, generate a final execution plan, and then coordinate the invocation of remote services as well as handle the data transfer between them [10].

EXPERIEMNTAL EVALUATION

For the evaluation of the semantics-enabled service discovery framework, two series of experiments are carried out to respectively evaluate the end user experience and the scalability of the SB. Except where stated otherwise, in the experiments, the SB is deployed on a workstation of AMD Opteron 252, 2.6GHz, with 1MB L2 cache and 2GB memory. As the SB is a Java application, Sun JDK 5.0 SE (Java Development Kit 5.0 Standard Edition) update 7 for Linux x64 platform is used, with the JVM (Java Virtual Machine) heap size set to 1GB. The SB knowledge base is built upon Sesame 1.2.4 (http://www.openrdf.org/), using OWLIM 2.8.3 as the SAIL (Storage and Inference Layer) to provide the OWL reasoning service. For experiments involving the Web service performance test, the SB is deployed in Apache Axis 1.3 final (http://ws.apache.org/axis/) and Apache Tomcat 5.5.17 (http://tomcat.apache.org/). Although the SB also has two WSRF interfaces respectively based on the Globus Toolkit 4.01 Grid middleware and the GRIA 5.1 middleware, the performance of SB's WSRF interfaces is not evaluated. Basically, the WSRF performance is expected to be similar to the Web service performance.

Evaluation of the End User Experience

The end user experience is evaluated based on the *IXodus* workflow with the focus on comparing the effort needed to perform a well-defined biological task using the SIMDAT Pharma Grid testbed with the conventional/manual way as it is currently being done in many institutes. For both manual analysis and workflow analysis, we choose at random a set of 20 *cDNA* sequences (*Ixodes ricinus*) from the research group on tick pathogenicity at the Laboratory of Applied Genetics of the ULB (Universite libre de Bruxelles). These sequences come from the automatic sequencing machine, among which about one-third have previously been characterized by the research team, and the other are considered as "unidentified".

In the manual analysis, expert intervention is required for each step throughout the whole procedure. As the procedure not only implies the treatment of huge amounts of output data, but also includes very repetitive and tedious tasks that are error-prone, the manual analysis necessitates two persons to accelerate the analysis process and minimize human errors. Because the time and effort involved in manual analysis are rather overwhelming, only eight sequences are analyzed. The experimental result of the manual analysis is:

- The average treatment time of a sequence is 322s; the average sequence per manpower time is 644s.

In the workflow analysis, the biologists need relatively less domain knowledge with the assistance of the SIMDAT workflow toolkit. Also the human intervention is greatly reduced. After the workflow is instantiated, the biologists need only copy/paste the sequences to the interface and click a button to start the whole procedure. A single person can easily manage this work with no risk of error. All 20 sequences are analyzed and the experimental result is:

- The average treatment time of a sequence is 40s; the average sequence per manpower time is 40s.

Though it can clearly be seen that the workflow analysis is more efficient, we cannot quantitatively show the reduced requirement on users' domain knowledge. This experiment is expected to be conducted in the next project phase when SIMDAT's usage is more widespread and more end users are involved in the testbed[8].

Scalability Evaluation of the Semantic Broker

As the number of services deployed in the current SIMDAT Pharma Grid testbed is still rather limited, a simulation experiment is conducted with the purpose of evaluating the scalability of the SB. In the experiment, up to 10,000 service annotations are generated based on a practical service annotation, each with over 1000 explicit triple statements. The SB repository is thus with a size up to over 10 million triple statements counting the SIMDAT domain ontology and various schema ontologies.

For the service publication/removal and query operation, the overheads of invoking the stand-alone SB and invoking SB via its Web service interface are respectively measured, as denoted in figure 12. The SB query overhead is evaluated based on two types of queries:

[8] For the whole evaluation report, we refer to the corresponding public project deliverable on the SIMDAT website (http://www.simdat.org).

- A query covering all annotated service features, which can retrieve all service instances ("the worst case" query); and
- A query covering only one annotated service feature, which can also retrieve all service instances ("the quasi-best case" query).

In either case, the ontology reasoning is included. In figure 12, we illustrate the experimental results.

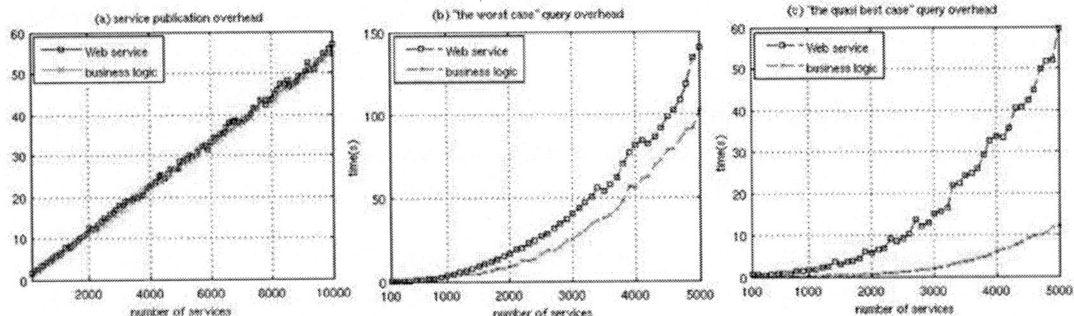

Figure 12. (a) service publication overhead. (b) "the worst case" query overhead. (c) "the quasi-best case" query overhead.

As the SB depends on the ontology reasoning, its scalability is basically determined by the OWLIM, i.e., the maximal scale achievable is limited by the available RAM (Random Access Memory) because the reasoning and query evaluation in the OWLIM are performed in-memory [32]. Figure 12(a) shows that the service publication overhead increases linearly with the repository size, which is in line with the "upload" operation overhead in the OWLIM [32] . For each new service published, the whole repository is reloaded excluding, thanks to the persistence policy of OWLIM, the domain ontology and the schema ontologies. The service removal overhead is basically the same as the service publication overhead; thus, it is omitted in figure 12. This is because that the "delete" operation in the OWLIM is relatively slow; thus, we simply reload the whole repository after a registered service has been removed.

Figure 12(b) and 12(c) show that the service query overhead also increases with the repository size, ranging from tens of milliseconds to tens of seconds. This can partly be explained by the fact that the query overhead in the OWLIM grows up in a linear dependency to the repository size and result set size [32]. However, comparing figure 12(b) and 12(c), we can clearly see that the key influential factor is actually the complexity of the SB queries, in particular, the number of query entries demanding the real-time ontology reasoning. In fact, with the increase of the semantic query entries, more ontology reasoning processes have to be initiated in the matchmaking algorithm. Thus, "the worse case" query overhead dramatically increases in comparison to "the quasi-best case" query overhead.

RELATED WORK

In recent years, several generic semantic matchmakers have been developed for Web service discovery and matchmaking, most of which are based on the OWL-S/DAML-S [12], e.g., the OWLS-UDDI matchmaker [41], OWLS-MX [44], DAMLS matchmaker [45], Mindswap Web service composer [46], and OWL-S matcher [47], etc. Whereas all these generic semantic matchmakers are mainly focused on service I/O-based matchmaking, the SB is more registry-alike, being able to include arbitrary number of interested service features into the matchmaking algorithm.

Besides the OWL-S-based ones, there are also other two series of semantic matchmakers, which are respectively based on the WSMO and the WSDL-S/SAWSDL, e.g., the WSMX (Web Service Execution Environment)[48] and METEOR-S [49]. In comparison to the OWL-S, the WSMO is a conceptually stronger approach for realizing semantic Web services, which "has a better separation of the service requester and provider point of view, includes the orchestration of a Web service enabling the static or dynamic reuse of Web services to achieve the functionality of a more complex Web service, provides formal semantics for the choreography of Web services, and allows multiple ways of interacting with a given service" [50]. Though most of WSMO-based products are still in prototypical phase, the WSMO seems rather promising for the future semantic Web services, in particular, as far as the service orchestration and choreography are concerned. Theoretically, the SB can be "upgraded" to become partly WSMO-compliant, as the OWL-S-based service annotation model can wholly be translated into the WSMO service description model, and both the OWL-DLP and the OWL-DL ontologies can be represented through corresponding subsets of the WSML (Web Service Modeling Language) [50]. However, the migration of SB from the OWL-S to the WSMO is deemed rather difficult, and the fully new design and implementation of the system needs to be done.

In contrast to the WSMO, the WSDL-S is conceptually much weaker than the OWL-S, which proposes a small set of extensions to the WSDL, by which semantic annotations may be associated with WSDL elements such as operations, input and output type schemas, and interfaces [35]. Since Aug. 2007, the WSDL-S has become a new W3C recommendation, namely, the SAWSDL [36]. As the only W3C recommendation for semantic annotation of Web services, the widespread acceptance of SAWSDL in the SWS community can be expected. In general, the SAWSDL is very similar to the OWL-S service grounding subset. The difference is in that the SAWSDL can only annotate service features contained in the WSDL, whereas the OWL-S service grounding proposes some WSDL extensions [51]. It is expected to be straightforward to "degrade" the SB from the OWL-S to the SAWSDL, and, in particular, the existing SB matchmaking algorithm can directly be reused. However, it is also clear that most of service annotations in the OWL-S will be lost, as only service I/O features can effectively be annotated through the SAWSDL. This implies that the SB's functionality will unavoidably be weakened.

In bioinformatics, a number of research projects have exploited semantic technologies in recent years for biological service discovery within Grid environments, principally including Semantic MOBY in the BioMoby [13], Feta in the myGrid [14], and Grimoires [15]. Restricted by the technology development at that time, these semantic service discovery frameworks are rather lightweight. "Semantically", Grimoires has no ontology support; Semantic MOBY uses a rather simple data model to describe services; and Feta leverages the RDF(S) (Resource Description Framework Schema) [52] reasoning instead of the more powerful OWL [16] reasoning. "Functionally", none of them can support semantic matchmaking, i.e., "matching queries and resource advertisements with reference to ontologies, and further returning results with a relevance ranking" [53]. Without matchmaking functionality, it is rather difficult to support advanced service discovery in workflow.

Conclusion

Though semantic technologies have been recognized as a key technology for biological service discovery, their potential is not yet fully revealed in previous work. Based on the state-of-the-art technology development, this chapter indicates several technology aspects, which are worth rethinking and further exploration.

1) The domain ontologies specifically designed for biological service discovery are not necessarily as complete and complex as generic biological/biomedical ontologies such as [26], [27], and [28]. As a consequence, some design principles for generic bioinformatics ontologies such as those proposed in [31] need to be revisited, in particular, when we only need a domain ontology to deal with lightweight service discovery requirements. Our practice in the SIMDAT evidences that it is possible to model a complex domain ontology for biological service discovery using lightweight ontology language OWL-DLP, which directly enables the usage of high-performance ontology reasoners. As the real-time ontology reasoning is always a "must" for semantic service discovery, our practice clearly points out that reducing ontology complexity is the key to system design.
2) The biological service discovery needs a more comprehensive, extendable, and semantic-rich data model to describe different service features. It may be beyond the capability of some pragmatic data models used in previous projects [13][14][15] to address more complex service discovery requirements, in particular, as far as their extendibility and interoperability are concerned. Our practice in the SIMDAT evidences that the OWL-S can efficiently be used as a standard data model to deal with various service features. OWL-S itself is also expected to have good interoperability with other mainstream standardization efforts in the SWS community such as the WSMO, the SAWSDL, and the SWSF (Semantic Web Services Framework) [50][51][54].
3) The service matchmaking is rather crucial for supporting (semi)automatic service discovery, selection, composition, and invocation, in particular, in a workflow centered Grid computing environment like the SIMDAT. Whereas this aspect was not fully recognized in previous projects, Our practice in the SIMDAT clearly shows its importance through (semi)automatic execution of the biological workflow *IXodus*. Additionally, we also highlight the significance of nonsemantic service features for biological service discovery and matchmaking, and demonstrate how to handle these features in the matchmaking algorithm in order to use them for advanced result ranking as well as possibly avoid costly computation for the ontology reasoning.
4) The OWL-S-based service annotation is not necessarily as complicated and cumbersome as described in [14]. Whereas some generic annotation toolkits are not really usable for biologists, Our practice in the SIMDAT shows that bioinformatics-specific annotation toolkits such as the TUAM and the Dynamo can greatly simplify

the service annotation process. Besides, we also demonstrate how domain ontologies can be engineered with the service annotation in mind. In general, these practices are different from those for generic biological/biomedical ontology design.

Last but not the least, our experimental evaluation indicates that the ontology reasoning, in particular, the OWL-DLP reasoning, is not necessarily as computationally expensive as imagined in previous projects [13][14][15][9]. For a middle-scale semantic application with a reasonable repository size, it proves possible to apply real-time ontology reasoning with acceptable performance. This may further inspire the exploration of more advanced semantic technologies, such as the full set of OWL-DL and semantic rules in the real-time reasoning. For this aspect, our practice indicates that Pellet [33] may deserve further investigation.

ACKNOWLEDGMENT

The authors are grateful to all project partners at the SIMDAT Pharma Activity Consortium. In particular, Jun Wang and Luigi Lo Lacono from the IT Research Division, NEC Laboratories Europe, NEC Europe Ltd., helped in setting up and operating the service discovery framework in the SIMDAT Pharma Grid testbed. Moustafa Ghanem from InforSense Ltd. kindly produced figure 10 to illustrate the SIMDAT workflow usage. The Dynamo was initially developed at the ULB (Universite libre de Bruxelles) by Aubin Rukera for his graduate thesis, and further improved by Joseph Mavor. The *IXodus* case study was inspired by Edmond Godfroid and Bernard Couvreur from the Laboratory of Applied Genetics of the ULB.

This research was supported in part by the EU IST FP6 project SIMDAT under Contract IST-2004-511438.

REFERENCES

[1] C. Upstill, and M. J. Boniface, "SIMDAT," *CTWatch Quarterly*, vol. 1, no. 4, pp. 16-24, Nov. 2005.

[2] M. Senger, P. Rice, and T. Oinn, "Soaplab - a Unified Sesame Door to Analysis Tools," in *Proc. UK e-Science, All Hands Meeting 2003*, Nottingham, UK, Sept. 2003.

[3] M.D. Wilkinson, and M. Links, "BioMOBY: an Open-Source Biological Web Services Proposal," *Briefings in Bioinformatics*, vol. 3, no. 4, pp. 331-341, 2002.

[4] T. Etzold, A. Ulyanov, and P. Argos, "SRS: information retrieval system for molecular biology data banks," *Methods Enzymol.*, vol. 266, pp.114-142, 1996.

[5] M.L. Hekkelman, and G. Vriend, "MRS: A fast and compact retrieval system for biological data," *Nucleic Acids Research*, vol. 33, pp.766-769, 2005.

[6] S.F. Altschul, W. Gish, W. Miller, E.W. Myers, and D.J. Lipman, "Basic local alignment search tool," *J. of Mol. Biology*, vol. 215, pp. 403-410, 1990.

[9] Our result is not directly comparable with Grimoires, as Grimoires does not make use of OWL-DL reasoning. It is comparable with Semantic MOBY and Feta. However, neither Semantic MOBY nor Feta has the performance evaluation report so far.

[7] E. M. Zdobnov, and R. Apweiler, "InterProScan - an integration platform for the signature-recognition methods in InterPro," *Bioinformatics*, vol. 17, no.9, pp.847-855, 2001.

[8] P. Rice, I. Longden, and A. Bleasby, "EMBOSS: the European molecular biology open software suite," *Trends Genet.*, vol. 16, no. 6, pp.276-283, 2000.

[9] M. Surridge, S. Taylor, D. De Roure, and E. Zaluska, "Experiences with GRIA-Industrial Applications on a Web Services Grid," in *Proc. 1st Int. Conf. on e-Science and Grid Computing*, Melbourne, Australia, Dec. 2005, pp. 98-105.

[10] M. Ghanem, N. Azam, M.Boniface, and J. Ferris, "Grid-Enabled Workflows for Industrial Product Design," in *Proc. 2nd IEEE International Conference on e-Science and Grid Computing (e-Science'06)*, Amsterdam, The Netherlands, Dec. 2006, pp.96.

[11] R. Stevens, C. Goble, P. Baker, and A. Brass, "A classification of tasks in bioinformatics," *Bioinformatics*, vol. 17, no. 2, pp.180-188, 2001.

[12] D. Martin, M. Burstein, J. Hobbs, O. Lassila, D. McDermott, S. McIlraith, S. Narayanan, M. Paolucci, B. Parsia, T. Payne, E. Sirin, N. Srinivasan, and K. Sycara, "OWL-S: Semantic Markup for Web Services." Available: http://www.daml.org/services/owl-s/1.2/overview/, March 2006.

[13] P. Lord, S. Bechhofer, M. D. Wilkinson, G. Schiltz, D. Gessler, D. Hull, C. Goble, and L. Stein, "Applying Semantic Web Services to Bioinformatics: Experiences Gained, Lessons Learnt," in *Proc. 3rd Int. Semantic Web Conference (ISWC2004)*, Hiroshima, Japan, Nov. 2004.

[14] P. Lord, P. Alper, C. Wroe, and C. Goble, "Feta: A Light-Weight Architecture for User Oriented Semantic Service Discovery," in *Proc. 2nd European Semantic Web Conference*, Crete, Greece, May 2005.

[15] W. Fang, S. C. Wong, V. Tan, S. Miles, and L. Moreau, "Performance analysis of a semantics enabled service registry," in *Proc. 4th All Hands Meeting (AHM'05)*, Nottingham, UK, Sept. 2005.

[16] D. L. McGuinness, and F. van Harmelen, "OWL Web Ontology Language Overview." Available: http://www.w3.org/TR/owl-features/, Feb. 2004.

[17] OASIS WSRF TC, "Web Service Resource Framework." Available: http://www.oasis-open.org/committees/tc_home.php?wg_abbrev=wsrf , Apr. 2006.

[18] The Globus Alliance, "Globus Toolkit." Available: http://www.globus.org/toolkit/, Aug. 2006.

[19] OASIS UDDI Specifications TC, "UDDI." Available: http://www.oasis-pen.org/committees/tc_home.php?wg_abbrev=uddi-spec, Feb. 2005.

[20] OASIS ebXML Registry TC, "ebXML registry." Available: http://www.oasis-open.org/committees/tc_home.php?wg_abbrev=regrep, Feb. 2007.

[21] A. Dogac, Y. Kabak, and G. Laleci, "Enriching ebXML Registries with OWL Ontologies for Efficient Service Discovery," in *Proc. 14th Int. Workshop on Research Issues on Data Engineering*, Boston, USA , March 2004.

[22] S. Schulze-Kremer, "Ontologies for Molecular Biology," *Pac. Symp. Biocomput.*, vol.3, pp.693-704, 1998.

[23] J. A. Blake, and C. J. Bult, "Beyond the Data Deluge: Data Integration and Bio-ontologies," *J Biomed Inform*, 39, pp. 314-320, 2006

[24] S. P. Gardner, "Ontologies and Semantic Data Integration," *Drug Discov Today*, vol. 10, pp.1001-1007, 2005.

[25] B. Smith, M. Ashburner, C. Rosse, J. Bard, W. Bug, W. Ceusters, L. J. Goldberg, K. Eilbeck, A. Ireland, C. J. Mungall, The OBI Consortium, N. Leontis, P. Rocca-Serra, A. Ruttenberg, S. A. Sansone, R. H. Scheuermann, N. Shah, P. L. Whetzel, and S. Lewis, "The OBO Foundry: Coordinated Evolution of Ontologies to Support Biomedical Data Integration," *Nat Biotechnol*, vol. 25, pp. 1251-1255, 2007.

[26] BioPax Group, "BioPax: Biological pathway exchange ontology." Available: http://www.biopax.org/, Dec. 2005.

[27] C. Rosse, and J.V.L. Mejino, "A reference ontology for biomedical informatics: the Foundational Model of Anatomy", *J. Biomed Inform.*, vol. 36, pp. 478-500, 2003.

[28] K. Eilbeck, S. E. Lewis, C. J. Mungall, M. Yandell, L. Stein, R. Durbin, and M. Ashburner, "The Sequence Ontology: A tool for the unification of genome annotations", *Genome Biology*, vol. 6, no. 5, Apr. 2005.

[29] D. L.Wheeler, C. Chappey, A. E. Lash, D. D. Leipe, T. L. Madden, G. D. Schuler, T. A. Tatusova, and B. A. Rapp, "Database resources of the National Center for Biotechnology Information," *Nucleic Acids Res*, vol. 28, no. 1, pp.10-14, 2000.

[30] B. Grosof, I. Horrocks, R. Volz, and S. Decker, "Description logic programs: combining logic programs with description logics," in *Proc. WWW 2003*, Budapest, Hungary, May 2003.

[31] C. Golbreich, "Web rules for Health Care and Life Sciences: use cases and requirements," in *Proc. Reasoning on the Web Workshop at WWW2006*, Edinburgh, UK, 2006.

[32] A. Kiryakov, D. Ognyanov, and D. Manov, "OWLIM – a Pragmatic Semantic Repository for OWL," in *Proc. Int. Workshop on Scalable Semantic Web Knowledge Base Systems (SSWS 2005)*, New York, USA, Nov. 2005.

[33] E. Sirin, B. Parsia, B. Cuenca Grau, A. Kalyanpur, and Y. Katz., "Pellet: A practical OWL-DL reasoner," *J. of Web Semantics*, vol. 5, no.2, pp.51-53, June 2007.

[34] D. Roman, U. Keller, H. Lausen, J. de Bruijn, R. Lara, M. Stollberg, A. Polleres, C. Feier, C. Bussler, and D. Fensel, "Web Service Modeling Ontology," *Applied Ontology*, vol. 1, no. 1, pp.77-106, 2005.

[35] R. Akkiraju, J. Farrell, J. Miller, M. Nagarajan, M. T. Schmidt, A. Sheth, and K. Verma, "Web Service Semantics - WSDL-S." Available: http://www.w3.org/Submission/WSDL-S/, Nov. 2005.

[36] J. Farrell, H. Lausen, Semantic Annotations for WSDL and XML Schema, http://www.w3.org/TR/sawsdl/, Aug. 2007.

[37] R. Stevens, C. Goble, P. Baker, and A. Brass, "A classification of tasks in bioinformatics," *Bioinformatics*, vol. 17, no. 2, pp.180-188, 2001.

[38] D. Elenius, G. Denker, D. Martin, F. Gilham, J. Khouri, S. Sadaati, and R. Senanayake, "The OWL-S Editor – A Development Tool for Semantic Web Services," in *Proc. 2nd European Semantic Web Conference*, Crete, Greece, May 2005.

[39] A. Arbona, S. Benkner, G. Engelbrecht, J. Fingberg, M. Hofmann, K. Kumpf, G. Lonsdale, and A. Woehrer, "A Service-oriented Grid Infrastructure for Biomedical Data and Compute Services," *IEEE Trans. Nanobioscience*, vol. 6, pp. 136-141, 2007.

[40] S. Bloehdorn, K. Petridis, C. Saathoff, N. Simou, V.Tzouvaras, Y. Avrithis, S. Handschuh, Y. Kompatsiaris, S. Staab, and M. G. Strintzis, "Semantic Annotation of Images and Videos for Multimedia Analysis", in *Proc. 2nd European Semantic Web Conference*, Heraklion, Greece, May 2005.

[41] M. Paolucci, T. Kawamura, T.R. Payne, and K. Sycara, "Semantic Matching of Web Services Capabilities," in *Proc. Int. Semantic Web Conference (ISWC)*, Sardinia, Italy, June 2002.

[42] D. Peterson, P. V. Biron, A. Malhotra, and C. M. Sperberg-McQueen, "XML Schema 1.1 Part 2: Datatypes." Available: http://www.w3.org/TR/xmlschema11-2/, Feb. 2006.

[43] J. A. M. Herveg,, F. Crazzolara, S. E. Middleton, D. Marvin, and Y. Poullet, "GEMSS: Privacy and security for a Medical Grid," in *Proc. HealthGRID 2004*, Clermont-Ferrand, France, Jan. 2004.

[44] M. Klusch, B. Fries, and K. Sycara, "Automated Semantic Web Service Discovery with OWLS-MX," in *Proc. 5th Int. Conference on Autonomous Agents and Multi-Agent Systems (AAMAS 2006)*, Hakodate, Japan, May 2006.

[45] L. Li, and I. Horrock, "A Software Framework for Matchmaking based on Semantic Web Technology," in *Proc. 12th Int. World Wide Web Conference Workshop on E-Services and the Semantic Web (ESSW 2003)*, Budapest, Hungary, May 2003.

[46] E. Sirin, J. Hendler, and B. Parsia, "Semi-automatic Composition of Web Services using Semantic Descriptions," in *Proc. of Web Services: Modeling, Architecture and Infrastructure. Workshop in Conjunction with ICEIS2003*, Angers, France, April, 2003.

[47] M. C. Jaeger, G. Rojec-Goldmann, C. Liebetruth, G. Mühl, and K. Geihs, "Ranked Matching for Service Descriptions Using OWL-S," in *Proc. KiVS 2005*, Kaiserslautern, Germany, 2005.

[48] A. Haller, E. Cimpian, A. Mocan, E. Oren, and C. Bussler, "WSMX - A Semantic Service-Oriented Architecture", in *Proc. International Conference on Web Service (ICWS 2005)*, Orlando, Florida, 2005.

[49] A. Patil, S. Oundhakar, A. Sheth, and K. Verma, "METEOR-S Web Service Annotation Framework," in *Proc. the 13th World Wide Web Conference (WWW2004)*, New York, USA, May 2004.

[50] R. Lara, A. Polleres, H. Lausen, D. Roman, J. de Bruijn, and D. Fensel, "A Conceptual Comparison between WSMO and OWL-S", in *Proc. European Conference on Web Services (ECOWS 2004)*, Erfurt, Germany, Sept. 2004.

[51] D. Martin, M. Paolucci, and M. Wagner, "Towards Semantic Annotations of Web Services: OWL-S from the SAWSDL Perspective," in *OWL-S Experiences and Future Developments Workshop at ESWC 2007*, Innsbruck, Austria, June 2007.

[52] D. Brickley, and R.V. Guha, "RDF Vocabulary Description Language 1.0: RDF Schema." Available: http://www.w3.org/TR/rdf-schema/, Feb. 2004.

[53] T. Di Noia, E. Di Sciascio, and F. M. "Donini, A non-monotonic approach to semantic matchmaking and request refinement in E-Marketplaces," in *Proc. 1st Int. Workshop on Semantic Matchmaking and Resource Retrieval: Issues and Perspectives*, Seoul, Korea, Sept. 2006, pp. 81-96.

[54] S. Battle, A. Bernstein, H. Boley, B. Grosof, M. Gruninger, R. Hull, M. Kifer, D. Martin, S. McIlraith, D. McGuinness, J. Su, and S. Tabet, "Semantic Web Services Framework (SWSF) Overview." Available: http://www.w3.org/Submission/SWSF/, Sept. 2005.

In: Cyberinfrastructure Technologies and Applications ISBN: 978-1-60692-063-3
Editor: Junwei Cao

Chapter 9

SERVICE COMPOSITION AUTOMATION WITH AI PLANNING

***Maozhen Li*[1], *Bin Yu*[2] *and Man Qi*[3]**
[1] Electronic and Computer Engineering,
School of Engineering and Design
Brunel University, Uxbridge, UB8 3PH, UK
[2] Level E Limited, Edinburgh, EH9 3JL, UK
[3] Department of Computing, Canterbury Christ Church University
Canterbury, Kent, CT1 1QU, UK

ABSTRACT

Grid computing is rapidly evolving into a service-oriented computing infrastructure that facilitates resource sharing and large-scale problem solving on the Internet. It is envisioned that many resources on the grid would be exposed as services for a wider use by the community. Service discovery and composition has thus become a vitally important component in utilizing grid facilities. This chapter focuses on service composition. One challenge in service composition is how to automate the composition process in terms of a large number of services (atomic services or component services) and a variety of user requests. A novel Hierarchical Two Directions Partial Order Planning (H2POP) algorithm is presented for discovery of composite services which are dynamically composed from component services. A use case is given to illustrate the application of the H2POP algorithm for travel planning service composition automation.

INTRODUCTION

With the development of Web services technologies [1], the computational grid is rapidly evolving into a service-oriented computing infrastructure that facilitates resource sharing and

[1] E-mail address: Maozhen.Li@brunel.ac.uk
[2] E-mail address: Bin.Yu@levelelimited.com
[3] E-mail address: mq4@canterbury.ac.uk

large-scale problem solving over the Internet [2]. Open Grid Services Architecture [3], promoted by the Open Grid Forum (OGF, http://www.ogf.org) as a standard architecture for developing the next generation service-oriented grid systems, has facilitated the evolution. It is expected that Web Service Resource Framework [4] will be acting as an enabling technology to drive this evolution further. Services are implemented as software components, the interfaces of which can be advertised in terms of their functional and non-functional properties. Advertising services in a grid environment means that service-associated properties are registered with a service registry, which can then be accessed by users or user applications to discover the services that meet user needs. In this way, users can utilize grid services without knowing their low level implementations, reducing the complexity of using grid facilities. Increasingly, grid environments host a large number of services -- exposing physical resources such as processors/CPUs, disk storage, network links, instrumentation and visualisation devices in addition to applications and software libraries. Service discovery has therefore become a vitally important component for utilising grid facilities.

ROSSE is a Rough sets based search engine for discovery of Grid or Web services that have WSDL interfaces or OWL-S interfaces [5, 6, 7]. The novelty of ROSSE lies in its capability to deal with uncertainty of properties when matching services. For a service query, ROSSE can find more relevant services than other service discovery mechanisms such as UDDI and OWL-S. It is worth noting that, in some cases, no individual services could satisfy a user specific query. A number of services may need to be composed together to answer a user query. ROSSE has been enhanced with the capability to discover composite services which is the focus of this chapter.

The rest of this chapter is organized as follows. Section 2 introduces AI planning for service composition. Section 3 presents a Hierarchical Two Directional Partial Order Planning (H2POP) algorithm for service composition automation. Section 4 gives a case study to illustrate the use of H2POP to discover a composite service for travel planning. Section 5 discusses some related work on service composition, and Section 6 concludes the chapter.

AI Planning for Service Composition

Service composition can be considered as a planning problem. AI planning has been widely used for service composition with an aim to automate the process in service composition. In the following sections, we give a brief overview of classic AI planning algorithms.

Classic AI Planning

Most AI planning approaches use a state-transition system. In a state-transition system, there are finite or recursively enumerable set of states, actions and events along with a transition function that maps a state, an action and an event to a set of states [8]. Classic AI planning mainly employs the initial modeling of the STRIPS [9] as a basic model to solve a planning problem.

The action of STRIPS can be represented as <P, A, D>, where

- The state P is the description of a precondition.
- The state A represents an "add" state, which adds a group of states in the current list of world states.
- The state D represents a "delete" state, which removes a group of states from the current list of world states.

Here, each state in STRIPS is defined by using a first-order language. Based on the STRIPS model, classic AI planning introduces an operator which can be represented as a triple, i.e. o= (*name(o), precond(o),effects(o)*), where

- The *name(o)* is the name of an operator.
- The *precond(o)* is the precondition of an action happened.
- The *effects (o)* can be positive or negative
- The $effect^{+}(o)$ is normally placed into an *add* list which represents a set of added states.
- Oppositely, the $effect^{-}(o)$ is normally placed into a *delete* list, which represents a set of removed states.

If the preconditions of an operator are satisfied with the state s (denoted by $precond(o) \subset s$), this operator *o* will act on the state s. However, the STRIPS model does not have expressions of conditions and functions. Then, the representation is insufficiently expressive for real domains. As a result, many language variations are developed such as the Action Description Language (ADL) [10]. In addition, many high-level planning models are developed to handle complex planning problems such as Hierarchical Task Network (HTN) planning [11]. The main idea of HTN planning is to decompose tasks to subtasks by using domain related pre-defined methods. However, the domain of grid services is very wide in terms of the variety of services, the number of relevant methods for decomposing user goals could be huge. As a result, for a user service request (goal), service composition using HTN planning would be hard as the mapping of user goals to decomposition methods is time consuming. Pre-defined decomposition methods are fixed for a particular domain problem using HTN planning, which is not flexible in terms of a variety of domain problems. Furthermore, a large space is needed to store decomposition methods. Therefore, a dynamic and flexible planning approach for service composition is needed to avoid the large scale of computing and storage.

Partial Order Planning

Partial Order Planning (POP) [12] only introduces ordering constraints as necessary (least commitment) in order to avoid unnecessarily searching through the space of possible orderings. POP has an advantage over HTN planning in terms of its least commitment.

Algorithm 1. A classic POP algorithm

```
function POP(initial, goal, operators) returns plan
Plan<-Make-Minimal-Plan(initial goal)
loop do
    if Solution?( plan) then return plan
    Sneed, c <-Select-Subgoal( plan)
    Choose-Operator( plan, operators, Sneed, c)
    Resolve-Threats( plan)
end
---------------------------------------------------------------
function Select-Subgoal( plan) returns Sneed, c
  pick a plan step Sneed from Steps( plan)
      with a precondition c that has not been achieved
  return Sneed, c
---------------------------------------------------------------
procedure Choose-Operator(plan, operators, Sneed, c)
  choose a step Sadd from operators or Steps(plan) that has c as an effect
  if there is no such step then fail
  add the causal link Sadd ---c--> Sneed to Links(plan)
  add Sadd < Sneed to Orderings(plan)
  if Sadd is a newly added step from operators then
          add Sadd to Steps(plan)
          add Start<Sadd<Finish to Orderings(plan)
---------------------------------------------------------------
procedure Resolve-Threats(plan)
  for each Sthreat that threatens a link Si --c->Sj in Links(plan) do
      choose either
              Demotion: Add Sthreat<Si to Orderings(plan)
              Promotion: Add Sj<Sthreat to Orderings(plan)
      if not Consistent(plan) then fail
  end
```

The key of POP is that, for a user request (goal), it searches partial plans rather than all possible situations. There are three main parts in classic POP algorithm, i.e. obtaining sub-tasks (sub-goals), searching and locating operators. When applying POP in service composition, a sub-task in POP can be considered as a component service in a composite service, an operator in POP can be considered as a candidate service (a concrete component service). Locating an operator can be considered as positioning a component service in a composite service. Thus, POP can be used for partially automating service compositions. A classic POP algorithm is depicted in Algorithm 1.

H2POP FOR SERVICE COMPOSITION AUTOMATION

H2POP Design

Based on classic POP algorithms, we have designed the H2POP algorithm for service composition automation. We extend POP from the following aspects.

Definition 1: A H2POP operator is defined as an expression of quaternion: (***Identity, Precondition, effect+, effect-***). There are two types of operators - ***operator+*** and ***operator-***, where

- ***Identity*** represents the identity of a candidate service.
- ***Precondition*** represents whether a candidate service is selected.
- ***effect+*** represents the H2POP operator's '*add*' list. The properties in the '*add*' list will be appended into an original property list after the service is selected.
- ***effect-*** represents the H2POP operator's 'delete' list. The properties in the 'delete' list will be removed from an original property list after the service is selected.
- ***Operator+*** represents ***add*** operator when a progressed search is executed.
- ***Operator-*** represents ***add*** operator when regressed search is executed.

Thus, a H2POP plan can be defined as follows:

$$\text{Plan} = \text{Operator+} \odot \text{Operator-}$$

where, the symbol $\odot$ combines all candidate services from their original service lists and workflow arrays (recording the positions of candidate services) of both ***Operator+*** and ***Operator-*** to form a uniform service list and a uniform workflow array. It should be stressed that the "+" and "-" of operators in H2POP algorithm represent progressed directions and regressed directions respectively.

The following definition defines how service composition can be considered as an AI planning problem.

Definition 2: A planning problem in H2POP is defined as a triple (***D, IN+, IN-***), where

- ***D*** is a domain
- *IN*+ is a positive goal
- *IN*- is a negative goal

The following definition describes a Hierarchical Partial Order Planning in one direction.

Definition 3: A Hierarchical Partial Order Planning (HPOP) is defined as a triple (***initial, goal, operators***), where

- ***initial*** represents a planning problem for services composition represented by (***D, IN+| IN-***).
- ***goal*** represents that the goal of HPOP is to gain ***effect-*** from selected ***operators*** to reduce ***IN+*** or ***IN-*** in ***initial.***
- ***operators*** represents selected operators, which are recorded in a service list and positioned in a workflow array.

HPOP can be a progressed planning (HPOP+) or a regressed planning (HPOP-).

Now the problem of service composition becomes an issue of AI planning. The H2POP algorithm provides a complete planning by recursively invoking partial planning processes. That is where a hierarchical POP comes from.

It should be noted that if a composite service is planned with one planning direction, be it a progressed planning or a regressed planning, the searching process may not be able to converge as shown in Figure 1. In Figure 1, the red line represents a progressed planning, and green line represents a regressed planning. Therefore, we add two directions in the HPOP to automate service composition.

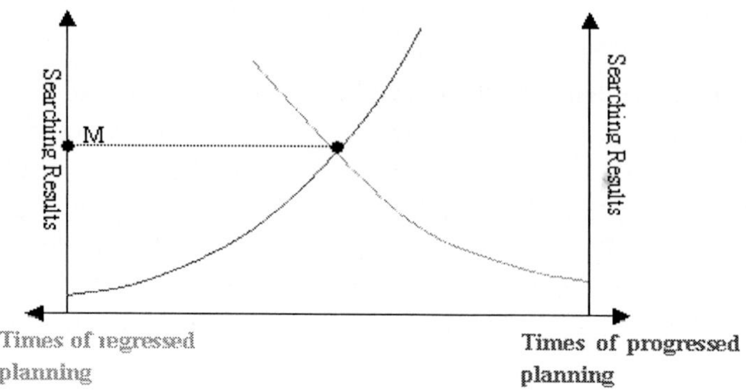

Figure 1. Progressed Planning and Regressed Planning.

The data flow in the H2POP algorithm is shown in Figure 2. The Progressed Planning unit and the Regressed Planning unit are used to compose a service in a positive direction and in a negative direction respectively. The Link unit aims to link a positive workflow array with a negative workflow array. The Judgement unit checks whether a Progressed Planning can be linked with a Regressed Planning.

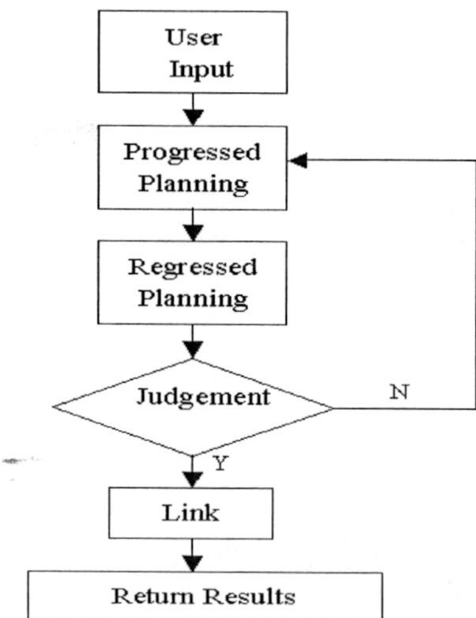

Figure 2. Data flow in the H2POP algorithm.

From Algorithm 2, only when all user inputs (goals) and outputs (goals) are satisfied and searching results using a progressed planning and a regressed planning are linked together, the algorithm returns correct plan. If user inputs and outputs are not satisfied or searching results of progressed and regressed cannot be linked together, the system will continue to search by using updated inputs and outputs. The function ***Available-Link*** checks whether outputs of ***operator+*** and inputs of ***operator-*** have any relationships. As defined in Definition

3, the goal of service composition is to satisfy all properties of user input and output. Algorithm 2 shows the H2POP algorithm, and Algorithm 3 shows the HPOP algorithm.

Algorithm 2. The HPOP function

```
   function H2POP(D, IN+ , IN-) returns plan
Plan <- Gain-Operator-From-ROSSE(initial value)
if Solution?( plan) then return plan
else plan <- null
loop do
 operator+ <- HPOP+(D \/IN+ \/IN-, goal, operator+)
 if goal? break
end
   IN+,IN- <- Renew(IN+, IN-) from HPOP+(-)
loop do
 operator- <- HPOP-(D \/IN+ \/IN-, goal, operator-)
 if goal? break
end
if Available-Link(operator+, operator-) and goal? then return plan
Else
   Renew(IN+, IN-) from Operator
   H2POP(D, IN+ , IN-)
   Return plan
   end if
end
```

Algorithm 3. The HPOP+(-) function

```
Function
HPOP+(-)( D \/IN- \/IN+, goal, operators+(-)) returns operators+(-)
 newoperators+(-) <-Choose-Operator+(-)( D \/IN+(-),operators+(-))
 IN-(+) <- Delete-Gain-Properties(newoperators+(-))
 operators+(-) <- Resolve-Threats( newoperators+(-),operators+(-))
 return operators+(-)
```

In HPOP, new operators are selected by using ***Choose-Operator*** function as shown in Algorithm 4, and then their locations are planned by using ***Resolve-Threats*** function as shown in Algorithm 5.

Algorithm 4. The Choose-Operator

```
   Function
   Choose-Operator+(-)(operators+(-)) return newoperator+(-)
    input(output)<- Gain-Property(operators+(-))
    newoperator+(-) <- Gain-Operator -From-ROSSE(input(output))
return newoperator+(-)
```

The ***Choose-Operator*** function builds on the ROSSE search engine and only uses input properties or output properties to search for related candidate services. Here, the new operator is different from the operator defined in Definition 1. It only includes the identity of a candidate service but does not include information related to a workflow array.

Algorithm 5. The Resolve-Threats function

```
Function
Resolve-Threats(newoperators+(-),operators+(-)) return operators+(-)
 for choose a step w_add from newoperators+(-) have eff as effects and pre as
preconditions do
 if there is no such step then fail
    choose either
        Demotion: Add w_add < w_0 to W(i,j )+(-)
        Promotion: Add w_j < w_add to W(i,j )+(-)
    if not Consistent(W(i,j )+(-)) then fail
    add newoperators+(-) to Service set S+(-)
    operators+(-) <- S+(-), W(i,j )+(-)
end
    return operators+(-)
```

In addition, the algorithm introduces a unit to handle loop states. The unit gives a flag for each service where the loop is possibly happened. The flag is also labelled into both workflow array and service set. In the Resolve-Threats function, both the progression approach and the regression approach in classical POP are used to solve ordering problem of workflow arrays. Then, generated operators are returned. These operators include related candidate services and their positions in a workflow array.

H2POP Implementation

This section focuses on the implementation of H2POP, which builds on the ROSSE search engine. Figure 3 shows the components related to H2POP.

The Service Selection unit is used to find possible individual component (atomic) services. Based on a service ontology, the unit reasons the relationships between properties of advertised services and properties used in a service query. For reasoning, the Service Selection unit uses Protégé OWL plug-in API (http://protege.stanford.edu/plugins/owl/api) and RACER [13]. In addition, the service property reduction of ROSSE is still used here to dynamically discern and reduce dependent properties.

Using selected component services, the Workflow Building unit is used to automatically generate workflows building on the H2POP algorithm. The process of building a workflow is alternately implemented between the Workflow Building unit and the Service Selection unit. In other words, if a possible component service is selected, then a workflow attempts to create a new step. Based on a created workflow, a new component service is selected. The process is

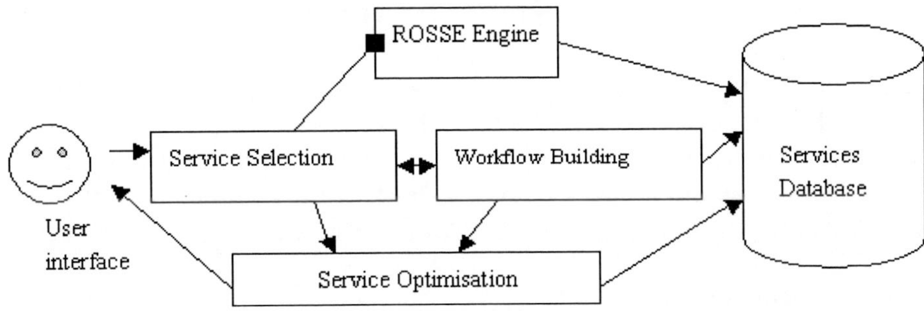

Figure 3. H2POP related components for discovery of composite services.

not completed until all component services are found and a whole workflow is built. There are three statuses after a composite service is built:

- Executable. It means that the search process is successful and a composite service is produced.
- Terminable. It means that searching process is a finite computation and a failure will be returned.
- Reusable. It means that a composite service can be directly discovered if user supplies an identical query.

The database in H2POP reuses and extends the structure of the ROSSE database as shown in Figure 4. In order to search services by input properties, output properties, preconditions and effects, we build input tables, output tables, precondition tables and effect tables. For services with OWL-S interfaces, the properties related to input, output, preconditions and effects are appended into relevant tables. For UDDI registries, only input and output properties are manually supplied to the system. In the mean time, H2POP also makes use of UDDI APIs to dynamically load services registered in UDDI registries. The OWL-S files mapping aims at mapping OWL-S files to their network access points, and the ontology table stores ontology information for semantic reasoning. The history table records previous search results.

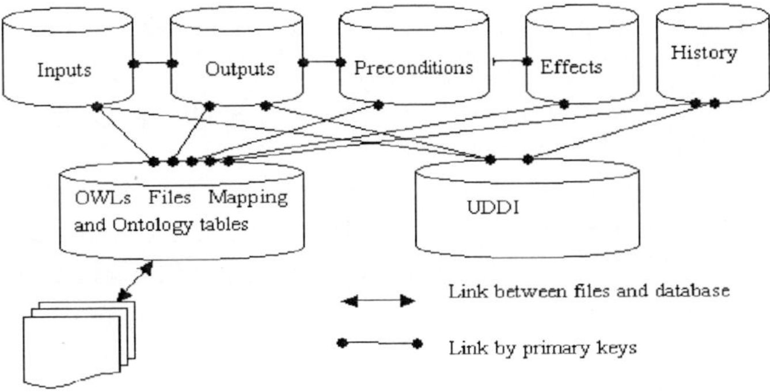

Figure 4. Service database management.

A CASE STUDY OF H2POP

In the section, a travel-planning example is used to illustrate how the H2POP algorithm works. To test the effectiveness of H2POP, we have designed a few hundred services in ROSSE including car rental, accommodation booking, money translator, flight tickets booking. The H2POP algorithm runs o top of the ROSSE environment. We designed a composite service using the following atomic services as shown in Table 1.

Table 1. Five atomic services for a composite service

Service Name	Input Properties			Output Properties			
Flight Ticket	Source Location	Destination	Dollar (price)	Availabilit y	City Location		
Car Rental	City Location			Charge (Day)	Cartype	Color	Payment
Accommodation	City Location			Price (Day)	Room no		Payment
Check Out	Payment			Totalprice			
Money Exchange	Pound			Dollar			

A service composition request is supplied to ROSSE using the user interface as shown in Figure 5. In the following sections, we describe how the service composition is performed.

Firstly, a service composition query which is provided in the form of *#composite#class:travel;input:pound,destination;output:cartype,price,color,charge,roomno,totalprice,*and is submitted to ROSSE, here,

- ***#composite*** means a service composition
- ***#class:*** means a searched domain
- ***input:*** means user inputs
- ***output:*** means user outputs
- *;* means the end of the request

Figure 6 (a), (b), (c) show the results of the service composition. The workflow diagram which is automatically drawn by ROSSE shows all the atomic services used in the composite service. It should be noted that service 3 and service 4 are used together in the service composition. The diagram also comes with a description on each service used in the composite service such as service number, service name, business name, business description, access point, overview document and all services ID of possible candidate services. Using the example as shown in Figure 5, we explain how the service is composed using H2POP.

Figure 5. ROSSE user interface for discovery of composite services.

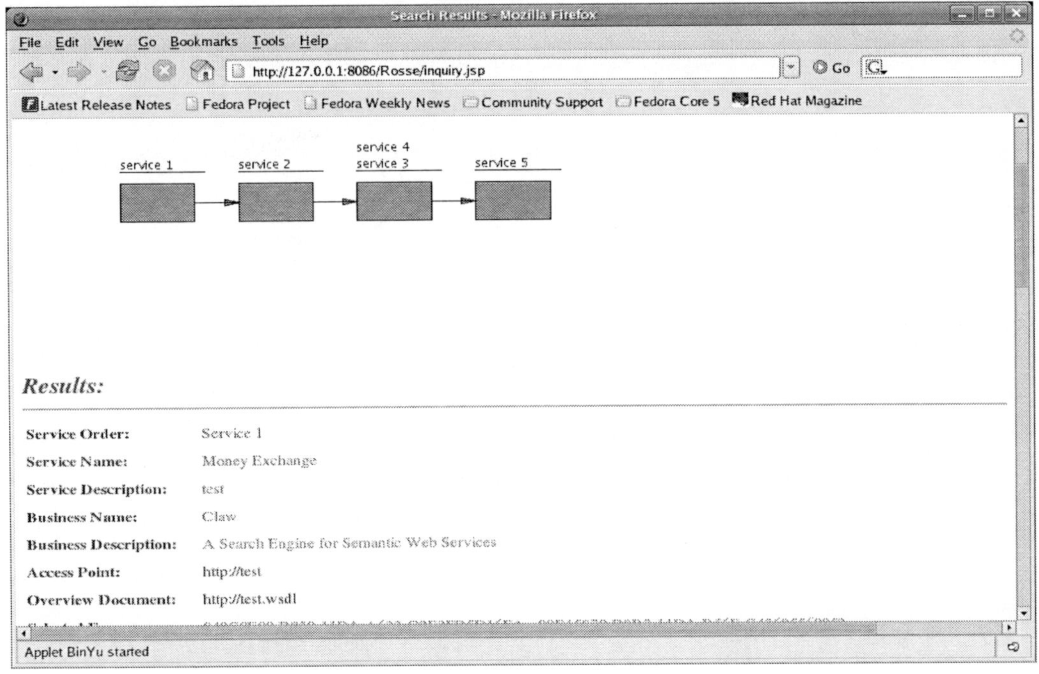

(a)

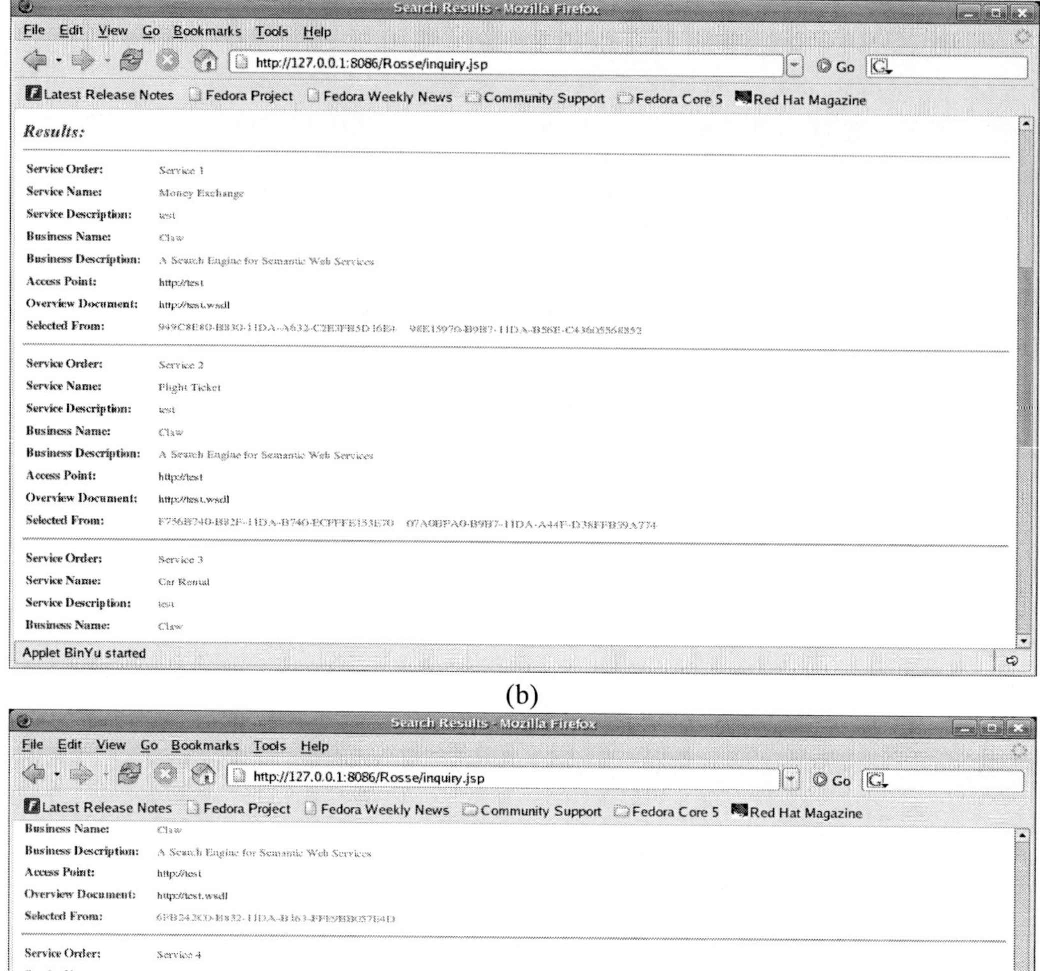

(b)

(c)

Figure 6. Service composition results.

- Step 1: the input proprieties of a user requirement were used to infer and discover the first service using the Progressed Planning as shown in Figure 7. The first service was descried as a quaternion

<flight ticket, add-this-service (precondition), Add (input: Dollar, Source Location output: Availability, City Location), Delete (input: Destination, output: null)>

Thus, Destinations were removed from current searching properties. Other properties were appended to the list of current searching properties. The goals of the user query are Pounds and Destinations.

Input: Pound, Destination

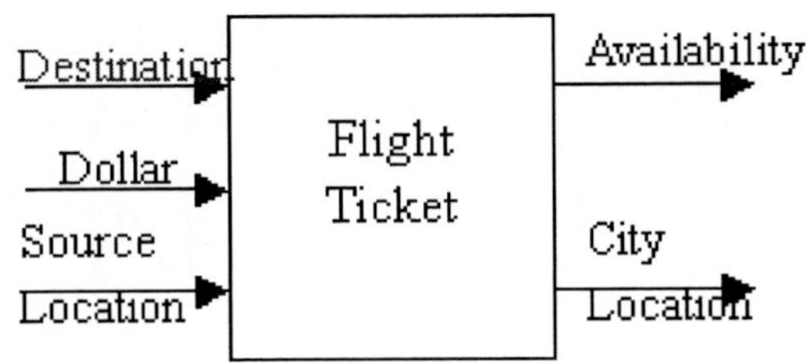

Figure 7. The 1st step of the service composition.

- Step 2: the inferred input proprieties in the second step were used to reason and match the second service in the goal of Progressed Composite Service as shown in Figure 8. The second service was descried as a quaternion

<Money Exchange, add-this-service, Add (input: null output: dollar), Delete (input: Pound output: null)>

Thus, Pound was removed from current searching properties. Other properties were appended to current searching properties. Then, the connection method was invoked to locate the relationship between the component service and services partially composed in previous steps. In the example, because the output of Money Exchange service was equal to an input of the service of Flight Ticket, the Money Exchange service was placed in front of the Flight Ticket service. Then, because the property of dollar belongs to input and output, the connected property Dollar was deleted from both of input and output property lists. In addition, all input properties of a user requirement are completed.

Input: Pound, Dollar, Source Location
Output: Availability, City Location

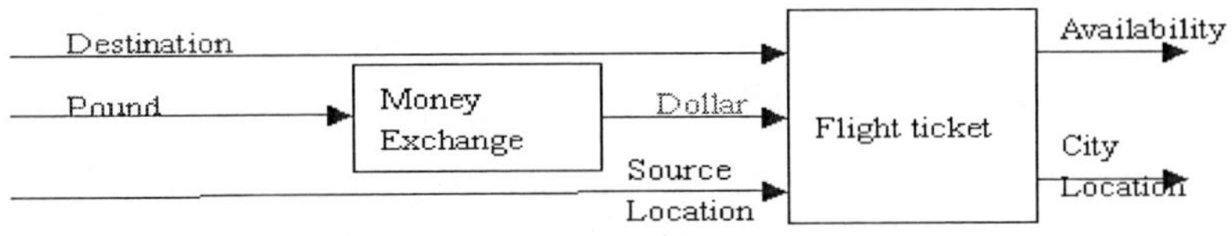

Figure 8. The 2nd step of the service composition.

- Step 3: the output proprieties of users' requirement were used to implement Regressed planning and discovered the first service in the goal of Regressed Composite Service as shown in Figure 9. The first service in the first Regressed reasoning was descried as a quaternion

<Car Rental, add-this-service, add-this-service, Add (input: City location output: Payment), Delete (input: null output: Charge, CarType, Color)>

Thus, Charge, CarType and Color properties were removed from current searching properties. The property of City Location was appended to current searching properties.

Output: Cartype, Price, Color, Charge, Room no, Totalprice

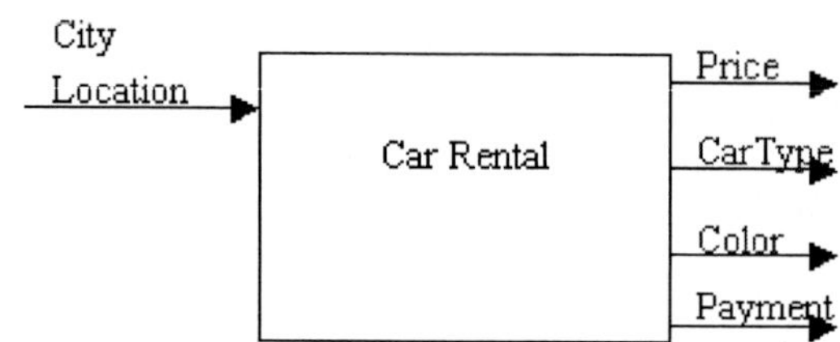

Figure 9. The 3rd step of the service composition.

- Step 4: the remainder of output proprieties of user requirement were used to implement Regressed reasoning and gain the second service in the goal Regressed Composite Service as shown in Figure 10. The second service in Regressed reasoning was descried as a quaternion

<Accommodation, Add-this-service, Add (input: City location output: Payment), Delete (input: null output: Price, Room no)>

Thus, the properties of Price and Room no were removed from current searching properties. Other properties were appended to current searching properties. Then, the connection method was invoked to locate the relationship between the component service and composed services in previous steps in process of Regressed reasoning. In the sample, because the input and output of the service of Car Rental was equal to an input of the service of Accommodation, the service of Car Rental and Accommodation were implemented in parallel. Then, because the property of City Location only belonged to input properties, one of two properties, Location, is deleted from input properties.

Input : City Location Output: price, room no, totalprice

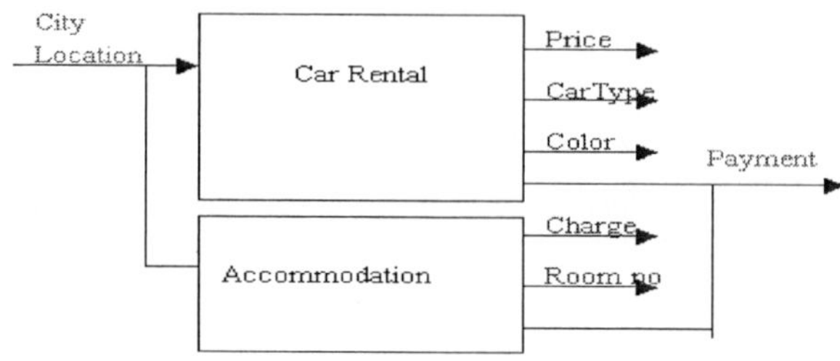

Figure 10. The 4th step of the service composition.

- Step 5: the remainder output propriety of user's requirement, totalprice, were used to implement Regressed reasoning and gain the third service in the goal Regressed Composite Service as shown in Figure 11. The third service in Regressed reasoning was descried as a quaternion

<Check out, add-this-service, Add (input: Payment output: null), Delete (input: null output: Totalprice)>

Thus, the output property of Totalprice was removed from current searching properties. Other properties were appended to current searching properties. Then, the connection method was invoked to locate the relationship between the component service and composed services in previous steps in process of Regressed reasoning. In the sample, because the input of the service of Check Out was equal to an output of the service of previous total Regressed composite service, the service of Check Out was located after the total Regressed composite service. Then, because the related property, Payment, belonged to input and output, it was deleted from both of input and output properties. In addition, all output properties of users' requirement had been completely gained. The first Regressed searching also had been completed. Then the Regressed searching and Progressed searching were attempted for connection.

Input : City Location Output: totalprice

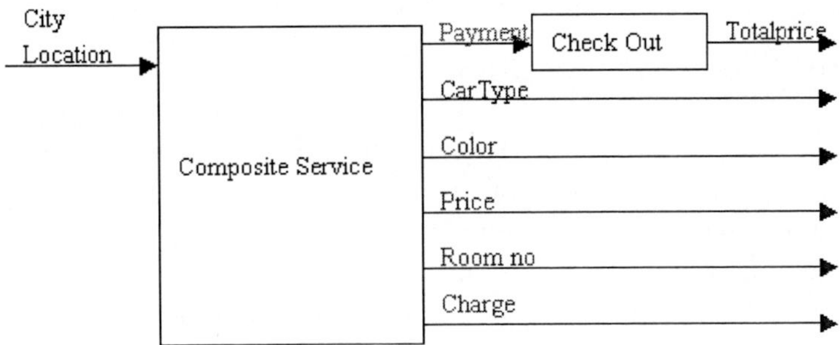

Figure 11. The 5th step of the service composition.

As shown in Figure 12, the Regressed searching and Progressed searching were attempted for connection. An output of the Progressed composite service was the same as an input of the Regressed composite service. The whole composite service was combined together. If the connection cannot be found, the output properties of Progressed composite service and the input of Regressed composite service will be implemented in a new round of service composite as new input properties and new output properties.

The whole service composition process was completed in 7413 ms. The composition process went through data transferring, service selection, workflow building, service optimisation, returning results and workflow diagram drawing. Hereinto, selecting services and building the workflow took 5034 ms.

Connection: Output of progressed planning: Availability, City Location
Input of regressed planning: City Location

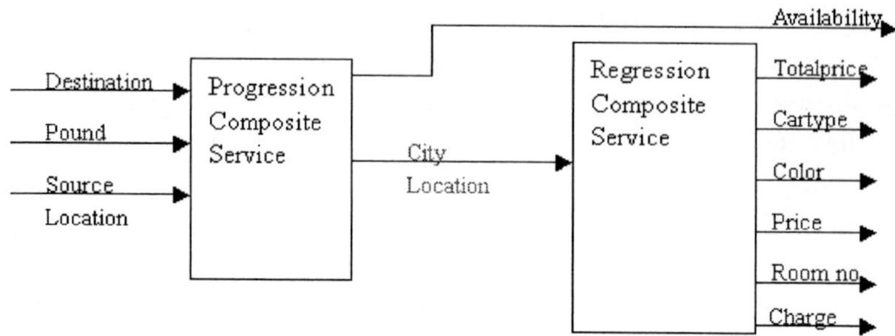

Figure 12. The 6th step of the service composition.

Related Work

In this section, we discuss some related work on service composition, from the aspects of semi-automatic service composition and automatic service composition.

Semi-Automatic Service Composition

Semi-automated service composition requires user involvement in a composition process. Currently, there are many successful work for semi-automated service composition such as OWL-S, Web Component [14], π-calculus [15] and Petri-nets [16]. OWL-S is a semantic service description langrage. OWL-S processes can be used to describe workflows of services. However, OWL-S files are very hard to generate and implement service binding, so many tools are developed for OWL-S to translate services from WSDL to OWL-S such as OWL-S Editor (http://owlseditor.semwebcentral.org/). The main idea of Web Component is to encapsulate composite-logic information inside a class definition and its public interface can be published and used for discovery and reuse. The composite-logic comprises service composition types and message dependencies. Similar to BPEL [17], Web component can be composed using Service Composition Specification Language (SCSL). However, the SCSL has to be manually edited, and its correctness has to be checked by using related tools. The π-calculus manually composes service components and automatically checks the correctness of composite services. However, the π-calculus is different form others, which is able to successfully compose large-scale composite services. In addition, Petri-nets is a well-established process-modeling approach. It can be used for modelling and checking composite services. However, it is not able to automatically compose services. Although the above-mentioned methods for service composition have strong analysis functions, they are only used for accessorial design.

Automated Service Compositions

The Finite-State Machines (FSM) [18] and Artificial Intelligence (AI) planning are two popular methods for automatic service compositions. The FSM checks the existence of a composite service. Therefore, if a system wants to discover composite services using FSM, it will have to require a fixed domain task structure or require an infinite storage space to locate all domain task structures. The AI planning includes situation calculus, Planning Domain Definition Language (PDDL) [19], Rule-based planning [20, 21] and Hierarchical Task Network (HTN) planning. The SWORD [22] is a tool for building composite services using rule-based plan generation. However, rule-base plan only build work flow in one direction. This method may lead to a large scale of searching space and a result which cannot trend to be constringency. And the HTN planning is an AI planning approach. SHOP2 [23, 24] uses the HTN planning to couple services. Its main idea is to decouple a composite task to many primitive tasks and to replace each primitive task with an existing operator. If the number of service in a system is large, SHOP2 will possibly introduce many intermediate services. Thus, the speed and correctness of service composition will be decreased. In addition, in order to decouple any complex tasks into primitive tasks, the system has to store a large number of primitive methods and hierarchical models.

Compared with the related work in service composition, the H2POP algorithm builds on a classic POP model, and extends POP with planning in two directions. One benefit of H2POP is its flexibility in that it does not need pre-defined methods to decompose user goals like HTN planning. H2POP uses the least commitments as constrains to compose services. The use of progressed planning and regressed planning ensures the composition process to be converged quickly instead of to be an infinite process. In addition, H2POP is a lightweight algorithm. The algorithm does not demand intensive resources when composing services.

CONCLUSION

This chapter presented H2POP for service composition automation. H2POP builds on Rough sets for dynamic service reduction, and extends a classic POP model for automatic service composition. A case study was presented to illustrate the use of H2POP for composition of travel services. H2POP can be improved in the following ways:

- *Distributed Services Composition.* The H2POP algorithm composes services in a centralised way. However, a service-oriented structure is typically used for a distributed system framework. Thus, services, which might be geographically distributed in a network environment, are more suitable to be coupled in a decentralised way. For this purpose, a distributed service registry is needed. UDDI 3, which provides a federated mechanism to couple services, can be explored for this work. Distributed service composition also entails structures for the distributed service registry to ensure the lookups of services in a distributed environment to be efficient.
- *Uncertainty of Service Properties.* The H2POP algorithm dynamically uses Rough sets to reduce irrelevant and dependent properties for a specific component service.

However, H2POP currently does not check the dependencies of service properties between component services. To speed up service composition process, and to achieve more accuracy in service composition, this issue should be considered in future work.

- *Service Composition with Loops.* H2POP deals with loops in service composition with pre-defined information. It is not flexible in this way as they can be a variety of services published in ROSSE. Future work should introduce new mechanisms to manage loops in service composition. There are two types of loops - self-loops and normal loops. For a self-loop service, the number of loops could be determined using historical execution records of the services. For a normal loop service, finite state machines (FSM) could be used to determine the number of loops and the services that are usually coupled by the service.

REFERENCES

[1] F. Curbera, M. Duftler, R. Khalaf, W. Nagy, N. Mukhi, and S. Weerawarana, "Unraveling the Web Services: An Introduction to SOAP, WSDL, and UDDI," *IEEE Internet Computing*, vol. 6, no. 2, pp. 86-93, 2002.

[2] M. P. Atkinson, D. De Roure, A. N. Dunlop, G. Fox, P. Henderson, A. J. G. Hey, N. W. Paton, S. Newhouse, S. Parastatidis, A. E. Trefethen, P. Watson, J. Webber, "Web Service Grids: An Evolutionary Approach," *Concurrency - Practice and Experience*, vol. 17, no. 2-4, pp. 377-389, 2005.

[3] Foster, C. Kesselman, J. M. Nick, S. Tuecke, "Grid Services for Distributed System Integration," *IEEE Computer*, vol. 35, no. 6, pp. 37-46, 2002.

[4] K. Czajkowski, D. F. Ferguson, I. Foster, J. Frey, S. Graham, I. Sedukhin, D. Snelling, S. Tuecke, W. Vambenepe, "The WS-Resource Framework," March 5, 2004. [online] http://www.globus.org/wsrf/specs/ws-wsrf.pdf. (last access: Dec. 2007)

[5] M. Li, B. Yu, C. Huang, and Y. H. Song, "Service Matchmaking with Rough Sets," *Proc. IEEE Int'l Symposium on Cluster Computing and the Grid (CCGrid)*, pp. 23-30, May 2006.

[6] B. Yu, W. Guo, M. Li, Y.H. Song, P. Hobson, M. Qi, "Service Matchmaking and Discovery with Rough Sets," *Proc. Int'l Conf. on Semantics, Knowledge and Grid (SKG)*, p. 80, Nov. 2006.

[7] Maozhen Li, Bin Yu, Omer Rana, Zidong Wang, Grid Service Discovery with Rough Sets, *IEEE Transactions on Knowledge and Data Engineering*, 19 Dec 2007.

[8] E Sirin, Automated composition of web services using AI planning Techniques, http://www.cs.umd.edu/Grad/scholarlypapers/papers/aiplanning.pdf.

[9] R. E. Fikes and N. J. Nilsson. "Strips: A new approach to the application of theorem proving to problem solving". In J. Allen, J. Hendler, and A. Tate, editors, *Readings in Planning*, pages 88–97. Kaufmann, San Mateo, CA, 1990.

[10] E Pednault, ADL and the state-transition model of action, *Journal of Logic and Computation* 4:467–512, 1994.

[11] K. Erol, J. Hendler and D. S. Nau, "Semantics for Hierarchical Task Network Planning", UMIACS Technical Report, 1994.

[12] Joachim Peer: A POP-Based Replanning Agent for Automatic Web Service Composition. *ESWC 2005*: 47-61.

[13] V. Haarslev and R. Möller, "Description of the RACER System and its Applications," *Proc. Int'l Workshop on Description Logics (DL-2001)*, Aug. 2001.

[14] Jian Yang, Mike P. Papazoglou: Web Component: A Substrate for Web Service Reuse and Composition. *CAiSE 2002*: 21-36.

[15] R. Milner, The ployadic pi-calculus: a tutorial, Technical Report, Laboratory for Foundations of Computer Science, University of Edinburgh, October 1991.

[16] R. Hamadi and B. Benatallah, "A Petri-Net-Based Model for Web Service Composition," *Proc. 14th Australasian Database Conf. Database Technologies*, ACM Press, 2003, pp. 191–200.

[17] F. Curbera, Y. Goland, J. Klein, F. Leyman, D. Roller, S. Thatte, and S. Weerawarana, Business Process Execution Language for Web Services(BPEL4WS) 1.0, August 2002.

[18] X. Fu, T. Bultan, and J. Su, "Formal Verification of Eservices and Workflows," *Proc. Workshop on Web Services, E-Business, and the Semantic Web (WES)*, LNCS 2512, Springer-Verlag, 2002, pp. 188–202.

[19] Ghallab, M., Howe, A., Knoblock, C., McDermott, D., Ram, A., Veloso, M., Weld, D., Wilkins, D.: PDDL—The Planning Domain Definition Language (1998)

[20] B. Medjahed, A. Bouguettaya, A. K. Elmagarmid, "Composing Web Services on the Semantic Web", *The VLDB journal*, September 2003

[21] S. A. Chun, V. Atluri, N. R. Adam, Using Semantics for Policy-Based Web Service Composition, Distributed and Parallel Databases, 2005.

[22] S. R. Ponnekanti and A. Fox. SWORD: A developer toolkit for Web service composition. In *Proceedings of the 11th World Wide Web Conference*, Honolulu, HI, USA, 2002.

[23] Dana S. Nau, Tsz-Chiu Au, Okhtay Ilghami, Ugur Kuter, Héctor Muñoz-Avila, J. William Murdock, Dan Wu, Fusun Yaman: Applications of SHOP and SHOP2. *IEEE Intelligent Systems* 20(2): 34-41 (2005).

[24] Dan Wu, Bijan Parsia, Evren Sirin, James A. Hendler, Dana S. Nau: Automating DAML-S Web Services Composition Using SHOP2. *International Semantic Web Conference 2003*: 195-210.

In: Cyberinfrastructure Technologies and Applications
Editor: Junwei Cao
ISBN: 978-1-60692-063-3

Chapter 10

WORKFLOW IN A SERVICE ORIENTED CYBERINFRASTRUCTURE/GRID ENVIRONMENT

Wei Tan[1], Yushun Fan[2], Ian Foster[3] and Ravi Madduri[4]

[1] Computation Institute, University of Chicago and
Argonne National Laboratory Chicago, IL, USA
[2] Department of Automation, Tsinghua University,
Beijing 100084, P. R. China
[3] Computation Institute, Argonne National Laboratory
and University of Chicago Chicago, IL, USA
[4] Mathematics and Computer Science Division,
Argonne National Laboratory, Argonne, IL, USA

ABSTRACT

With the emergence of Service Oriented Computing, workflow has become an important method to compose services and reuse existing resources in the Cyberinfrastructure (CI) and the Grid. In this chapter, we first summarize research activities in the field of workflow in service oriented computing. We discuss five major research topics, i.e., languages and tools for service orchestration, automatic service composition, mediation-aided service composition, verification of service workflow, and decentralized execution of workflow. Although some of this work was originally targeted at the area of business process management, they can be adopted by the CI community with some modification or enhancement. In the second part of this chapter, we introduce a service-oriented workflow execution system, WS-Workflow, and explain the key modules in this system, including the data-driven composition module, the mediation-aided composition module, and the model fragmentation module. This system has been used in many projects to facilitate the flexible workflow composition and decentralized execution.

---

[1] E-mail address: wtan@mcs.anl.gov
[2] E-mail address: fanyus@tsinghua.edu.cn
[3] E-mail address: foster@mcs.anl.gov
[4] E-mail address: madduri@mcs.anl.gov

INTRODUCTION

From a technical point of view, with the emergence and maturation of Service Oriented Architecture (SOA), Web Services have begun to play a key role in enterprise information systems, the Grid, and cyberinfrastructure (CI) [1]. CI and Grids are becoming service oriented, i.e., using Web Services to provide a uniform interface to various resources, data and appliances. For example, grid middleware such as Globus [2] are adopting Web Services standards. From economic aspect, Web Services are now gaining momentum in the so-called network economics. Companies such as eBay, Google, and Amazon all use Web Services to extend their business and make money out of it.

In the scope of this chapter, CI is regarded as a collaborative platform to involve more users (typically scientific researchers) and resources (including data, applications and appliances), enabling greater capabilities in science and engineering research across multiple institutions and disciplines. Moreover, the term Grid is used to represent a concept similar as CI. However, this focus does not preclude the application of the techniques described in the following sections to overlapping fields like high performance computing, enterprise computing, etc.

Service not only provides interoperability. More important, service orchestration, or service workflow, offers enhanced reusability, agility and dynamicity in solution engineering. Nowadays, many resources in CI, including data, computational resources and others are wrapped with service interfaces to provide a uniform access. At the same time, many applications in CI are achieved through complex and distributed procedures. Given this requirement, many studies have been conducted on how to tailor the traditional workflow technology in order to address the new challenges brought about by such a service oriented computing (SOC) paradigm. Workflow systems are increasingly being developed to enable users in CI to integrate, orchestrate, and monitor various local or remote service resources to perform tasks like parallel data processing, batch job execution, and scientific discovery. Because we are talking about workflow in SOC, we use terms like service workflow, service composition, and service orchestration interchangeably.

Workflow technology can be traced back to late 1970s. It is originated from Office Automation (OA) and flourishes in Business Process Management (BPM) area. The requirements and characteristics of workflows in CI are partially overlapping those of business workflows. For example, they both need to define tasks and dependencies, and the structures of workflow systems in both areas are similar to the workflow reference model [3] proposed by WfMC (See figure 1). On the other hand, workflow in CI, especially in a service oriented paradigm, has its unique features, and therefore raises new challenges to be addressed by academia as well as industry. Here we classify these features and challenges into five categories which correspond to five interfaces defined in the workflow reference model. Some of these challenges have their counterparts in business process management domain, others do not. We will emphasize these different features and challenges in the following summary.

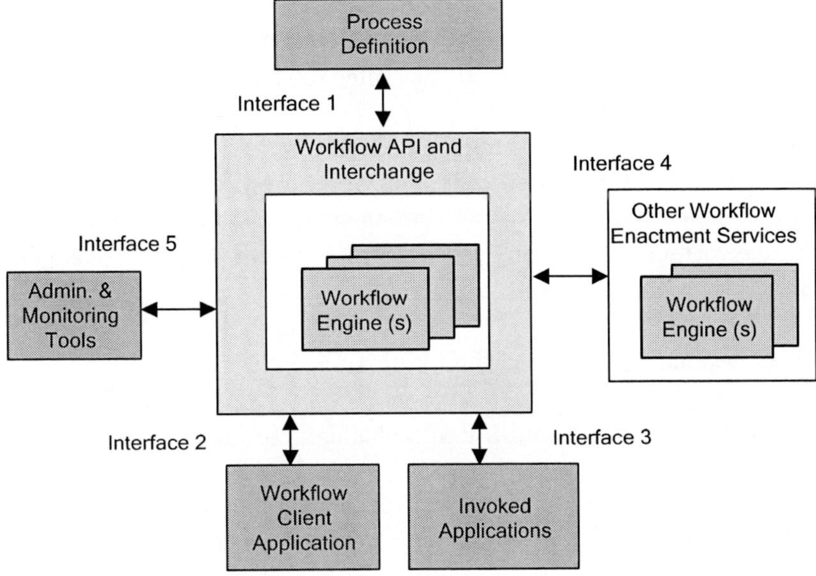

Figure 1. The workflow reference model (by WfMC [3]).

Interface 1: Process Definition

- Business workflows are usually constructed in a control-flow pattern. In contrast, data flow is considered to be the first-class citizen in a workflow in CI [4]. In a data flow, tasks and links represent data processing and data transport, respectively. A challenge is how to define a meta-model for the workflow in CI given this data-oriented nature. Whether and how service orchestration languages like BPEL can be used in CI still needs more exploration.
- Another challenge is how to derive a composite workflow that fulfills a given requirement. Because in CI typically there are many resources that provide similar functionality, techniques to facilitate the automatic service matchmaking and composition is welcomed. Given the heterogeneity of various service resources, an approach to glue them together with least cost and engineering effort is also desired.
- When multiple workflows are involved in an interaction, some technique is needed to guarantee the correctness of workflow(s).

Interface 2: Workflow Client Application

The Workflow Client Application let the users retrieve the tasks together with related data, and submit the execution results. In CI, user authentication and authorization is of great importance to ensure the integrity of workflows that span multiple organizations.

Interface 3: Invoked Applications

- The applications in CI workflow are often data-intensive. Data stage in/out are necessary before and after application execution.

- Fault handling is also crucial because the applications in CI are often long-running, computation-intensive, and expensive. The workflow system should be capable of identifying and handling failures to guarantee reliable execution in such cases.
-

Interface 4: Other Workflow Enactment Services

To take part in the collaboration with the workflows controlled by other enactment services, the flow engine should overcome heterogeneity at both process level (for example, in data format, data semantics, and process logic) and system level (for example, in security mechanisms).

Interface 5: Administration and Monitoring Tools

- In a CI without central control, it is a challenge to guarantee QoS commitments in workflow execution.
- Data provenance, i.e., to track the steps by which the data was derived, can provide significant value addition in data-intensive workflows in CI.

This chapter tries to give an overview of the research activities in service workflow, with a concentration on its application to CI. Section 2 provides a summary of the studies in service-oriented workflow area. Besides the languages and tools which have been widely used in this community, we also discuss some key technologies enabling the flexible and efficient modeling and execution of a service workflow. These technologies include automatic service composition, mediation-aided composition, verifications of workflow, and decentralized execution of workflows. Section 3 introduces the implementation of a service workflow system called WS-Workflow, including its framework and the major supporting techniques inside it. In Section 4 a summary is given and directions in future research are pointed out.

RESEARCH OF WORKFLOW IN SOC: STATE OF THE ART

SOA can be regarded as the backbone for CI, just like TCP/IP as the backbone for the Internet. In this section we summarize the research of workflow in SOC into five topics, i.e., languages and tools for service orchestration, automatic service composition, mediation-aided service composition, verification of service workflow, and decentralized execution of workflow. Although some of the studies are not originally targeted at CI, they can be used by the CI community without difficulty

Languages and Tools for Service Orchestration

Industrial community has proposed many service orchestration specifications, including Web Service Business Process Execution Language (WS-BPEL, or BPEL for short) [5], Web Service Choreography Interface (WSCI) [6], Web Service Choreography Description Language (WS-CDL) [7], and Ontology Web Language-Service (OWL-S) [8]. Among these specification languages, BPEL is becoming dominant not only because it has been proposed by OASIS as an industry standard, but also because it is execution-oriented and supported by major software companies such as IBM, Oracle and SAP, as well as the open source community.

BPEL

BPEL defines a meta-model and a XML-based grammar for describing the behavior of a business process that are composed of Web services as well as exposed as Web services. The BPEL process defines the execution structure on how multiple Web Service invocations are coordinated to achieve a business goal. BPEL also include other model elements like data manipulation, fault/exception handling, event processing, etc. One favorite feature of BPEL is that, it utilizes many XML specifications like XML Schema, WSDL, XPath, XSLT, WS-Addressing, etc. WSDL messages and XML Schemas are used as data type definitions; XPath and XSLT provide support for data assignment. WSDL is used to model all external partner services; WS-Addressing is used to access stateful resources which are especially relevant in CI.

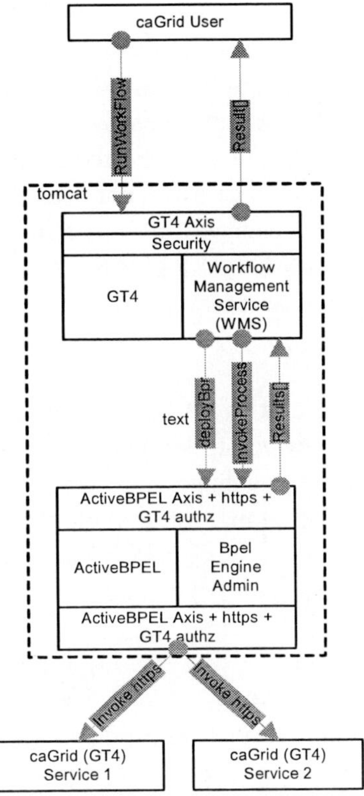

Figure 2. BPEL workflow in caGrid.

BPEL version 1.0 is proposed by IBM, BEA, Microsoft, SAP, etc, in 2002. BPEL 1.0 inherits two former specifications, XLANG by Microsoft and WSFL by IBM. In 2007 BPEL version 2.0 became an OASIS standard. BPEL specification can be used to model both abstract process and executable process. An abstract process is to define the externally visible message exchange sequence of a process, and this abstraction is used to depict interaction protocol without revealing its internal implementation. Whereas an abstract process provides a public view of the process logic, an executable one provides an internal view which is fully specified and thus can directly be executed by a workflow engine. There are many tools provided by commercial and open source community to support the modeling and execution of BPEL processes. These tools include IBM WebSphere Process Server, Oracle BPEL Process Manager, Active Endpoints ActiveBPEL, Eclipse BPEL Designer, etc.

Although BPEL is originally designed for business workflows, it also attracts attention from the CI community [9-11]. Because BPEL is built on top of the XML standard stack, standards that are particularly relevant in CI, like those in WSRF, can be easily integrated into BPEL. For example, BPEL is used in caGrid [12]. caGrid is the service-based Grid software infrastructure of the National Cancer Institute (NCI) cancer Biomedical Informatics Grid (caBIG™) program. The caBIG™ program was established to implement the enabling informatics technologies so that researchers can more efficiently share, discover, integrate, and process disparate clinical and research data resources with the overarching goal of accelerating cancer research. It implements a workflow service for orchestration of Grid services. The workflow management service uses BPEL, enabling the execution and monitoring of BPEL-defined workflows in a secure Grid environment. The general architecture is shown in figure 2. The Workflow component leverages an infrastructure stack that consists of Globus Toolkit, Apache Tomcat, Java, and Ant, with the addition of the ActiveBPEL workflow engine. The WorkflowFactoryService is a standard Globus Toolkit based grid service that allows a workflow to be created from a BPEL workflow document. An EndPointReference is returned to a WorkflowManagementService resource that can be used to start, stop, pause, resume, cancel, and destroy the created workflow. The WorkflowManagementService is layered on top of the ActiveBPEL workflow engine, which provides the primary functionality for running the BPEL-defined workflow.

WSCI and WS-CDL

Different from BPEL, WSCI and WS-CDL are not targeted at describing executable workflows. They, from a global viewpoint, define the peer-to-peer collaborations of services by specifying the ordered message exchanges between them. These two specifications are both abstract, and describing the global observable behavior of multiple parties in a collaboration. While BPEL is called a service orchestration language, WSCI and WS-CDL are usually referred as service choreography languages.

DAGMan, Kepler, Taverna and Other Scientific Workflow Systems

Besides the specifications proposed by the Web Service community, there are also some systems developed by the CI community. These tools mainly support scientific data analysis and therefore are called scientific workflow systems. Some of these systems, like DAGMan, Pegasus and Swift, are sometimes called "workflow" systems, but they really have nothing to do with "workflow" in the traditional sense--they are parallel programming systems. Here we briefly introduce DAGMan, Kepler, and Taverna.

DAGMan is the flow manager of the well-known Condor job scheduling system [13]. In DAGMan a directed acyclic graph (DAG) which is made up of nodes and edges, is used to represent a set of programs and the dependencies between them. The nodes in a DAG represent programs, and the edges identify the dependencies between nodes. DAGMan submits jobs to Condor in an order represented by a DAG. For robustness, DAGman provides fault-handling mechanisms like task-level retry and checkpointing, and flow-level re-submission (rescue workflow).

Kepler [4] is a scientific workflow system which is developed for the biology, ecology, and astronomy research community, for the analysis and modeling of scientific data. Kepler workflow consists of processing steps called actors that perform computations such as signal processing, statistical operations, etc. Kepler has an extensible library of actors such that in-house or third-party software like Web and Grid services can be added to this library by scientists.

Taverna [14], developed by the UK myGrid project (www.mygrid.org.uk), aims to provide a open-source tool to facilitate the use of workflow and distributed resources within the e-Science community. The Taverna workbench provides a GUI to model a workflow, using Taverna's proprietary workflow language called Scufl, and then the model can be executed by a workflow engine called Freefluo. Taverna has build-in support to access many biological databases and analytical services (like Biomart, Soaplab, etc), which is welcomed by the biology scientists. Taverna also provides an extensible framework so that additional applications, like grid services, can be populated to it.

Other scientific workflow systems include Chimera [15], Pegasus [16], Swift [17], Triana [18], BioOpera [19], and Askalon [20].

Automatic Service Composition

Automatic service composition methods can be further classified into three categories, i.e., semantic based [21-27], automata based [28-30], and optimization based [31, 32]. Semantic based and automata based methods concern the functional properties (i.e., the behavior) of the composite service, while the optimization based approach concerns the non-functional properties.

Semantic-based approaches assume that service providers give semantic descriptions of the service, like input, output, pre-condition, effect, etc. Based on these semantic descriptions, various methods can be used to automatically generate composite service, or service workflow. A common problem description is given in [33]: the automatic service composition problem can be described as a five tuple $< S, S_0, G, A, \Gamma >$, in which S is the set of all possible system states; S_0 and G are the initial and goal states, respectively, for the service composition; A is the set of all possible actions (i.e., all service operations); and $\Gamma = S \times A \times S$ defines the pre-condition and effect for each action. Then, if we start from the initial (or goal) state, with all the possible actions in Γ, various AI-planning methods can be used to derive a feasible plan to the goal (or initiate) state. Research in this category usually uses an ontology modeling language like OWL-S [8] to add semantic annotations to Web Service descriptions. Then situation calculus [23, 24], PDDL [22], or rule based approaches such as SWORD [25] and HTN [27] can be used to derive feasible solution(s).

Automata-based approaches use automata to describe the behavior of target composite service as well as participant services. The key of this approach is computing a delegator which coordinates the activities from those participant services so that the overall behavior corresponds to the desired composition. If a delegator exists, the behavior of the target composite service can be achieved by synthesizing the participant services and delegating their individual behaviors [29, 30, 34].

Optimization-based approaches are used to derive composite services which satisfy certain Quality of Service (QoS) index. Usually the QoS constraints include local constraints (constraints imposed on single service) and global ones (constraints imposed on multiple services). Optimization methods like linear programming [31] and genetic algorithm [32] are used to solve the QoS constraints and yield feasible solution(s).

Mediation-Aided Service Composition

The service orchestration and choreography specifications, and most of the automatic service composition methods, assume the direction composition between services. Direct composition is made based on the assumptions that:

- The incoming messages of one service are the exact ones provided by its partner(s); the outgoing messages of one service are the exact ones consumed by its partner(s), and
- Two services in composition consent to the message format and exchange sequence, such that their composition process always executes in a logically correct way (e.g., terminates properly).

It is observed that partial compatibility is common in real-life service composition. Partial compatibility refers to the situation that, two (or more) Web Services provide complementary functionality and can be linked together in principle; however, their interfaces and interaction patterns do not fit each other exactly. Hence, they cannot be directly composed. The problem of partial compatibility arises mainly because services have to interact with one another in the ways not necessarily foreseen when they are separately developed.

Briefly, there are two methods to make partially compatible services work with each other, i.e., configuration [35] and mediation [36]. Configuration is a heavyweight approach that changes the original service within some predefined variable points to make it work smoothly with other services. Casati and Shan [35] propose a configuration-based approach that enables the dynamic and adaptive composition of e-services. Its adaptability is achieved through dynamic service discovery, dynamic conversation selection, multi-service nodes, and dynamic service node creation.

Compared to configuration, mediation is considered as a more economic and labor-saving approach to address the challenge of partial compatibility in real-world Web Service composition. The basic idea of mediation-aided composition is similar to the concept of an adapter to make two pieces of hardware compatible. Mediator wraps the various services (which are in general heterogeneous, e.g., have different interfaces, impose different message sequences, and support different data structures) such that they can appear homogeneous and are therefore easier to be integrated.

Figure 3 illustrates several cases of partial compatibility and corresponding (message) mediators. For example, in figure 3 (a), messages A and B are exchanged between Services 1 and 2. In figure 3(a), Service 1 first sends A and then sends B; Service 2 waits for A and B to arrive simultaneously. Services 1 and 2 cannot be directly composed by simply adding links, because Service 1 has two interfaces while Service 2 has only one. In this case we can add a module (i.e., a mediator) between them. Its function is to combine messages A and B from Service 1 and forward it to Service 2. With the aid of this mediator, Services 1 and 2 can be correctly composed. Figure 3 (b) illustrates a similar case where message needs to be spitted by a mediator and dispatched to the receiver. In figure 3 (c) the message sequence needs to be altered by a mediator.

The examples in figure 3 are elementary and intuitive, but in real cases several challenges have to be addressed:

- A rigorous method, to check the existence of a mediator, should be used before any effort is taken to actually generate the mediation model or code. For example, in figure 3 (d), Service 1 first expects *A* and then sends *B*, Service 2 first expects *B* and then sends *A*, and in this circumstance no message mediator exists between them.
- Mediation code should be generated when the existence of a mediator is determined. For example, if two BPEL services need to be glued and is checked to be compatible, a mediator, which should also be a Web Service, needs to be generated in order to communicate with the two BPEL services.

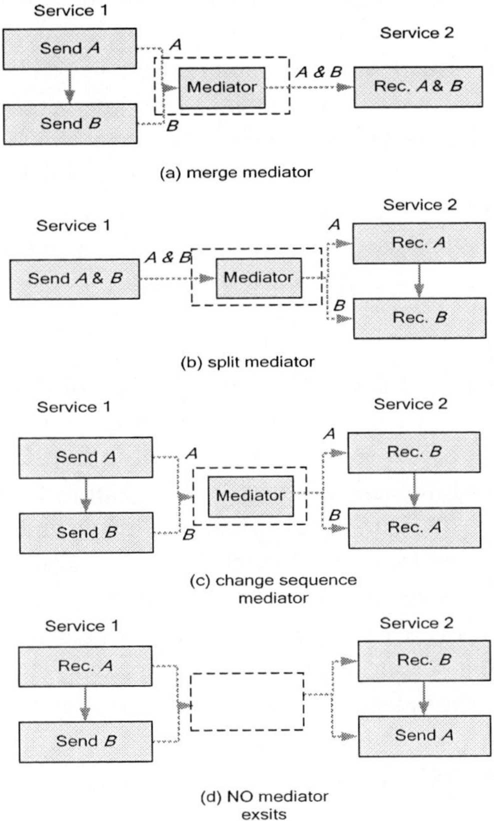

Figure 3. Partial compatibility and mediation-aided composition.

Recently the mediation-aided approach is attracting more attention. Benatallah et al. [37] provide a summary of the mismatch patterns in Web Service composition. These patterns include Message Ordering Mismatch, Extra Message Mismatch, Missing Message Mismatch, Message Split Mismatch, and Message Merge Mismatch. Based on the work in [37], Kongdenfha et al. [38] propose the use of an aspect-oriented programming (AOP) approach to weave adaptor code into the partial compatible services. Brogi and Popescu [39] present a method to automatically generate a mediator between two BPEL services. They translate a BPEL process into YAWL (which is a workflow specification language with formal semantics), and do mediator existence checking and adapter generation afterwards. Nezhad, et al. [40] also propose an automata-based method to model the protocol of service interaction and to identify mismatches automatically. They use a schema matchmaking technique to handle the issue of message mapping, and they also propose some heuristic methods for deadlock resolution between two services.

Verification of Service Workflow

The verification of workflows in an SOC paradigm has its unique features, compared to the verification of traditional workflows. First, the model elements in specifications like BPEL, are much more complicated than those in former workflow specifications like WfMC's XPDL [37]. BPEL concepts such as correlation set, death path elimination, compensation and fault handling are unique, which brings complexity in verification. Second, because workflows in SOC usually interact and interrelate to each other by message exchange, the correctness of a workflow not only relies on its own internal logic, but also relies on how its partners collaborate with it. Even if a workflow is correct from a traditional single-process point of view, its composition with another workflow may still fail: for example, if the two workflows do not agree on message exchange sequence, as figure 3 (d) shows.

Based on the formal method used, the researches in this area can be classified into several categories, i.e., Petri net based, automata based, and process algebra based methods.

Petri Net Based Methods

Ouyang et al. [38] build up a comprehensive Petri net formalism for various BPEL model elements, including basic activities, structured activities, event handler, control link, fault handling, etc. This model covers major part of BPEL specification, and is ideal for the verification of a single BPEL process. Martens, et al. [39] tries to verify the choreography of multiple BPEL processes. The properties been verified include usability (give a BPEL process, whether a proper environment exists to interact with it), equivalence (whether one BPEL process can safely replace another in a choreography), and compatibility (whether the composition of BPEL processes can be used by an environment). Because Martens' work concentrates on interactions among multiple workflows, its Petri net model for BPEL are more compact, omitting some features like event, fault handling, etc. Hinz, et al. [40] transform BPEL into Petri net, and then use CTL (Computational Tree Logic) and model-checking tool to verify various properties.

Automata Based Methods

Su et al. [41, 42] focus on the automata model for services and apply model checking via LTL (Linear Temporal Logic). A special point of their research is a technique called synchronizability analysis to tackle the problem of state space explosion brought about by asynchronous messaging. Their result shows that, if a composite Web service is synchronizable, its conversation set remains the same when asynchronous communication is replaced with synchronous communication, so a synchronous communication model can be used in LTL model checking. Kazhamiakin et al. [55] develop a set of parametric communication models in service compositions. These models range from synchronous communications to asynchronous communications with complex buffer structures. In addition, they develop a technique to associate with a service composition the most adequate communication model that is sufficient to capture all the behaviors of the composition. Using this model, the analysis before the actual verification can bring about an improved performance in verification.

Process Algebra Based Methods

Process algebra [43] is an algebraic approach to the modeling and analysis of concurrent processes. Its advantage is that it provides not only temporal logic model checking, but also bisimulation analysis through which whether two processes have equivalent behaviors can be determined. Foster et al. transform BPEL into a kind of process algebra called Finite State Process (FSP), and then use a model checking tool to verify properties like whether the implementation satisfies the abstract design specifications [44], whether the composition of two services are compatible [45], and whether the composition of BPEL services satisfies the properties defined in WS-CDL [46]. Pi-Calculus [47] is a kind of process algebra developed based on Calculus of Communicating system (CCS). This model takes into account the notion of mobility which can describe the dynamic behaviour like late binding in service choreography. A formal BPEL model based on Pi-Calculus can be find at [48].

Decentralized Execution of Workflow

Workflow systems are often built on the client/server architecture in which a single workflow engine takes the responsibility for the operation of a whole process. However, this sort of centralized systems may not fully meet the requirements in CI. The services involved in a grid workflow come from different organizations, and probably these services can only be controlled by the workflow engines insides their own domain boundary; therefore sometimes a cross-organizational workflow cannot be orchestrated by one single engine. Secondly, the reliability and performance of a workflow can be increased when multiple workflow engines collaborates to execute it.

Partitioning an integrated workflow into small fragments each of which is orchestrated by one engine is a preliminary requirement for decentralized execution. A team from IBM India Research Lab has conducted a series of studies in the decentralized execution of composite BPEL services [49-52]. The research include: how to partition a BPEL into multiple parts, especially the partition of fault-handling code [50]; the model partition policies to improve execution performance, like to reduce communication between engines and increase execution parallelism [51, 52]; how to partition the model when dataflow is constrained [49].

Recently a similar distributed execution approach has been used in grid workflow to optimize the non-functional properties [53].

WS-Workflow: a Workflow System in SOC

To assist services composition and orchestration in SOC, a workflow system, WS-Workflow [54] is designed and implemented. We do not mean to offer a comprehensive system that tackles all the challenges given in Section 1. However, our system provides an integrated solution with some nice features that other approaches do not provide. These features include:

- A data-driven composition module providing guidance for service composition with a given service portfolio.
- A mediation-aided composition module providing a rigorous approach to analyze the compatibility of two BPEL services. The method to generate a mediator which glues two partial compatible BPEL services is also given to provide run-time support.
- A model fragmentation module partitioning a BPEL process into multiple parts each of which can be orchestrated by one engine.

This section first introduces the WS-Workflow framework and then discusses the major supporting techniques inside it.

WS-Workflow: the Framework

The WS-Workflow framework shown in figure 4 consists of three layers, the data layer, the logic layer and the user layer. Here we briefly introduce the modules in each layer. In the next subsection a more detailed explanation on the techniques needed by each module will be given.

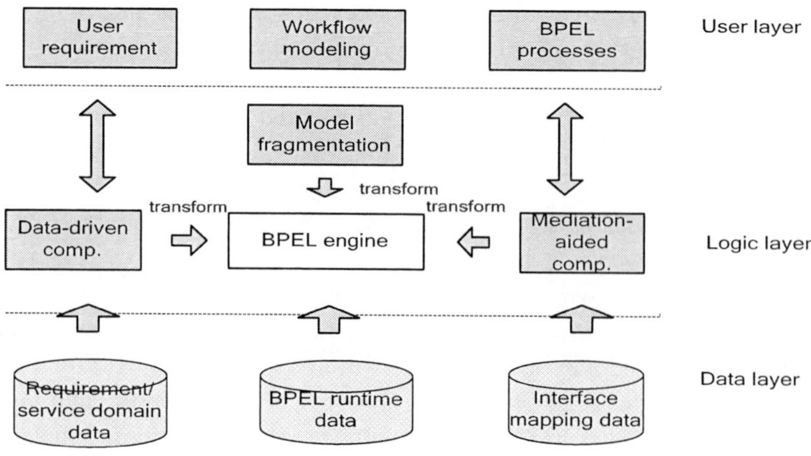

Figure 4. WS-Workflow framework.

The data layer provides persistence storage for all the other modules in WS-Workflow. It includes a requirement/service domain database, a BPEL run-time database, and an interface mapping database. The requirement/service domain database stores the data elements in requirement domain and service domain, and the relations between them; BPEL run-time database stores the instance data used by a BPEL engine in workflow execution; interface mapping database stores the mapping relations of the data elements in BPEL services which are to be composed.

The logic layer contains the core functionalities offered by the WS-Workflow system. The BPEL engine coordinates the execution of BPEL processes input by other modules in this layer. It navigates a process according to BPEL definition, invoke Web Services, and monitor the execution trace of the process. The other three modules around the BPEL engine are key components in WS-Workflow system, and embody the features offered by WS-Workflow. The functions of these modules are:

- Data-driven service module. This module receives the service composition requirement from *user requirement* module in the user layer, checks the requirement/service domain data, and provides a composition of the operations from services in the given service portfolio. The service composition method is based on a Colored Petri nets formalism and Petri net decomposition method. The derived service composition can be transformed into a BPEL process and feed into the BPEL engine for execution.
- Mediation-aided composition module. This module receives two BPEL services and the interface mappings between them from the *BPEL process* module in the user layer. In this module BPEL processes are transformed to Colored Petri nets. A rigorous method is used on these colored Petri nets to check their compatibility. If the two nets are compatible, mediation code is generated to glue these two BPEL processes; if not, detailed reasons are given back to the users.
- Model fragmentation module. This module partitions a BPEL process into multiple parts each of which can be orchestrated by one engine.

The user layer provides user interfaces to access the functions provided by WS-Workflow: user requirement module allows users to model new service composition requirements represented by abstract activities in the form of I→O in which I and O are input and output data types of desired service composition; workflow modeling module allows users to set up an integrated workflow model (in BPEL or Petri net format) to be later partitioned and executed in a decentralized manner; BPEL process module allows users to input two BPEL processes to be later analyzed and glued by a mediator.

The layered and modularized design of WS-Workflow has two advantages: first, the individual modules are relatively independent and can easily integrate with other systems or solutions; second, the interfaces of these modules are well-defined so that additional building blocks can be added into the system without any difficulty.

Key Technologies in WS-Workflow

Data Driven Service Composition

Reuse of existing services is a key issue in SOC. When new requirements emerge, solution designers should devise a composite process that makes the best use of available services. There are a number of good studies on automatic Web Service composition, but there are not many tackle the problem from the perspective of data. As mentioned in Section 1, data flow is considered to be the first-class citizen in the workflow in CI. Therefore we believe it is worthy of investigating a data-driven service composition method that can be used in both business and grid workflow.

There are two domains involved in service composition scenario, i.e., requirement and service domains. In requirement domain a requirement is represented by a process that consists of abstract activities with input/output data. In service domain, a service is modeled by a set of operations with input/output data defined in WSDL.

To devise a data-driven composition approach, first the data relations and composition rules are to be explored. The taxonomy we use to define data relations is rather similar to the Unified Modeling Language (UML). The semantics of aggregation, generalization and realization relation is identical to those in UML. (Later on in this chapter, the data types in requirement domain are denoted with uppercase strings, and the data types in service domain are denoted with lowercase strings. One's name in lower case and the other in upper case has realization relation between them.)

Other relations are:

- *Generation* describes the relations between input and output data types of an abstract activity or a service operation.
- *Partial realization* describes the situation that a service data realizes one part of a requirement data.
- *Specialized realization* describes the situation that a service data realizes one special kind of a requirement data.

Given the defined data relations, three data-driven composition rules, i.e., sequential, parallel, and choice composition rules, are given in figure 5 (a), (b) and (c), respectively. These rules are represented with colored Petri nets. Here we explain these three rules.

Sequential Composition Rule

For a requirement A→C, if there are two operations in service portfolio, i.e., a→ b and b→ c, we can refine the requirement into a composite service realized by existing services, i.e., a→b→c.

Parallel Composition Rule

If $a=a_1 \times a_2$ (a is composed of a_1 and a_2), $b=b_1 \times b_2$, and we have two operations $\{a_1 \to b_1, a_2 \to b_2\}$. Then requirement $A \to B$ can be realized by the AND collection of $\{a_1 \to b_1, a_2 \to b_2\}$, with two additional mediation transitions (which are represented with black rectangles in figure 5). With mediation transitions, a is decomposed into a_1 and a_2, and b is decomposed into b_1, b_2.

Choice Composition Rule

In the requirement domain, we have a requirement $A \rightarrow B$; in the service domain, we have two operations $\{a_1 \rightarrow b, a_2 \rightarrow b\}$; $a = a_1 \cup a_2$ (a can be specialized into two categories, i.e., a_1 and a_2), $a_1 \cap a_2 = \emptyset$. Then requirement $A \rightarrow B$ can be realized by the XOR (exclusive OR) collection of $\{a_1 \rightarrow b, a_2 \rightarrow b\}$, with two additional mediation transitions (which are also represented with black rectangles in figure 5). With mediation transitions, data type a is specified to either a_1 or a_2.

Based on the data relations and composition rules we propose a formal method to derive all the possible composition candidates given a service portfolio. First we get a connected service net from the service portfolio; then we reduce the service net with respect to the given requirement (i.e. the input/output data types); afterwards we decompose the reduced service net into subnets each of which represents a composition candidate. More details regarding the service composition algorithm can be found in [55].

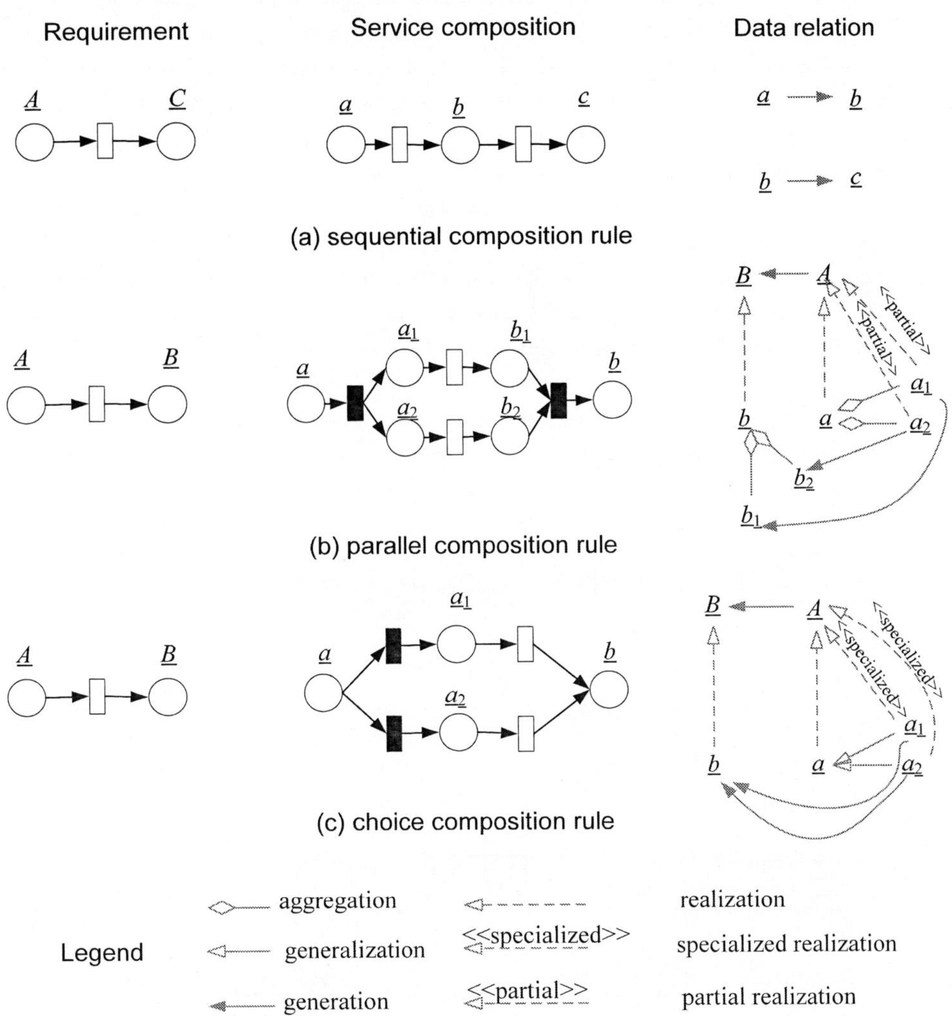

Figure 5. Data relation and data driven composition rules.

Compatibility Analysis and Mediation-Aided BPEL Service Composition

In Section 2.3 we have claimed that, when two BPEL services provide complementary functionality and can collaborate in principle, but their interfaces and internal logic do not fit each other exactly, they cannot be directly composed. The mediation-aided composition module in WS-Workflow embedded an approach to analyze inter-BPEL compatibility and automatically generate mediation to compose services. The steps of the proposed approach are as follows (see figure 6):

1. Take two BPEL processes as the input, and check whether these two processes are directly composable. If *yes*, existing methods can be used to compose them. If *no*, continue with step 2.
2. Translate two BPEL processes into Service Workflow net (SWF-net) which is a kind of colored Petri nets, acting as a unified formalism to describe services, composition and mediator.
3. Read the data mapping information that will be used in later steps.
4. Check mediation existence. We formally define the compatibility between services, two services are *compatible* iff 1) Once the interaction begins, it will complete successfully; 2) When interaction completes, both services must reach their ending state, and possibly there are remaining messages sent out by one service but are not consumed by the other.

We introduce the concept of Communicating Reachability Graph (CRG) whose function is to concurrently construct the reachability graph of two services, using data mapping as the communication mechanism. In this step, CRG is generated and based on it the existence of mediation, i.e., the compatibility, can be determined.

5. If mediation exists, generate mediation code to glue two services.
6. If no mediation exists, provide the reasons (for example, message missing or message sequence collision) to the user.

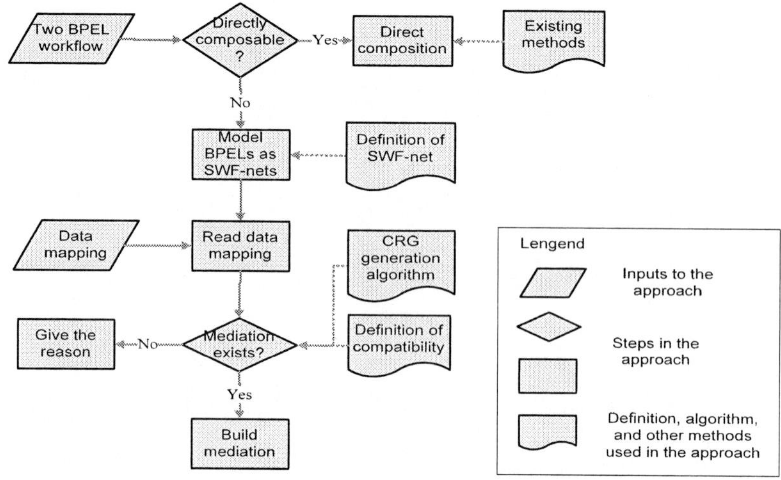

Figure 6. Compatibility analysis and mediation-aided BPEL service composition.

Figure 7 gives a sample CRG (left) and the corresponding SWF-nets as well as the mediation model (right). The CRG is a basically a state-space model. The two SWF-nets are in two rectangles respectively and the mediator in between glue them together. Mediator code is also generated to actually compose two BPEL services.

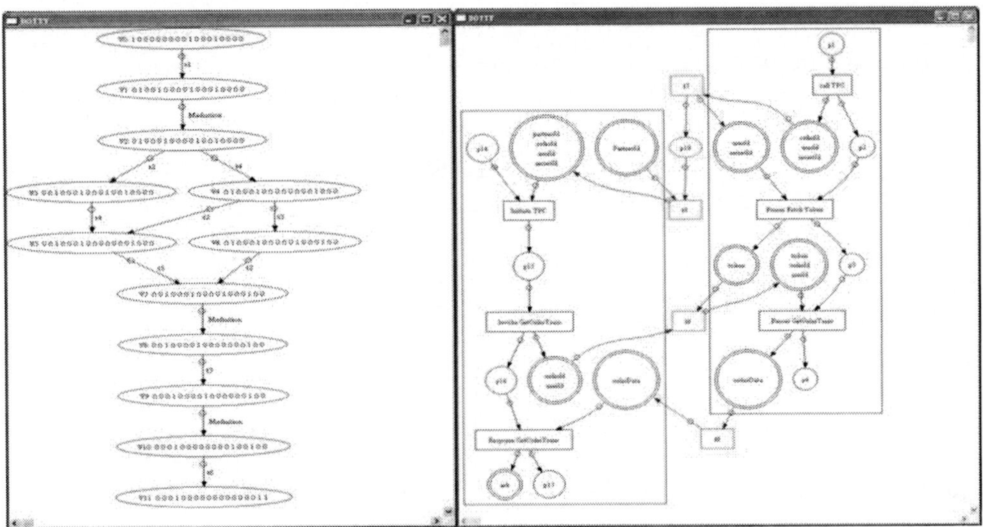

Figure 7. Communicating Reachability Graph and mediation generation.

The advantages of our approach are:

- We take the de facto standard, i.e., BPEL, as input, and adopt service workflow nets as a model to formally represent service, composition and mediator.
- We formally define the compatibility between services, and this definition is more reasonable in stating the correctness of a service composition, compared with the related definitions in business process integration field.
- We propose a reachability based method to check the existence of mediators. The concept of Communicating Reachability Graph (CRG) is introduced to concurrently construct a reduced reachability graph of two services. We prove that the CRG contains all needed properties in mediator existence checking. Moreover, the state-space exploration is efficiently performed by using mediation transitions as stubborn sets.
- We define an algorithm to generate a mediator to glue two services, based on data mapping and derived CRG.

More details regarding the background, algorithms and application of this approach can be found in our paper [56].

Decentralized Execution

In the model fragmentation module in WS-Workflow, two kinds of fragmentation method, i.e., static and dynamic fragmentation, are embedded to support decentralized execution of workflows. Static fragmentation means that the integrated workflow model is partitioned before process instantiation, while the dynamic one means that the centralized model is partitioned step by step during process execution. The static fragmentation method is used when the execution engine (site) of each task can be determined when a workflow instance starts; while the dynamic one is used in a more flexible way, when the execution site can to be determined at run-time.

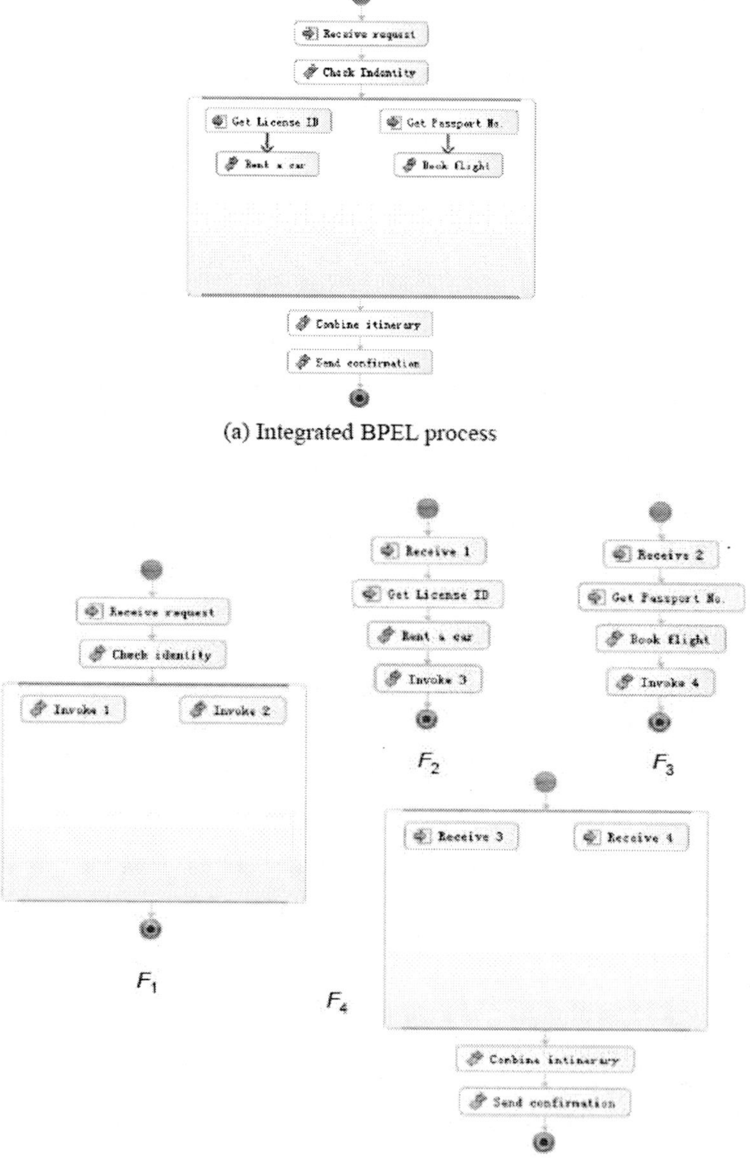

(a) Integrated BPEL process

Figure 8. Static model fragmentation for decentralized execution.

Figure 8 is an example of static model fragmentation. Figure 8 (a) is an integrated BPEL process that is to be partitioned. This itinerary booking workflow first receives the user's request and then checks the user's identity. Afterwards the workflow splits into two parallel flows: rent a car and book flight, after the two flows both complete they are merged and itinerary confirmation is sent back to the user. There are three parties involved in this workflow, i.e., the travel agency, the car rental company and the airline agency. Each party has its own workflow engine. Figure 8 (b) shows the four fragments after model fragmentation. The fragments' codes come from the integrated BPEL, with additional receive and invoke activities as communication blocks between fragments. Fragment F1, F2, F3 and F4 are to be executed at the site of travel agency, car rental company, airline agency and travel agency, respectively.

We also devised a dynamic fragmentation method to partition an integrated workflow model into fragments step by step while the workflow is in execution [57]. The fragments created can migrate to proper engines, where tasks are performed and new fragments are created and forwarded to other engines to be executed in succession. The advantages of this approach include the enhanced scalability by outsourcing the required functionalities, the increased flexibility by designating execution sites on-the-fly, the avoidance of redundant information transfer by judging their pre-conditions before forwarding fragments, etc. Currently the dynamic model fragmentation and execution is based on colored Petri nets model, we are now working to make it work with BPEL model.

The WS-Workflow system has been used in several projects to facilitate flexible workflow composition and decentralized execution in an SOC paradigm.

CONCLUSION

With the emergence of the SOC paradigm, Web services are gaining momentum as key elements not only in the World Wide Web, but also in the cyberinfrastructure and the Grid. Building workflows using service composition has become an important method to build composite applications and reuse existing resources. Therefore, workflow-based service composition and orchestration is now a hot topic in both academia and industry.

This chapter consisted of two parts. The first part summarized research activities in the field of workflow in SOC. Five major research topics, i.e., languages and tools for service orchestration, automatic service composition, mediation-aided service composition, verification of service workflow, and decentralized execution of workflow were discussed. We emphasize that although some of the studies are originally targeted at the area of business process management, they can be adopted by the CI community with some modification or enhancement. The second part introduced a service-oriented workflow execution system, called WS-Workflow. The key modules in this system, including the data-driven composition module, the mediation-aided composition module, and the model fragmentation module were explained. This system has been used in many projects to facilitate flexible workflow composition and decentralized execution.

There are still many challenging issues to tackle when bringing workflow and service composition technology to CI. Currently there is not a well-accepted workflow specification to address grid-specific features such as large data transport, authorization and authentication, and stateful resource access. The various home-grown meta-models hamper workflow interoperability. The run-time issue is even more challenging. Techniques supporting large data transport, fault handling and recovery, real-time resource discovery and composition, are in urgent need to enable cyberinfrastructure-wide collaboration.

REFERENCES

[1] D. E. Atkins, K. K. Droegemeier, S. I. Feldman, H. Garcia-Molina, M. L. Klein, D. G. Messerschmitt, P. Messina, J. P. Ostriker, and M. H. Wright, "Revolutionizing Science and Engineering Through Cyberinfrastructure," National Science Foundation Blue-Ribbon Advisory Panel on Cyberinfrastructure, January, 2003.

[2] I. Foster and C. Kesselman, "Globus: a Metacomputing Infrastructure Toolkit," *International Journal of High Performance Computing Applications*, vol. 11, pp. 115-128, 1997.

[3] WfMC,"The Workflow Reference Model", http://www.wfmc.org/standards/docs/tc003v11.pdf, 2005.

[4] B. Ludäscher, I. Altintas, C. Berkley, D. Higgins, E. Jaeger-Frank, M. Jones, E. Lee, J. Tao, and Y. Zhao, "Scientific Workflow Management and the Kepler System," *Concurrency and Computation: Practice and Experience*, 2005.

[5] OASIS,"Web Services Business Process Execution Language Version 2.0", http://docs.oasis-open.org/wsbpel/2.0/CS01/wsbpel-v2.0-CS01.html, 2007.

[6] W3C,"Web Service Choreography Interface (WSCI) 1.0", http://www.w3.org/TR/wsci/, 2002.

[7] W3C,"Web Services Choreography Description Language Version 1.0", http://www.w3.org/TR/2004/WD-ws-cdl-10-20041217/, 2004.

[8] W3C,"OWL-S: Semantic Markup for Web Services", http://www.w3.org/Submission/OWL-S/, 2004.

[9] F. Leymann, "Choreography for the Grid: towards fitting BPEL to the resource framework," *Concurrency and Computation-Practice and Experience*, vol. 18, pp. 1201-1217, Aug 2006.

[10] A. Slominski, "Adapting BPEL to Scientific Workflows," in *Workflows for E-science: Scientific Workflows for Grids*, I. J. Taylor, E. Deelman, D. B. Gannon, and M. Shields, Eds.: Springer Press, 2007, pp. 212-230.

[11] B. Wassermann, W. Emmerich, B. Butchart, N. Cameron, L. Chen, and J. Patel, "Sedna: A BPEL-Based Environment for Visual Scientific Workflow Modeling," in *Workflows for E-science: Scientific Workflows for Grids*, I. J. Taylor, E. Deelman, D. B. Gannon, and M. Shields, Eds.: Springer Press, 2007, pp. 428-449.

[12] J. Saltz, S. Oster, S. Hastings, S. Langella, T. Kurc, W. Sanchez, M. Kher, A. Manisundaram, K. Shanbhag, and P. Covitz, "caGrid: design and implementation of the core architecture of the cancer biomedical informatics grid," *Bioinformatics*, vol. 22, pp. 1910-1916, 2006.

[13] P. Couvares, T. Kosar, A. Roy, J. Weber, and K. Wenger, "Workflow Management in Condor," in *Workflows for e-Science: Scientific Workflows for Grids*, I. J. Taylor, E. Deelman, D. B. Gannon, and M. Shields, Eds.: Springer Press, 2007.

[14] T. Oinn, P. Li, D. B. Kell, C. Goble, A. Goderis, M. Greenwood, D. Hull, R. Stevens, D. Turi, and J. Zhao, "Taverna/myGrid: aligning a workflow system with the life sciences community," in *Workflows for E-science: Scientific Workflows for Grids*, I. J. Taylor, E. Deelman, D. B. Gannon, and M. Shields, Eds. Guildford: Springer, 2007, pp. 300–319.

[15] I. Foster, J. Vockler, M. Wilde, and Y. Zhao, "Chimera: a virtual data system for representing, querying, and automating data derivation," *Proc. 14th International Conference on Scientific and Statistical Database Management*, Edinburgh, Scotland, 2002, pp. 37-46.

[16] E. Deelman, J. Blythe, Y. Gil, C. Kesselman, G. Mehta, S. Patil, M. H. Su, K. Vahi, and M. Livny, "Pegasus: Mapping Scientific Workflows onto the Grid," *Proc. 2nd European Across-Grids Conference*, Nicosia, Cyprus, 2004.

[17] Y. Zhao, M. Hategan, B. Clifford, I. Foster, G. von Laszewski, I. Raicu, T. Stef-Praun, and M. Wilde, "Swift: Fast, Reliable, Loosely Coupled Parallel Computation," in *2007 IEEE Congress on Services*, 2007, pp. 199-206.

[18] I. Taylor, M. Shields, I. Wang, and A. Harrison, "The Triana Workflow Environment: Architecture and Applications," in *Workflows for E-science: Scientific Workflows for Grids*, I. J. Taylor, E. Deelman, D. B. Gannon, and M. Shields, Eds. Guildford: Springer, 2007, pp. 320–339.

[19] W. Bausch, C. Pautasso, R. Schaeppi, and G. A. A. G. Alonso, "BioOpera: cluster-aware computing," *Proc. 2002 IEEE International Conference on Cluster Computing*, Chicago, IL, USA, 2002, pp. 99-106.

[20] T. Fahringer, R. Prodan, R. Duan, F. Nerieri, S. Podlipnig, J. Qin, M. Siddiqui, H. L. Truong, A. Villazon, and M. Wieczorek, "ASKALON: A Grid Application Development and Computing Environment," *Proc. The 6th IEEE/ACM International Workshop on Grid Computing*, Seattle, Washington, USA, 2005, pp. 122-131.

[21] J. Cardoso and A. Sheth, "Semantic e-workflow composition," *Journal of Intelligent Information Systems*, vol. 21, pp. 191-225, Nov 2003.

[22] D. McDermott, "Estimated-Regression Planning for Interactions with Web Services," *Proc. AI Planning Systems Conference (AIPS'02)*, 2002.

[23] S. McIlraith and T. Son, "Adapting Golog for Composition of Semantic Web Services," *Proc. Eights International Conference on Principles and Knowledge Representation and Reasoning (KR-02)*, Toulouse, France, 2002, pp. 482-493.

[24] S. Narayanan and S. A. McIlraith, "Simulation, verification and automated composition of web services," *Proc. Eleventh International World Wide Web Conference (WWW2002)*, Honolulu, Hawaii, USA, 2002, pp. 77-88.

[25] S. R. Ponnekanti and A. Fox, "SWORD: A Developer Toolkit for Web Service Composition," *Proc. Proc. of the Eleventh International World Wide Web Conference*, Honolulu, HI, 2002.

[26] E. Sirin, J. Hendler, and B. Parsia, "Semi-automatic composition of web services using semantic descriptions," *Web Services: Modeling, Architecture and Infrastructure* workshop in conjunction with ICEIS2003, 2002.

[27] E. Sirin, B. Parsia, D. Wu, J. Hendler, and D. Nau, "HTN Planning for Web Service Composition Using SHOP2," *Journal of Web Semantics*, vol. 1, pp. 377-396, 2004.

[28] D. Berardi, D. Calvanese, G. De Giacomo, M. Lenzerini, and M. Mecella, "Automatic composition of e-services that export their behavior," *Proc. 1st Int. Conf. on Service Oriented Computing (ICSOC)*, 2003, pp. 43-58.

[29] Z. Dang, O. H. Ibarra, and J. W. Su, "On composition and lookahead delegation of e-services modeled by automata," *Theoretical Computer Science*, vol. 341, pp. 344-363, Sep 2005.

[30] E. Gerede, R. Hull, O. H. Ibarra, and J. Su, "Automated composition of e-services: lookaheads," *Proc. 2nd International Conference on Service Oriented Computing*, New York City, NY, USA, 2004, pp. 252-262.

[31] D. Ardagna and B. Pernici, "Dynamic web service composition with QoS constraints," *International Journal of Business Process Integration and Management*, vol. 1, pp. 233 - 243, 2006.

[32] G. Canfora, M. D. Penta, R. Esposito, and M. L. Villani, "An approach for QoS-aware service composition based on genetic algorithms," *Proc. Conference on Genetic and evolutionary computation* Washington DC, USA, 2005, pp. 1069 - 1075

[33] J. Rao and X. Su, "A Survey of Automated Web Service Composition Methods," *Proc. Semantic Web Services and Web Process Composition*, San Diego, California, USA, 2004, pp. 43-54.

[34] D. Berardi, D. Calvanese, G. De Giacomo, M. Lenzerini, and M. Mecella, "Automatic composition of e-services that export their behavior," *Proc. 1st Int. Conf. on Service Oriented Computing (ICSOC)*, Trento, Italy, 2003, pp. 43-58.

[35] F. Casati and M. C. Shan, "Dynamic and adaptive composition of e-services," *Information Systems*, vol. 26, pp. 143-163, 2001.

[36] D. Fensel and C. Bussler, "The Web Service Modeling Framework WSMF," *Proc. Electronic Commerce Research and Applications*, 2002, pp. 113–137.

[37] WfMC,"Process Definition Interface-- XML Process Definition Language", http://www.wfmc.org/standards/docs/TC-1025_xpdl_2_2005-10-03.pdf, 2005.

[38] H. M. W. Verbeek and W. M. P. v. d. Aalst, "Analyzing BPEL Processes using Petri Nets," *Proc. Second International Workshop on Applications of Petri Nets to Coordination*, Workflow and Business Process Management, Miami, Florida, USA, 2005, pp. 59-78.

[39] A. Martens, "Analyzing Web Service Based Business Processes," *Proc. 8th International Conference on Fundamental Approaches to Software Engineering (FASE 2005)*, Edinburgh, UK, 2005, pp. 19-33.

[40] S. Hinz, K. Schmidt, and C. Stahl, "Transforming BPEL to Petri nets," *Proc. 3rd International Conference on Business Process Management*, Nancy, France, 2005, pp. 220-235.

[41] T. Bultan, J. Su, and X. Fu, "Analyzing conversations of Web services," *IEEE Internet Computing*, vol. 10, pp. 18-25, 2006.

[42] X. Fu, T. Bultan, and J. W. Su, "Synchronizability of conversations among Web services," *IEEE Transactions on Software Engineering*, vol. 31, pp. 1042-1055, Dec 2005.

[43] J. C. M. Baeten and W. P. Weijland, *Process algebra*: Cambridge University Press New York, NY, USA, 1991.

[44] H. Foster, S. Uchitel, J. Magee, and J. Kramer, "Model-based verification of Web service compositions," *Proc. 18th IEEE International Conference on Automated Software Engineering*, 2003, pp. 152-161.

[45] H. Foster, S. Uchitel, J. Magee, and J. Kramer, "Compatibility verification for Web service choreography," *Proc. IEEE International Conference on Web Services* 2004, 2004, pp. 738-741.

[46] H. Foster, S. Uchitel, J. Magee, and J. Kramer, "Model-Based Analysis of Obligations in Web Service Choreography," *Proc. Advanced International Conference on Telecommunications and International Conference on Internet and Web Applications and Services (AICT/ICIW 2006)*, Guadeloupe, French 2006, p. 149.

[47] R. R. Milner, Communicating and Mobile Systems: The Pi Calculus: Cambridge University Press, 1999.

[48] R. Lucchi and M. Mazzara, "A pi-calculus based semantics for WS-BPEL," *Journal of Logic and Algebraic Programming*, vol. 70, pp. 96-118, 2007.

[49] G. Chafle, S. Chandra, V. Mann, and M. G. Nanda, "Orchestrating composite Web services under data flow constraints," *Proc. 2005 IEEE International Conference onWeb Services* 2005, pp. 211-218 vol.1.

[50] G. B. Chafle, S. Chandra, V. Mann, and M. G. Nanda, "Decentralized orchestration of composite web services," *Proc. 13th international World Wide Web conference*, New York, NY, USA 2004, pp. 134-143.

[51] M. G. Nanda, S. Chandra, and V. Sarkar, "Decentralizing execution of composite web services," *Proc. 19th Annual ACM SIGPLAN conference on Object-oriented Programming, Systems, Languages, and Applications*, Vancouver, BC, Canada 2004, pp. 170-187.

[52] M. G. Nanda and N. Karnik, "Synchronization analysis for decentralizing composite Web services," *Proc. 2003 ACM Symposium on Applied Computing*, Melbourne, Florida 2003, pp. 407-414.

[53] W. Binder, I. Constantinescu, B. Faltings, and N. Heterd, "Optimized, decentralized workflow execution in grid environments," *Multiagent and Grid Systems*, vol. 3, 2007.

[54] W. Tan, "Research on the Composition and Fragmentation of Service-oriented Workflow Model," Department of Automation, Tsinghua University, Beijing, PhD Thesis 2007.

[55] W. Tan, Z. Tian, F. Rao, L. Wang, and R. Fang, "Process Guided Service Composition in Building SoA Solutions: A Data Driven Approach," *Proc. IEEE International Conference on Web Services (ICWS'06)*, 2006, pp. 558-568.

[56] W. Tan, Y. Fan, and M. Zhou, "A Petri Net-based Method for Compatibility Analysis and Composition of Web Services in Business Process Execution Language," *IEEE Transactions on Automation Science and Engineering*, 2007 (Accepted).

[57] W. Tan and Y. Fan, "Dynamic workflow model fragmentation for distributed execution," *Computers in Industry*, vol. 58, pp. 381-391, 2007.

In: Cyberinfrastructure Technologies and Applications ISBN: 978-1-60692-063-3
Editor: Junwei Cao

Chapter 11

FEDERAL MANAGEMENT OF VIRTUAL ORGANIZATIONS WITH TRUST EVALUATION

***Zhen Wang*[1] *and Junwei Cao*[2]**
Research Institute of Information Technology
Tsinghua National Laboratory for Information Science and Technology
Tsinghua University, Beijing 100084, P. R. China

ABSTRACT

Dynamical and flexible resource aggregation tools are required in 21st century research. Scientists need to aggregate various digital equipments and cooperate with each other in different organizations through Virtual Organizations (VO) on the Internet in a flexible and dynamical way. In this cooperation and resource sharing process, trust evaluation is of great importance for flexible VO management. Traditional tools such as VOMS for grids are short in dynamism and trust evaluation. In this chapter, we propose a new scheme providing federal VO membership management based on trust evaluation, with which researchers can achieve appropriate trust relationships with each other and establish a particular VO dynamically to aggregate resources for their own purposes.

INTRODUCTION

Background

Modern science research has great requirement for experimental instruments, computational and storage capability, and cooperation across organizations and discinplines [1]. However, grid technology, which is considered as a traditional solution enabling resource integration and sharing within a virtual organization (VO), can not meet current requirements

[1] E-mail address: zhen-wang07@mails.tsinghua.edu.cn
[2] E-mail address: jcao@tsinghua.edu.cn

for multiple VO management. Scientists and researchers require a more general and flexible digital platform for data analysis, information integration, instruments sharing and so on.

In 2003, CI [1], short for Cyberinfrastructure, was proposed as a new infrastructure in Cyberspace in future by National Science Foundation in USA. In the year of 2006, Office of Cyberinfrastructure [2] was founded for CI implementation and a detailed plan [3] was established. Compared with grid or other similar technologies, CI is a more general and flexible platform, with which everyone can contribute their resources or obtain sufficient distributed resources to meet their own requiurments. In a CI environment, resource providers contribute their resources (Cyberresources) and users benefit from these resources via a Cyberenvironment. A member of CI may be a *RP* and a *User* at same time, depending on his requirement. CI knocks down the barriers between different grids VOs and makes it possible to share resources and cooperate across them.

The implementation of CI will be a large distributed system and brings a lot of challeages. The problem we are trying to address in this chapter is how to provide a mechanism to support trustable cooperation and resource sharing dynamically and flexibly, as we called Trustable Federal VO Management (TFVOM). TFVOM help Users or RPs to realize effective and trustable resource sharing and access control. There are already some traditional mechanisms which achieve similar functions: Mandatory Access Control (MAC) [4][5], Role-Based Access Control (RBAC) [6][7], Discretionary Access Control (DAC) [8], Virtual Organization Membership Service (VOMS) [9][10][11], Grid User Management System (GUMS) [12], PRIvilege Management and Authorization (PRIMA) [13], Privilege and Role Management Infrastructure Standards Validation (PERMIS) [14] and so on. DAC is implemented by maintaining an account list. In CI environment, the number of CI members is huge and they also change dynamically. On another hand, Users or RPs require flexible resource privilege management, which cannot be implemented using DAC. RBAC and MAC cannot be adopted either because of the different jurisdiction distribution. These two mechanisms are more suitable for centralized organizations with fixed architecture, where there is an account with the highest privilege to all the resources. However, in the CI environment, all the resources are owned by RPs who have the highest privilege over their own resources. This means CI environment is an incompact system and the privilege locates on terminals. Other mechanisms, including VOMS, GUMS, PRIMA and PERMIS, are all used in grids without any trust management mechanism since Users have already built trust relationships to some extent before a grid is enalbed. How to make cross-domain users and resource providers achieve appropriate trust relationships is the main purpose of TFVOM.

In this chapter, we will introduce TFVOM mechanism in details with corresponding implementation. How to deploy the TFVOM mechanism in the CI environment is also described.

Challenges

Compared with other resource aggregation environments, such as grids, a CI environment is more dynamic, complicated, and open. This is why CI is regarded as the future advanced infrastructure for 21st scientific discovery, but this also imposes significant challenges.

The openness of CI leads to the complexity and wideness of origins of Users and RPs in a CI environment, which is quite different from that of the grid. Before a grid is established,

Users or RPs have already achieved an agreement about the purpose of the grid and how to contribute and share the resources. This agreement or protocol is established by non-technological manner. A grid usually has more fixed organization architecture and members of corresponding VO are within a specific domain. A grid is used to enable resource sharing between RPs and Users that already form a VO beforehand. Members of a grid believe that all other members are trustable and resources are shared following the existing agreement. Members of a CI environment do not have any agreement with each other on sharing their resources before they join in the CI. The CI provides an environment to enable members to build such agreement and thus fosters VOs and grids. In this process, RPs want to make sure that they have full control on their resources, and meanwhile Users also have requirements to ensure quality of services (QoS) when using these resources. TFVOM is designed to provide RPs and Users with a negotiation mechanism to achieve agreement on resource sharing with trust evaluation supports. This process should be implemented using advanced computing technologies that can adapte to various situations.

Compared with centralized organizations, the privilege of resources in a CI environment is distributed to each RPs, as mentioned before. No matter what has happened, RPs, resource owners, always have full control of their resources. Each RP has specific and various policies on how to share his resource, which makes traditional aggregation and access control mechanisms not feasible in the CI environment, since most of them are suitable for centralized organizations. To implement a CI environment, a new mechanism should be proposed that can ensure that RPs has the highest privilege on their resources.

Another challenge sources from the variety of resources. Any resource that can be connected through Cyberspace can join in the CI resource pool. Cyberresources include hardware facilities, e.g. a computer, a sensor or an astronomical observatory, driven by different middlewares running on different operating systems. In general, a VO is founded usually for some specific research purpose, used for a certain discipline and enabled using grid technologies, e.g. Network for Earthquake Engineering Simulation (NEES) [16], National Ecological Observatory Network (NEON) [18], The Geosciences Network (GEON)[19], National Center for Atmospheric Research (NCAR) [20], US National Virtual Observatory (NVO) [21] and TeraGrid [15]. CI provides an environment to operate on multiple VOs or grids, e.g., Open Science Grid (OSG) [17], where cross-VO resource sharing becomes available.

In the CI environment, researchers and scientists can easily form a VO, aggregate sufficient and appropriate resources, and collaborate together to work for a specific project. After the project is finished, these resources are released again and can be used for other VOs. Existing VOs should also be able to share resources with each other. From this point of view, a CI could be a platform with many grids. The access control mechanism must be robust enough to handle various situations. Moreover, this mechanism should not be centralized. The primary challenges faced by the new access control mechanism can be summarized as follows:

- Trust evaluation mechanism: With the help of certificates, members can know identity of each other. However, members still can not trust each other based on pure authentication. We must provide a trust evaluate mechanism to help them establish appropriate trust relationships.
- Dynamic and flexible VO membership management: requirements of Users and RPs are various and dynamically changing over time, so we must provide appropriate

tools help them change VO membership easily and dynamically when their purpose and requirements change.

- Communication with other middlewares seamlessly. The new access control implementation must be deployed in different environment and communicate with other CI components seamlessly.
- Reliability: as this mechanism is one of essential tools in a CI environment, we must ensure high reliability since it will influence all Users and RPs of a CI.

To address to these challenges, we propose TFVOM as the solution for the access control mechanism in CI. The details will be introduced in the following. In Section 2, we will introduce some related and similar technologies used in grids or other applications currently. In Section 3, TFVOM mechanism is introduced in details. We will demonstrate the implementation architecture and several typcial applications in Section 4. Then we will evaluate this mechanism, make conclusion and indicate future work in Section 5.

Related Work

In this section, we will introduce several mechanisms widely being used in the current systems. These implementations are developed for some specific purposes. For example, VOMS is developed by European DataGrid (EDG) [22] and Data TransAtlantic Grid (DataTAG) [23] to knock down the barrier between the two grids originally, and be accepted and used by other grids. Privilege and Role Management Infrastructure Standards Validation (PERMIS), supporting authentication of the personal idnode and determination of the role, status, entitlements, or other socio-economic attributes, is developed by ISIS, Institute for Science and International Security. These different schemes all support access control but focus on different aspects. In this section, we will introduce these schemes to show current technology landscape.

VOMS

VOMS, short for the Virtual Organization Membership Service, is one of the most famous and widely used implementation about authorization and authentication of access privilege over resources to the members in grids. The project is supported and developed by European DataGrid (EDG) and Data TransAtlantic Grid (DataTAG) and used by many other grids [24] such as Laser Interferometer Gravitational-wave Observatory (LIGO) [25], Structural Biology Grid (SBGrid) [26], Open Science Grid (OSG), Georgetown University Grid (GUGrid) [27] and so on.

VOMS is developed for the purpose of authorization and authentication on the organization level. VOMS maintains a database to manage and store the information of user roles and capabilities and provides user a set of tools for accessing and manipulating the database. Then VOMS can generate Grid credentials for users through the database contents when needed. The VO is established by the administrator, who is in charge of managing the VO, e.g. adding new member, creating new group, changing roles. Every member in VO is assigned with specific role, with which the member has the privilege corresponding to the role assigned in VO. The role assignment is described in the certificate in local sites and stored and managed by the administrator. Access control over the resource is achieved by the role definition and assignment. In the VO, there are two important facts: administrator and user. Administrator is the one who has the responsibility to manage and maintain the VO and is in charge of role assignment and maintaining the membership information, while the User is a part of the VO and can request information onVO memberships when needed.

VOMS consists of four modules, User Server, User Client, Administration Server, Administration Client, which have different functions, respectively:

- User Client: contact to the Server with certificate and obtain the information of membership in the VO after confirmation, e.g., member lists, role assignments, sub-groups, and capability of a User.
- User Server: receive the request from the User Client, and return the results for user requests.
- Administrator Client: this client is used by the administrator who is in charge of management tasks, e.g., adding new user, creating sub group, role assignment.
- Administrator Server: this server is mainly a data server, which is used to maintain the database and response to the request for the membership information from the client.

VOMS adopts the GSI security control mechanism provided by Globus Toolkit package. User can use the command "voms-proxy-init" to get the certificate generated in the VOMS server. This certificate adopts RFC 3281[29] format and signed by the VOMS server. In order to make sure user can be a member of several VO and may have communication with other non-VOMS GateKeepers, this certificate is extendable and can be an aggregation of several certificates. The VOMS combines two different mechanisms: RBAC and VO. The policy on how the User uses the resources is defined by two aspects: which VO the User belongs to and which role he plays.

GUMS

The Grid User Management System (GUMS) is a system running in the local site in the grid. The major function of GUMS is to manage the mapping process from User's grid certificate or credential to the local site-specific certificate or credential. In the grid, User shares the distributed resources through mapping User grid account to the RP local account, which is similar with remote access to the resources using the account RP assigned to the User. One RP may belong to many VOs or grids, so it is a big challenge how to map the grid certificate to the local account according to the access policy. The accounts provided for the remote access may be different according to different Users jobs. For example, a RP who owns a computer joins in two different VOs. He may provide two different kinds of accounts with different privileges locally for the two VOs. The accounts may even be different because of the job recieved.

GUMS can be configured to map the grid certificate to a local account in two manners: 1) generate statistic map-files according to the Users 2) or map the grid certificate to the local account dynamically according to the job submitted. For example, a user wants to submit some jobs to a certain resource. When the job arrives at the resource with the grid certificate and be passed to the job manager, the gatekeeper must obtain a local account for this job. The gatekeeper will either consults with the map-file generated by the GUMS or pass a request to the local GUMS for a local site, depending on the GUMS configuration. If the GUMS is configured to map the grid certificate dynamically according to the job, the gatekeeper will act as later case, or just consult with the map-file.

The function of the GUMS can be summarized as follows [12]:

- Retrieve membership information from a VO server such as LDAP or VOMS.
- Maintain a manual group of people, and stored in the GUMS database (this is useful to handle special cases).
- Map groups of users to the same account (a group account).
- Map groups of users to an account pool, in which one account will be forever assigned to each user.
- Map groups of users according to the information present in NIS or LDAP.
- Map groups of users according to a manual mapping, stored in the GUMS database.

However, GUMS do not perform authentication but provide information to the gatekeeper. In this view of point, GUMS is just a Policy Decision Point (PDP) not a Policy Enforcement Point (PEP). It must cooperate with other middlewares.

PRIMA

PRIMA, short for PRIvilege Management and Authorization, is a system combining the resource access request with the appropriate privilege. In PRIMA model, a privilege is independent and self-contained, taking files access privilege for example. The privilege of access files are configured by the administrator and stored in the database. In order to ensure the seamless communication, PRIMA describe this privilege in XML-based language.

DAC, MAC and RBAC

DAC (Discretional Access Control), proposed in the 1970s, is based on the access control matrix. In this mechanism, any member in the system can empower other member's the privilege that is the subset of the privilege he has. And this information is stored and managed by the access control matrix. In the access control matrix, in which the line factors represent the User, the row factors represent the RP, and the elements represent the access privilege. If a User wants to use a certain resource, the DAC monitor will check the element on the intersection of the User and RP who own the resource. If this kind of access is allowed according to the element record, the access to the resource can be established, or be forbidden. However, the advantage of discretion of DAC also brings a big problem the DAC can not deal with: security. In DAC, the information and privilege always flows and be empowered from on member to another member. A User U_1 forbidden to the resource R_1 may get the privilege over it because another User who has the privilege over the R_1 empowers U_1 the privilege accessing to the R_l. Another problem is that this scheme is too complicated to maintain for the members, Users and RPs, and system managers.

MAC, short for the Mandatory Access Control, determines the privilege of access by the security level between Users and RPs. All the members, Users and RPs, are assigned with security tags recording the security level by the system security administrator and these tags can not be changed by other members. User can only access to the resources whose security level is not higher than him. Multiple privilege management can be achieved based on the security tags. This mechanism is suitable for the centralized organizations and cannot be adopted a CI environment.

Role-based Access Control (RBAC), proposed by National Institute of Standards and Technology (NIST) in the 90's, is another access control mechanism that is widely used. RBAC binds the privilege with roles defined by the administrator. User can get the privilege only through the roles assigned to him. Administrator in charge of assigning the roles to the right Users according to the functions they have in the organizations. A role may be assigned to several Users, and meanwhile a User may have different roles in different departments. RBAC cut the direct relationships between privileges with specific Users, which makes the access control system more secure and easy to maintain. This scheme is only suitable for centralized organization for it has an administrator with the highest privilege over all the resources to assign the roles.

PERMIS

PERMIS [29], founded by ISIS, is designed to address the issues of an authorization of the personal idnode and determination of socio-economic attributes. Based on the RBAC mechanism, this scheme provides an authorization system that complements an existing authentication system. PERMIS is a privilege management with two major functions: provide policy editor for the owners to construct policies and assign appropriate privilege to the remote users. There are two kinds of policies: authorization policies define how to empower a remote user an appropriate privilege on local sites; delegation policies determine how to delegate to a trustable member the power to assign roles to other users in the same group. All these policies are in XML format. PERMIS also provides the Attribute Certificate Manager (ACM) and the Bulk Loader for managers to allocate privilege to users. The generated privilege information is stored in X.509 Attribute Certificate format [30]. With this policies and privilege information, PERMIS can provide following services:

- When users request access to your resources, PERMIS makes the access control decisions for you based on your access control policies and the roles of the users.
- Edit policies according to the requirement by the owner.
- It allows you to delegate to trusted individuals the ability to assign roles to users on your behalf.

PERMIS is kept in the local site which RP controls. So if being deployed in a dynamical environment, it is hard for PERMIS to maintain consistency among various repositories. Besides, PERMIS is more likely a policy engine without negotiation mechanism.

Federal VO Management

In a CI environment, everyone, including scientists, researchers, institutions and common PC users, are potential RPs. They can contribute their digital equipments, computers, sensors, instruments and other resources, to others and benefit from sharing resources through the CI platform. All Cyberresources are various from access policies defined by the RPs. On the other hand, Users have different requests for resource sharing. TFVOM is designed to achieve trustable cooperation and resource sharing based on the agreement accepted by both RPs and Users. TFVOM helps Users to achieve agreements and trust relationships with RPs and then aggregates sufficient resources by establishing a suitable VO. In a VO, there can be three different kinds of members: Users, RPs and sub-VOs and two types of policies: resource policy and VO policy. Sub-VO is both a User and a RP in a VO. Resource policy, formulated by the RP, is the description on how much privilege the User can have to use the resource. VO policy, formulated by the VO administrator, describes what resources can join in and what privilege the User should have over the resources in the VO. VO policy is used to balance User requirements with RP interests. In this section, we will introduce the details of the TFVOM.

VO Architecture

In a CI environment, VO is established for the purpose of cross-domain resource sharing and member collaboration. A VO consists of two different types of members: RPs and Users. A member can be both a RP and a User in one VO. A sub-VO can be considered as a RP and a User simultaneously.

All Users, RPs and VOs are regarded as Nodes in a CI environment. If node A directly belongs to node B, we call node B is a *Father Node* of A and node A is a *Child Node* of B. If node A belongs to node B indirectly, we call node B is an *Ancestor Node* of A. If nodes A and B belong to the same Node directly, we call they are *Brother Nodes*.

As shown in figure 1, VO2 is a Child Node of VO1 and VO1 is a Father Node of VO2. VO1 is an Ancestor Node of RP3. One VO, RP or User can belong to multiple VOs. As shown in figure 1, VO2 belongs to both VO1 and VO4. RP3 is both a member of VO3 and VO4. In a CI environment, Users, RPs and VOs are same logically, so we manage them using a uniform abstract: *Node*. VOs in CI consist of nodes and has a hierarchical architecture.

Resource and VO Policies

In a CI environment, there are two types of nodes with policies: RP and VO. The policy defined by the RP is different from that defined by a VO administrator.

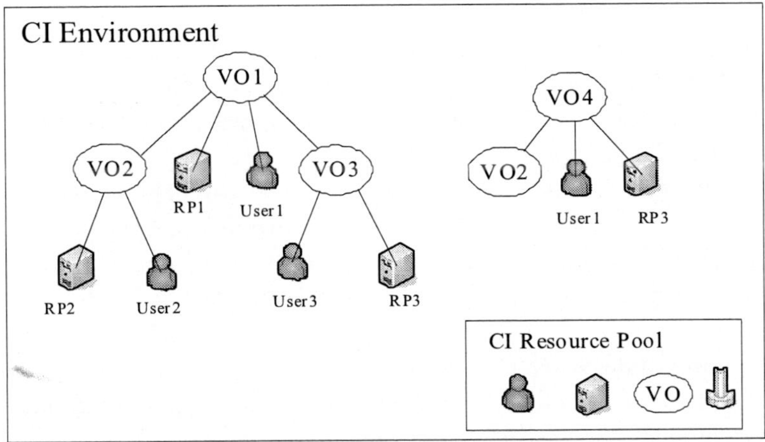

Figure 1. Hierarchical VO architecture.

Resource Policy: This policy represents RP configuration about how to share their resources. A PC owner may only allow sharing his PC when the CPU utilization ratio is lower than 10 percents, providing no more than 30 percents EMS memory of the PC and reading on the hard disk is not allowed. All these are configured by the RP according to his preferences. Cyberresources consist of various types of instruments and services, policies of which are quite different from each other. For scientific computing, some jobs are computation intensive and others may be data intensive. More over, Cyberresources do not just include physical instruments, but also Web Services developed by the scientists. There may be more limitations and definitions on how to share their services.

VO Policy: This type of policy defines rules all members in the VO should obey and rules other nodes out of the VO should obey if they want to cooperate with nodes in this VO. Former rules are use policy, representing Users' reqirements about what kind of resources and collaborators they need. These include instrument status such as memory usage, CPU frequency, core number, instrument type, and bandwidth, and the way to use the resource such as available time, memory limitation and cost. A program for data analysis would collect computers with large memories and bandwidth, which can be defined via the VO policy. Another rule type, share policy, is proposed for the nodes out of the VO. As mentioned before, VO has two roles in function: User and RP. So a VO may join in another VO as RP. Share policy, having the same function with the RP policy, are the policy for other nodes on how to combine and share this VO. If a scientist in Bioinformatics established a VO to analyze data in biology, he may not be happy to include any other VOs which have nothing to do with bioinformatics. He can write this policy as VO policy to avoid such a situation.

In summary, there are only two types of policies in function: policy for outer nodes and that for inner nodes. First policy, stipulating outer nodes how to share local resources, is regarded as use policy. Second type of policy, stipulating members how to share resources in the VO, is regarded as share policy. RP only has use policy while VO has both of them.

Federal Cooperation and Sharing Mechanisms

Organizations are usually formed to achieve aggregation and cooperation using two different types of mechanisms: centralized mechanism and federal mechanism. In the centralized mechanism, power is distributed in a pyramid way, centralized to the top level organizations. Most of organizations with fixed architecture apply this mechanism. If two organizations, organization 1 and organization 2, all have privilege on an instrument and organization 2 is a sub-organization and belongs to organization 1, organization 1 has higher privilege than organization 2 over the instrument when there are some conflicts. But in the federal model, a big organization, consisting of small organizations and individuals, has smaller privilege over resources of sub organizations. Organizations at bottom of the hierarchy have highest privileges over resources.

CI is an open environment with high freedom and flexibility, in which cooperation and aggregation between nodes happens frequently. The reason we adopt a federal mechanism to deal with cooperation and sharing in CI can be summarized in two aspects. 1) In CI, all the resources owned by various RPs who have the highest privilege over their resources. VO can not have higher privilege than the RP for security problems. Besides, VO can join to a higher-level VO for the cooperation. Joining a VO does not mean that the sub VO or resources will be controlled by the father VO. Node only collaborate based on the common goals. This feature is quite different from organizations with a centralized architecture. 2) In CI, VO is established, removed and combined in other VOs dynamically according to the various requirements. When a node needs to aggregate resources or collaborate with other User or VO, it will establish a VO or join in the established VO. Such actives happen frequently. In this case, the federal mechanism is more suitable for the CI environment than the centralized mechanism.

In CI environment, privilege is defined by the policies. VOs at different levels have different privileges over a specific resource, though this resource belongs to all these VOs. Figure 2 shows the difference of privilege scopes of VOs at different levels. We denote the rectangle as a node and the context covered by the rectangle as the privilege the node has over the appointed resource. RP has the highest privilege and can control the resource completely, and contributes part of his privilege for the members in VO2 based on its use policy. The same situation happens between VO1and VO2: VO2 has higher privilege over the RP than the VO1, for VO2 is one of the RPs of VO1 in this case.

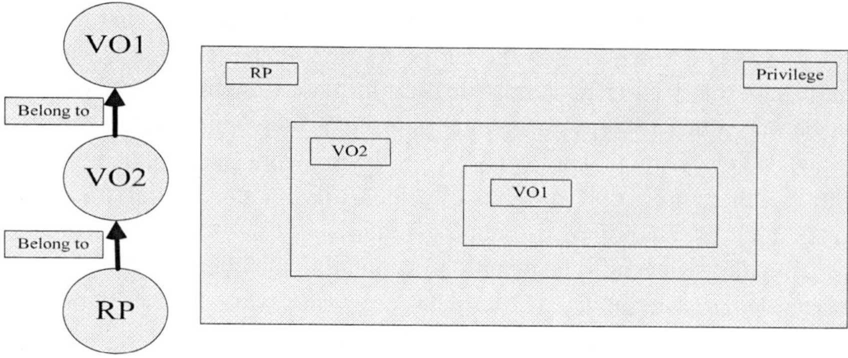

Figure 2. Privilege architecture in the federal mechanism.

TRUST MANAGEMENT

Federal VO architecture can ensure the platform flexibility and dynamism to satisfy the various requirements, however, it is not so easy to be implemented, since privileges are quite different from one to another. A computer owner may just want to only contribute his computer to certain users or just provides computational capability without data storage. Because policies are defined by common users and RPs individually according to their specific requirements, it is hard to describe all these policies in a uniform way. A lot of privilege search and manage scheme which are all based on semantic analysis are proposed, such as SIMDAT (semantics-enabled service discovery framework in a pan-European pharmaceutical Grid) [30].

In fact, RPs define various resource policies just because they have different trust levels for different users. Resource policies will transform different trust levels to appropriate privileges on the local site. Since trust levels can be somehow charaterized in a uniform and quantfatitative way, we can just map different trust evaluation to different levels of privileges. This support can help RPs to assign appropriate privileges to the Users.

Schemes to manage and evaluate trust values are wildly used in E-commerce and P2P applications [31, 32, 33]. Considering characteristics of a CI environment, a trust management model is proposed that can help Users and RPs achieve appropriate, flexible and dynamic trust relationships automatically.

Current Trust Models

Trust is defined in different ways: In [34], trust was defined as “a particular level of the subjective probability with which an agent assesses that another agent or group of agents will perform a particular action, both before he can monitor such action and in a context in which it affects his own action.” In [35] , the trust that Device A places in Device B is defined as: the level that A believes B will implement the desired operations and will not initiate or transfer attacks on Device A or a system that runs on Device A. In CI environment, trust value A places in B is the level of risk A would take to empower privileges to B. More privileges always lead to high risk. When a RP receives request from a User, he must first calculate the value of trust he places on the User, and then the RP will empower appropriate privileges to the User according to the trust values.

In the CI environment, trust value A places on B is not the same as that B places on A, for the trust value is not symmetrical. We use $T_{A\text{-}B}$ to denote trust value A places on B.

In an E-commerce environment and other virtual communities on the Internet, trust management schemes originate from real world scenarios. In real world, people get trust information in two ways: recommendation from others and their own judgment. Recommendation from others is usually called as reputation in real world. On the other hand, ones own judgment comes from his individual experience. So in a virtual environment on the Internet, researchers also adopt similar ways to evaluate and manage trust values via both *direct trust* and *indirect trust* [34, 36].

Direct trust: Direct trust value can be calculated from historic information Node A observes from Node B. When Node B wants to establish trust relationship with Node A, it will present its trust related information to Node A, such as certificate, membership, etc. Node A also has historic access records. Then A will calculate trust values according to these two aspects of information: trust related information Node B presents and historic records maintained by Node A:

D_{A-B}: Direct trust value Node A places on Node B;

Inf_B : Information Node B presents to Node A;

Rc_{A-B} : Historic records between Nodes A and B.

Direct trust values will be calculated as $D_{A-B} = f(Inf_B, Rc_{A-B})$. This trust value includes the individual judgment of Node A to Node B.

Indirect trust/reputation: Indirect trust values are provided by a third party, which is regarded as a trustable node for Node A. Assume Node B wants to establish trust relationship with Node A. Node B should send a request to Node A. Then Node A needs to make a decision on if Node B is trustable. Besides direct trust values, it is also important to consider other trustable nodes' opinion about Node B. For example, Node C is trusted by Node A as a third party node. How Node C evaluates Node B influence the trust value Node A evaluate on Node B. The importance of the recommendation depends on the trust level Node A places on Node C. The recommendation from the third node is more likely regarded as reputation of Node B in real world:

I_{C-B}: Indirect trust values Node A places on Node B from Node C (recommendation from Node C);

Indirect trust values represent global judgment on a certain node. In E-commerce or distributed scenarios, final trust value is combined with two types of trust values with different weights and calculated as follows:

$$T_{A-B} = \omega D_{A-B} + (1-\omega) I_{C-B} \quad 0 \le \omega \le 1;$$

where ω presents the level a node trusts its own judgment. $\omega=1$ means Node A judges Node B totally according to its own experience and does not trust any recommendation from other nodes; $\omega=0$ means Node A totally trusts the recommendation from others and do not adds any self experiences in it.

Trust Modeling in CI Environment

CI is a distributed environment in which all resources are aggregated and managed through VOs. VO is an aggregation of nodes in CI based on the agreed policies and trust levels. If Users or RPs evaluate trust values every time they want to share or contribute resources with others, the platform is not only complicated to use but also hard to support a large scope of cooperation and resource sharing. Modern science research always need large scope of cooperation across disciplines and huge number of various instruments. CI should support dynamism, usability and large scope of cooperation of VOs. In fact, nodes in a VO already establish a certain level of trust relationships between each other when they agreed on the VO policies and joined the VO. There are three different types of trust values in the trust model of CI:

- *Global trust value:* This trust value is managed and calculated by the CI Management Center (CIMC). It represents the reputation of a node in the whole environment, for it is an accumulated value which is calculated from all historical records of activities in the environment. CIMC is a trustable node which is in charge of managing all nodes' global trust values as recommendation values.
- *Local trust value:* This trust value is based on the specific relationship. When Node A receives request from Node B, Node A will examine historic records of Node B and idnode information presented by Node B. And then Node A will calculate local trust value according to all these information. Local trust value, which makes sense only for specific relationship, represents individual opinion. Nodes can establish flexible and individual trust relationship through local trust values.
- *VO trust value:* This trust value defines the basic trust level of a VO and assigned by the VO administrator. All members of a VO are trustable at the level of the VO trust value. In default situation, all members can establish trust relationship among each other without any authorization and negotiation processes. This trust value represents the agreement of nodes in a certain VO.

The first two types of trust values are calculated while VO trust value is assigned by the VO administrator according to the VO purpose.

Global Trust Value

When a task is submitted by a User to a specific RP, there are two main properties to describe the completion of the submitted task: duration time and quality of execution. Duration time of a task represents task complexity. Quality of execution of a task can be characterized in three grades: success in time, success but delayed and failure. Longer duration time and higher quality contribute positively to trust evaluation.

When a task is finished and the task result is returned to the User, User will send a task report to the CIMC, which records the duration time and quality of execution of the task. CIMC will calculate contribution value from the report. And then we need an appropriate scheme to assign this contribution value to the related nodes: User, RP and VO (if necessary).

The trust contribution of task execution is assigned to Users and RPs with different weights. Users and RPs are designed to take the responsibility of the task together to avoid spite activities from Users and RPs. This can also encourage Users submit suitable tasks and configure sufficient expected time while RPs provide QoS services.

Final global trust value is an accumulated value from contribution of many tasks. Last task status has highest influence and trustability as it can more accurately reflect current situation.

If the task is submitted and finished across VOs, VOs which the User and the RP directly belongs to take part in this trust relationship, because a high trustable VO is a guarantor aided to provide credit situation of nodes thatbelong to the VO. Further more, scientists or researchers usually collaborate with each other on the VO level to establish resource sharing among all members of the two VOs. This cooperation is always the result of negotiation between administrators of the two VOs. In this case, VO plays a key role in the process of establishing trust relationship between members from different VOs. As VO is a guarantor in this process, it also takes responsibility in the resource sharing.

Global trust values represent nodes' reputation and recommendation of CIMC. It is basic trust evaluation in CI which provides three functions:

- Determine whether or not a node can join in a VO, for the global trust value of the node must satisfy the VO requirement. Generally speaking, global trust value of a node must higher than the trust value of VO it belongs to.
- This value is also one part of idnode information which helps node to calculate local trust value.
- If a node wants to establish a VO for his applications, the VO trust value configured by him is limited by his global trust value. VO trust value must be lower than founder's global trust value.

Local Trust Value

Local trust value represents the individual judgment on specific relationships. This type of trust values is calculated from two types of information: idnode information from the applicant and historical records in the local site. Assuming Node B sends a request to Node A to establish a trust relationship. The local trust value is calculated using historic data.

Historic records are items that record former activities between Nodes A and B. Every item consists of several properties: task duration, completion time and subjective judgment. Completion time is the time when a task is finished. Recent items always are more important and influential. Task duration time for a task is also of importance, the longer duration is, the more influential this task for the local trust is. Subjective judgment comes from the local site: the RP will give a rough judgment on the User who submits his tasks while the User will also make a judgment if the RP provide high quality of services.

VO Trust Value

The VO trust value is initially configured by the VO founder, namely the VO administrator, under the limitation of administrator global trust value. After all, a node can not establish a VO with trust value higher than its global trust value. Generally speaking, administrator's global trust value should higher than his VO trust value in a certain number for he must be surplus to handle unexpected global trust value changing. When a VO trust value is configured, this parameter should keep stable in its entire life. There is only one limitation: administrator should have sufficient global trust value to ensure the VO trust value.

Establishment of Trust Relationships

In a CI environment, all nodes are organized through VOs. The trust relationship between two nodes can be only established inner a VO directly or indirectly. If two nodes belong to the same VO directly, they can establish a trust relationship. If they do not belong to the same VO directly, they can establish trust relationship only if they have at least one common ancestor VO. Nodes many have many common ancestor VOs. In this case, we just choose the VO which has the highest VO trust value as the smallest common VO.

IMPLEMENTATION

There are three components in a CI environemnt: CI Management Center (CIMC), VO Management Center (VOMC) and Clients. CIMC has three functions: authentication, trust management and node management. VOMC (VO Management Center) manages VO membership and monitors member status in real time. VOMC and CIMC are all implemented as Web Service to provide flexible and security information service. There is also a small optional client installed at client sides (Users and RPs). This client can help User manage local historic records and aid to make decision on the access control and privilege management. Common Users or RPs can also access to the VOMC or CIMC through a web browser.

Compared with VOMS, membership in a CI environment is recorded and authenticated by VOMC and CIMC rather than in the certificate. When two nodes want to establish a trust relationship, they must check the other node membership and global trust value from CIMC and VOMC. This scheme can ensure the dynamism and flexibility of VO memberships. Figure 3 shows how these three components work together.

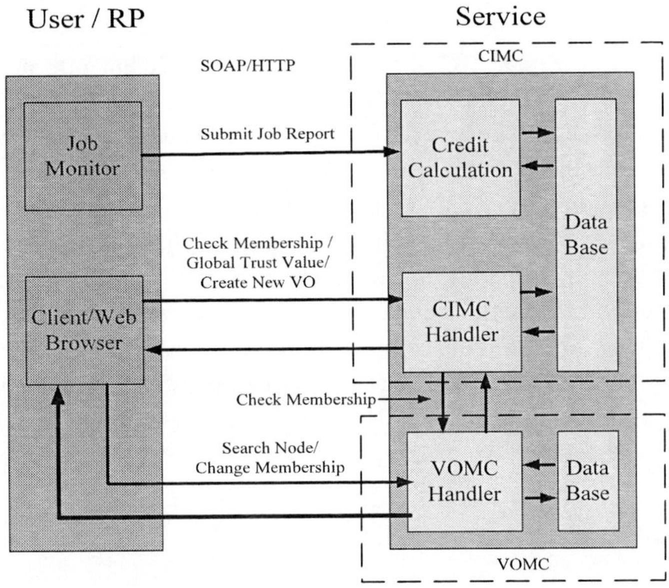

Figure 3. Architecture of three Components.

CIMC

CIMC is composed by three function modules: Certificate Authority Module, Trust Management Module and Node Management Module and CI Database maintain all concerned information:

- *Certificate Authority Module*. This module is the authority center for all the nodes in the CI, signing and sending certificates. There are three sections in a certificate: a section for node basic information, an extendable section and digital signature.
- *Trust Management Module*. This component is mainly in charge of calculating global trust values. Input data is task reports from Users and RPs. When a new report is received, new global trust values are calculated and databased is updated.
- *Node Management Module*. Node Management Module is designed to manage and monitor all nodes in CI, including VOs, RPs and Users. This module is in charge of registering of VOs, Users and RPs and checking their qualifications. It is also used to provide identity information, e.g. membership of a node, global trust value of a node, etc.

VOMC

A VO is always established by a certain individual or another VO. Though this establishment may be the negotiation result among several individuals and organizations, there is only one administrator to found a VO. The administrator maintains a VO Management Center (VOMC) and undertakes the following responsibilities: managing VO membership, monitoring members status in real time, communicating with other VOs on VO level collaboration. VOMC has two components: VO Monitoring Module and VO Membership Module. There is also a VO database to store and manage all this information.

- *VO Membership Module*. This module records sufficient information about all the members in the VO and deals with dynamical changing of these information. Information on father nodes or brother nodes is also maintained. As mentioned before, nodes achieve resource sharing only when they belong to the same VO directly or indirectly. This module provides membership management and information service. Membership management handle requests related to the membership such as joining or leaving the local VO, and collaborating with other VOs and so on. Information service provides membership search and lookup services.
- *VO Monitoring Module*. This component is designed to monitor the member status in the VO in real time, especially the member global trust value, since every member should meet the requirement for the node global trust value from the VO and global trust value is dynamically changing.

Clients

The client is an optional component for Users or RPs. It is used to store individual historic records about past activities and evaluation. The client provides a tool to guide Users or RPs to record integrated information about a task and store them locally.

The client receives applications and provides processed information to help Users or RPs implement access control and privilege management. It calculates local trust value from past records, and final trust value as mentioned before. The client also retrieves identity information of applicants from CIMC and VOMC if necessary. All these information is provided to Users/RPs to help them make a decision. Users or RPs can also run a small component like GUMS to help them assign appropriate privilege to the applicant according to the information the client provides following the policies they are configured with.

CONCLUSION

TFVOM, as proposed in this chapter, provides federal VO management mechanism to achieve trusted collaboration. Compared with traditional technologies, TFVOM has the following features:

- *Trust Evaluation Supports.* This is one of core functions of TFVOM. Traditional resource aggregation or authority/authentication mechanisms can not provide this function. Members' credits cannot be evaluated only with certificates. The Grid, which is based on the public key infrastructure, is widely used as a resource aggregation and cooperation platform. However, all the members of a grid have to achieve agreement on resource sharing beforehand. TFVOM provides an environment for members who know nothing about each other before to build credits, gain trust relationships with each other and form a VO together if required.
- *Federal VO Management.* Most of related mechanisms, such as RAC and RDAC, are suitable for centralized organizations. RAC and RDAC requires a center with highest privilege managing all members in the organization, e.g. assigning roles, empowering privileges, determine security levels and so on. The federal mechanism makes TFVOM suitable for incompact organizations.
- *Portability and Extendibility.* TFVOM can handle various resources and dynamic membership changes. Any RPs or Users can join the environment. Besides, the extendibility also indicates that TFVOM can meet various requirements and requests since policies are extendable.
- *On-the-fly Collaboration.* VOMS is another toolkit to enable cooperation and resource sharing across VOs. VOMS is more suitable for stable cooperation and resource sharing. TFVOM provides a series of tools that facilitates the process of dynamic VO operations, e.g. creating, joining, leaving, removing or merging VOs.

REFERENCES

[1] D. E. Atkins et al. Revolutionizing Science and Engineering through Cyberinfrastructure. National Science Foundation Blue – Ribbon Advisory Panel on Cyberinfrastructure, January 2003.

[2] OCI – Office of Cyberinfrastructure. http://www.nsf.gov/oci.

[3] NSF Cyberinfrastructure Council. NSF's Cyberinfrastructure Vision for 21st Century Discovery, Version 7.1. National Science Foundation, July 2006.

[4] Thomas, T.; A mandatory access control mechanism for the Unix file system, *Aerospace Computer Security Applications Conference*, 1988.

[5] Yixin Jiang; Chuang Lin; Hao Yin; Zhangxi Tan, Security analysis of mandatory access control model, 2004 IEEE International Conference on Systems, Man and Cybernetics, Vol. 6, 10-13 Oct. 2004.

[6] Andreas Schaad, Jonathan Moffett, Jeremy Jacob. The role-based access control system of a European bank : a case study and discussion, *Proceedings of the sixth ACM symposium on Access control models and technologies*, May 2001.

[7] Ninghui Li; JiWon Byun; Bertino, E, A Critique of the ANSI Standard on Role-Based Access Control, *IEEE Security and Privacy*, Vol. 5, No. 6, 2007.

[8] Sandhu, R.S. Samarati, P. Access control: Principles and Practice. *IEEE Communications Magazine*, Vol. 32 , No. 9 pp. 40 – 48, 1994.

[9] http://vdt.cs.wisc.edu/VOMS-documentation.html

[10] http://hep-project-grid-scg.web.cern.ch/hep-project-grid-scg/voms.html

[11] Dongguk Univ. Grid Information Retrieval Management System for Dynamically Reconfigurable Virtual Organization, *Fifth International Conference on Grid and Cooperative Computing (GCC 2006)*, Oct. 2006.

[12] https://www.racf.bnl.gov/Facility/GUMS/1.2/index.html

[13] http://computing.fnal.gov/docs/products/voprivilege/prima/prima.html

[14] Privilege and Role Management Infrastructure Standards Validation: http://www.permis.org

[15] TeraGrid. http://www.teragrid.org.

[16] NEES – Network for Earthquake Engineering Simulation. http://www.nees.org.

[17] OSG – Open Science Grid. http://www.opensciencegrid.org.

[18] NEON – National Ecological Observatory Network. http://www.neoninc.org

[19] GEON – The Geosciences Network. http://www.geongrid.org

[20] NCAR – National Center for Atmospheric Research. http://www.ncar.ucar.edu

[21] NVO – US National Virtual Observatory. http://www.us-vo.org

[22] EDG – European DataGrid http://eu-datagrid.web.cern.ch/

[23] DataTAG – Data TransAtlantic Grid http://datatag.web.cern.ch/datatag/

[24] VOMS Monitoring Documentation http://voms-monitor.grid.iu.edu/cgi-bin/index.cgi

[25] LIGO – Laser Interferometer Gravitational-wave Observatory. http://www.ligo.caltech.edu

[26] SBGrid – Structural Biology Grid http://www.sbgrid.org/

[27] GUGrid – Georgetown University Grid http://gugrid.arc.georgetown.edu/

[28] I. Foster, C. Kesselman and S. Tuecke, The Anatomy of the Grid, *International Journal of High performance Computing Applications*, 15, 3 (2001).

[29] Permis http://sec.cs.kent.ac.uk/permis/

[30] C. Upstill, and M. J. Boniface, "SIMDAT," *CTWatch Quarterly*, vol. 1, no. 4, pp. 16-24, Nov. 2005.

[31] Alexandria, Virginia, Trust Management for Trusted Computing Platforms in Web Services, Conference on Computer and Communications Security, *Proceedings of the 2007 ACM workshop on Scalable trusted computing*, Nov. 2007 - Nov. 2007.

[32] Guangwei Zhang, Jianchu Kang, Rui He, Towards a Trust Model with Uncertainty for e-Commerce Systems, *Proceedings of the 2005 IEEE International Conference on e-Business Engineering (ICEBE'05).*

[33] Alireza Pourshahid, Thomas Tran, Modeling Trust in E-Commerce: An Approach Based on User Requirements,

[34] D. Gambetta, "Can We Trust Trust?" in *Trust: Making and Breaking Cooperative Relations*, Basil Blackwell, New York, 1988, pp. 213-237.

[35] Tao Sun, Mieso K. Denko. A Distributed Trust Management Scheme in the Pervasive Computing Environment, *Canadian Conference on Electrical and Computer Engineering (CCECE 2007),* pp. 1219-1222, 22-26 April 2007.

[36] Pavlou, P. A., Yao-Hua Tan, and Gefen, D. The transitional role of institutional trust in online interorganizational relationships, *Proceedings of the 36th Annual Hawaii International Conference on System Sciences*, 2003.

In: Cyberinfrastructure Technologies and Applications
Editor: Junwei Cao
ISBN: 978-1-60692-063-3

Chapter 12

COMMUNITY-SCALE CYBERINFRASTRUCTURE FOR EXPLORATORY SCIENCE

Peter Bajcsy[1], Rob Kooper[2], Luigi Marini[3] and Jim Myers[4]
National Center for Supercomputing Applications (NCSA)
University of Illinois at Urbana-Champaign (UIUC)

ABSTRACT

This chapter presents some of the key aspects of Cyberinfrastructure (CI) research and development targeting community-scale exploratory science. The motivation comes from the fact that successful software for CI is increasing scientific productivity of a single investigator, small groups of scientists as well as dispersed teams spanning multiple institutions. Community scale scientific activities and their informatics requirements are driving the development of new CI solutions. It becomes critical to follow CI design principles based on past, present and future efforts. In addition, data- and hypothesis-driven explorations are fundamental scientific activities leading to discoveries. In this work, our focus is on informatics requirements and CI design principles behind existing software. We have included our experiences and described several prototype CI solutions to support exploratory science.

INTRODUCTION

The word Cyberinfrastructure (CI) has emerged as the latest in a series of terms to describe the potentially revolutionary impact of computational technologies on scientific and technological progress. There is a plethora of definitions, community specific interpretations and reports describing the future of CI [1-4], but it are generally accepted that "cyberinfrastructure refers to infrastructure based upon distributed computer, information and

[1] E-mail address: pbajcsy@ncsa.uiuc.edu
[2] E-mail address: kooper@ncsa.uiuc.edu
[3] E-mail address: lmarini@ncsa.uiuc.edu
[4] E-mail address: jimmyers@ncsa.uiuc.edu

communication technology. If infrastructure is required for an industrial economy, then we could say that cyberinfrastructure is required for a knowledge economy" [3]. A critical distinction between CI and earlier terms, such as collaboratories and grids, is the recognition that CI is infrastructure that will become vital in addressing the key research and development challenges and increasing scientific productivity of the 21st century. Despite its promise, CI is as yet far from ubiquitous and is still considered by many to be difficult to use productively. Thus, in this chapter we present some key challenges related to the need for CI to be deployed as broad community infrastructure while enabling ubiquitous adoption and productive use by independent exploratory research efforts.

In some sense, CI has been in development since the advent of the computer and the Internet and has been benefiting from the exponential increases in computing, storage, and networking capabilities ever since. Inflection points such as the development of the Mosaic Web browser in 1993 at NCSA and the emergence of the World Wide Web have helped shift our perception of CI from simply providing cycles and storage to understanding that its true power includes organization of information and coordination of efforts. Over the past decade, there have been developments such as grids, portals and social networking sites, and community databases that have helped to drive these concepts of community scale sharing and organization of computation, data, and expertise. The developments have shown that CI will enable qualitatively new approaches to scientific research in addition to the quantitative improvements leveraging hardware advances. Most recently, the emergence of Web 2.0 technologies has shown how core capabilities created by service providers can rapidly be 'mashed-up' and customized to support the needs of specific projects and communities.

In general, one could capture two concepts pertinent to CI design. First, it is the concept of scientific communities planning and designing CI based on the expectation of both quantitatively and qualitatively new CI capabilities. Second, it is the concept of designing and building CI from the ground up to support quantitative and qualitative change in how research communities work. Following from these CI design considerations, NCSA has coined the term "Cyberenvironments" and worked with a number of communities to explore the transformative potential of CI within their domains and to enable cross-disciplinary work. Based on this analysis, we recognized that both technological and scientific progress will be rapid in comparison to the timescale at which infrastructure can be developed and deployed. Thus, NCSA worked on identifying key design patterns somewhat analogous to those behind Web 2.0 that will enable core CI to quickly and cost-effectively be 'mashed-up' to create environments tailored to specific scientific needs. Building on these design patterns, we have begun to create a new generation of middleware and components that implement them. Furthermore, we are working with pilot communities to begin deployment of capabilities that will help catalyze transformative changes in the scope, scale, and impact of their research over the coming decade.

In this chapter, we begin with a discussion of the changes CI is expected to enable in various communities. We focus on Earth science applications, where a number of large environmental observatory projects are being supported by the National Science Foundation (NSF) with the goal of transforming Earth systems research. A discussion of the requirements related to these observatories is then used to motivate discussion of the necessary design patterns. We introduce specific work at NCSA that has begun to apply the design patterns to create new capabilities for earth systems research. The chapter concludes with the thought that, as at the beginning of the WWW, we are entering a phase of rapid progress in the use of

CI to power a next-generation of scientific research and development that will drive new discoveries and help to provide a scientific basis for addressing the grand challenges facing humanity.

INFORMATICS REQUIREMENTS FOR CYBERENVIRONMENTS

The word "informatics" has been used widely in a multitude of application areas. Informatics refers to the increasing amounts of often highly complex data that have to be analyzed, interactively explored, and transformed from raw form to information and to knowledge [5]. In all application areas, scientists desire to learn from their data about a spectrum of complex phenomena surrounding us. From this perspective, any software for CI becomes a part of an X-informatics system, where X stands for a specific application domain. Some common uses for X are bio, hydro, medical, document, astro, or sensor [6, 7]. In all these instances, classes of the X-informatics systems are typically characterized by methodologies of steps or activities that have to take place to conduct experiments and to arrive to knowledge. It is understood that knowledge is gained regardless of the experimental outcome. Thus, the research and development (R and D) of software for CI is therefore driven by a set of requirements for building X-informatics solutions following common methodologies. Broad community infrastructure can be built to support specific research areas by providing common informatics frameworks. The frameworks are decoupled into specific components developed by the individual researchers and provide simple mechanisms for component inclusion, no matter what the specific area of study might be.

One of the scientific fields that have benefited the most from the current CI is Earth science. Earth science includes domains ranging from geosciences, atmospheric sciences, hydrology, environmental engineering to ecology, geochemistry, earthquake engineering, oceanography, ground water or water quality sciences. Scientists and practitioners from multiple sub-domains have formed communities and consortiums funded primarily by NSF to articulate their requirements for CI development and deploy prototypes of CI that could support community level science. One should mention a few of these communities such as WATERS (environmental engineers), CUAHSI (hydrologists), LTER, SEEK and NEON (ecologists), CZEN (geochemists), GEON (mineralogists) and OOI (oceangraphers). While working with multiple Earth science communities, it has become apparent that the research and development of Cyberenvironments has to lead to community-scale CI for exploratory science.

In Earth sciences and many other domains, the R and D requirements of software CI typically vary depending on the scientific activities of individuals alone and the activities of collaborating teams. Thus, the software components in community-scale Cyberenvironments need to scale from the activities of individual scientists to groups of scientists and eventually to very large communities. In addition, community scale research involves sharing not only data but also scientific analyses and the trails of exploratory processes as documented in the NSF workshop on Challenges in Scientific Workflows [8]. Sharing of scientific analyses is important for reproducibility of results, validation of each other's hypotheses and comparisons of results obtained using multiple techniques. It also implies to put in place gathering of rich provenance information and leads to a concept of self-describing software execution so that scientists do not have to be burdened with the provenance gathering. Based on our observations, the software scalability requirement with the number of users and the concept of self-describing software execution are rarely imposed on software solutions designed by a single investigator and his students.

Whether at a community scale or at the individual scientist scale, most scientific activities deal with data. Scientific data sets can come from sensors, instruments or visual observations, as well as from simulations and modeling efforts. When working with scientific data, common challenges are encountered and would be usually related to (a) sharing of potentially large volumes of data, (b) data volume and data rate, and associated computational and bandwidth requirements, (c) heterogeneity of data, software and hardware technologies, (d) management of time-critical data exchanges and analysis execution, (e) curation of data and preservation of scientific analyses, and (f) complexity of data-software-hardware interfaces during data sharing, interactive data manipulations and configurations of data-driven analyses. This list of common data-centric challenges outlines what a scientist has to cope with today in order to execute a data-driven scientific study. Thus, the requirements on software for CI are clearly defined by the need to remove the burden from a single scientist or a team of scientists. For example, given sufficient computational, storage and networking resources, scientists should not be concerned with data volume and data rate since the software for CI would dynamically allocate resources on demand [9, 10]. In general, Cyberenvironments must address the above data-centric challenges by automation in order to increase the productivity of scientists using such Cyberenvironments.

As the requirements are presented at a high level from domain specific X-informatics methodologies thru community scale activities to data-centric challenges, one might neglect the requirements coming from conducting Exploratory Science. It is very common in Earth sciences to explore historical data, global geospatial changes over time from spatially and temporally partitioned data, indirect geospatial variables, or data with heterogeneous geospatial, temporal and spectral sampling rates. In the majority of these exploratory studies, there is an inherent uncertainty about the source of data, acquisition parameters, metadata information (e.g., geo-referencing information), variable transformations (e.g., indirect variables or un-calibrated variables) or approximating transformations (e.g., interpolations or extrapolations in space, time and spectrum). Thus, exploratory studies frequently involve verification, validation and confidence evaluations of the obtained results. In addition, many of the exploratory studies are about comparing multiple implementations of methodologies, a range of parameters associated with software execution or several data products. One could also view these activities as computer-assisted decisions and the associated software as computer-assisted decision support. Driven by the exploratory science, one of the requirements for building software for CI is the easy-to-use user interface to support computer-assisted scientific explorations and discoveries.

In a summary, the informatics requirements for building Cyberenvironments should include considerations about (a) components of domain specific X-informatics methodologies, (b) the spectrum of community scale activities associated with data and science discovery process sharing, (c) the automation benefits and costs of solutions addressing data-centric challenges, and finally (d) the user interfaces to support scientific explorations and discoveries. In order to meet these requirements with limited resources, one has to identify common components of science domain methodologies, decouple specific domain science characteristics from generic functionality of software for CI, and then pursue research and development of Cyberenvironments with inter-disciplinary teams.

Meeting Informatics Requirements in Cyberenvironments

Software for CI has been evolving along the historical course of the cyberinfrastructure development from basic communication protocols and secure data transfers to advanced cyberinfrastructure-enabled environments. According to the Atkins report [3], "generic names for such cyberinfrastructure-enabled environments include collaboratory, co-laboratory, grid community/network, virtual science community, and e-science community." Several NSF, NIH, and DOE funded projects have contributed to prototyping examples of cyberinfrastructure-enabled environments, for instance, the NSF Network for Earthquake Engineering Simulation (NEES), the NSF National Virtual Observatory (NVO), the NSF National Ecological Observatory Network (NEON), NIH Biomedical Informatics Research Network (BIRN) or the DOE Scientific Discovery Through Advanced Computing (SciDAC) projects just to mentioned a few. While the communities building software for CI are numerous, the human resources are limited. Thus, sharing human resources, establishing and following technology standards, and coordination of efforts in meeting informatics requirements have become eminent.

In the Earth science domains, the communities have adopted the term "virtual observatories" (VO) as their end-goal of cyberinfrastructure developments. For instance, the hydrologic community has been building digital hydrologic observatory or Hydrologic Information System (HIS). It is is designed as "a combination of hydrologic data, tools and simulation models that supports hydrologic science, education and practice."[2]. Similarly, the environmental engineers envisioned "an observatory system that will engender a community-wide shift towards systems-level, CI-enabled research, global research coordination, and bi-directional integration of experiment and simulation." [4]. These communities of hydrologists and environmental engineers together with the geochemists building the Critical Zone Observatories (CZEN) have been coordinating their efforts to leverage their cyberinfrastructure investments by holding joint workshops, designing cyberinfrastructure for sharing data (measurements) and metadata, as well as by leveraging developed technologies. One of the challenges for such large communities has been to identify the priorities for meeting the informatics requirements given the existing software technologies. For instance, according to the community survey [4], Excel spread sheets are the most common data formats and software tools used by the communities. Therefore supporting conversions of data entries from Excel to community standards (e.g., Observational Data Model[5]) and the ability to execute Excel macros on demand by the cyberinfrastructure are of a high priority for the domain users.

Within the framework of shared development of software for CI, it is critical to understand what design principles should govern developments of new software for CI, what

[5] http://water.usu.edu/cuahsi/odm/

software technologies are currently available for building virtual observatories, and how the features of these technologies meet the informatics requirements.

Design Principles

A large number of researchers and developers are contributing to building software for CI. Software teams follow basic software design principles that come from software engineering. These principles include modularity of software components, computational scalability, standard application programming interfaces, user interfaces, and ability to use common data formats. With the current fast pace of information technology, new technologies are being invented and existing technologies are failing. Any new design of software for CI has to take into consideration the fault tolerance of distributed cyberinfrastructure systems whether due to a failure of existing components or due to an introduction of new components. Following the basic software engineering design principals is the key to interoperability, low cost maintenance and long term sustainability of software for CI. Along with basic design principles of software engineering, there is a category of design principles that assumes particular importance when it comes to software for science and engineering. The following list is by no means comprehensive, but should receive some attention when designing CI: strong community involvement, open distributed systems, active curation, data-centric strategies for dealing with heterogeneous data and software, and adaptable interfaces.

The most successful and currently available software for CI has a community of scientific users behind it. It is the support of the X-informatics methodologies by the software for CI that attracts domain users since the step-by-step execution of those methodologies is a part of their daily scientific activities. "Careful technologists will take the time needed to understand fully how users currently work, and why, rather than simply assuming that the innovations they propose are an inevitable improvement." [11] Thus, the design and the development of Cyberenvironments supporting informatics processes of a community is an inter-disciplinary effort with each expertise being represented. One has to take into an account technology adoption within a community and across communities, incentives and rewards for migrating legacy data and technologies to new Cyberinfrastructure and social aspects of engagements in multi-disciplinary efforts that are needed for building Cyberenvironments supporting scientific explorations. Needless to say, meeting this design requirement is typically one of the most challenging one since typically the computer science team members are focused on decoupling specific domain science characteristics from generic functionality of software for CI, while domain science team members are seeking specific features supporting an individual informatics process of their interest. Thus, design principles for Cyberenvironments have to be derived from community surveys and community workshops in order to identify and support community informatics methodologies.

The Cyberinfrastructure must be built not only with multiple communities in mind, but also with the considerations of distributed research teams and distributed resources. In such a distributed world, spanning both digital constructs and people, software system must be open, distributed and architected with no master control in mind. The Cyberinfrastructure development landscape is filled with relatively small teams that need simple ways to leverage what other groups have already done. This is a similar problem to what the business community is facing, with a cultural shift added to the mix. In a culture of "publish or perish" careful attention must be placed in privacy and recognition of individual contributions.

Another design principle comes from the spectrum of community scale activities associated with data and science discovery process sharing. It is many times hard to foresee what would be the scale of active users of Cyberenvironments and their activities. In addition, it is unknown in exploratory scientific experiments what parts of the activities would be of interest later when a discovery has been made. All activities and data sets created should typically be actively curated during the scientific end-to-end cycle. We define active curation as the use of cyberinfrastructure to integrate curation and preservation directly with data collection, analysis, modeling, referencing and scholarly publication. Thus, one of the design principles for Cyberenvironments is the capability to record data provenance (data history) as data moves from sensors to reference stores, as integrated and derived data products are created, and as researchers perform analysis and modeling activities at their desktop and on the computational Grid. Cyberenvironments will capture additional annotations, usage information, and associations with papers and other publications. Significant new functionality becomes available when this information is captured across the full data lifecycle [12]. Incorporating this active curation principle in response to the wide spectrum of community scale activities, researchers will be able to better manage complex, multidisciplinary synthesis science activities, and curators can make informed decisions.

The design principles addressing the automation of basic data-centric operations draw from the fact that the explorations could be much more productive if software for CI would (a) automatically resolve heterogeneities of data and software, (b) seamlessly allocate computational resources and (c) effortlessly enable curation of results and preservation of scientific analyses. For example, it has been known that data cleaning and data preparation takes the majority of time in many data-driven analyses and there is an abundance of literature about techniques available to address these problems [13, 14]. When it comes to community scale software for CI, data preparation techniques have to be readily available and the cost of software development must be clearly counter-balanced by increased productivity of all users. In order to minimize the cost of software development and still automate the data-centric operations, one of the design principles is not only to establish and follow existing data and software interface standards, but also to develop limited number of modules for easy translations of historical data and easy integration of legacy software interfaces into new Cyberenvironments. We could list a few particular open specification standards for data sets, such as GeoTIFF and HDF file formats for raster data, ESRI Shapefile for vector data, separator delimited tabular data and Adobe PDF for office documents. The web software interfaces are primarily defined by the World Wide Web Consortium[6] (W3C) and the standards include for publishing language (HTML), Service Modeling Language (SML), Synchronized Multimedia Integration Language (SMIL 3.0), query language for the Semantic Web (SparQL), Ontology Language (OWL), Simple Knowledge Organization System (SKOS), etc.

Depending on the community targeted, the level of technical knowledge can greatly vary amongst its members. For example, in a survey on technology user adoption done for the WATERS Cyberinfrastructure plan [4], Excel was the most popular software (88%) with ArcGIS and Matlab being the most commonly used software package. Despite the long list of

[6] http://www.w3.org/

scientific workflow management system, no software system belonging to this area has make it to the WATERS list of the community most commonly used software packages. We believe that one of the reasons for poor adoption lies in using inappropriate metaphors for workflow composition that do not actively support the end user [15]. In general, much more care must go towards interfaces that support scientists in the way they are used to do their science. When new interfaces must be introduced, it is important to do so with a level of complexity that supports both basic and advanced users. A novice user should be guided from simple and illustrative capabilities to more advanced and still clear ones. The ability to easily create new interfaces besides the ones available needs to be supported as well. Therefore, a clear decoupling between features and graphical user interfaces (GUIs) should be built into the software for CI so that new software developers and even researchers can easily create new GUIs or aggregations of the existing GUIs.

Technologies Available for Building Cyberenvironments

There is a stack of exciting technologies available for building virtual observatories that range according to their maturity level. The software landscape of cyberinfrastructure is still evolving and adapting, trying to learn from and contribute to some of the software changes in both the open source world and the commercial world. The software includes components such as science portals/gateways, context management systems (CMS), process management systems, security layers and communication mechanisms for remote access to data, software and computational resources. These technologies address certain requirements described in the previous two sections, and are outlined next. We omit describing database technologies that are typically connected to various components of Cyberenvironments but are not viewed as a part of the on-going Cyberenvironment research and development.

Software Technology Components

Several broad categories of software have emerged over the years to support scientific and engineering communities. A short discussion of some of the most fundamental ones follows, starting from portal technologies, to content management systems, process management systems, and security and data transfer protocols.

One could view science portals as the points of access on the Internet through which information and services are delivered to a user (a client) from central or distributed computational resources (servers). Although there exist multiple definitions of portals in the literature, [16, 17] the definition above includes multiple purposes of portals. For example, portals could have the purpose of sharing supercomputing resources like the tera-grid resources, or the purpose of sharing scientific publications, presentations, data and metadata like in many digital library systems, or the purpose of communication via email, blog, chat room, or Skype (video and audio). Several commercial portal solutions are available on the market, for instance, Liferay, Blackboard, Campus Ads, Jenzabar, ORACLE PeopleSoft, or uPortal. In the realm of scientific portals, MyExperiment.org[7] adopted some of the successful models developed by popular social networking web site such as Facebook.com and MySpace.com, to provide a way for scientists to share personal profiles, create collaborations on the fly and share workflow representations. Nanohub.org[8] mixes delivery of teaching material with access to desktop applications inside the browser via the use of virtual machines and VNC. The Nanohub users can easily access existing applications without having to worry about setting up the software environment. Although science portals do not specifically consider informatics methodologies in exploratory science, the technologies have matured to provide scalability with the number of users, support for limited data-centric activities and customizable user interfaces for data sharing and browsing.

Content management systems (CMS) are designed for content creation, editing, and version control. In the case of cyberinfrastructure, content includes any data and metadata about data, and activities of communities represented in-silico regardless of whether the content came from local or remote resources. Open source projects, such as Joomla!, focus on providing extensible interfaces to organize, create and publish content. Tools, such as the Tupelo semantic content repository[9], enable distributed management of datasets and Resources Description Framework (RDF) descriptions. The management is supported by a variety of storage implementations, including file systems, relational databases, and RDF triple stores such as Mulgara[10]and Sesame[11]. A key concept in semantic content management is that, at the level of CMS's operation, all information about any kind of entity is simply a combination of an opaque blob of bits and metadata associated with a globally unique identifier. Thus, at the repository level, people, scientific instruments, data, workflows, documents, etc. are all first-class, co-equal entities that can be managed and annotated by any application.

Process management systems and specifically scientific workflow management systems have been originally designed for automation of procedures, linking stand-alone codes and creating flows of data governed by rules. The simplest workflows are scripts of batch files that describe a sequence of tool executions. In the context of Cyberenvironments, workflow systems aim at providing complex problem-solving environments from heterogeneous tools. Driven by systems-science use cases and complex informatics problems, there are several dimensions along which current workflow technologies have grown to become a robust cyber-infrastructure capable of scaling to meet the national needs [5]. These dimensions include (1) hierarchical structure and organization of software, (2) heterogeneity of software tools and computational resources, (3) usability of tool and workflow interfaces (e.g., workflow by example), (4) community sharing of fragments and publications, (5) user friendly security and provenance, (6) built-in fault-tolerance, etc. It is apparent that workflow environments address immediately several of the informatics requirements described in the Informatics Requirement for CI section. There is a plethora of existing workflow technologies although the features of workflows designed for scientific communities are different from those designed for business communities. The primary difference is in responding to the large

---

[7] http://myexperiment.org
[8] http://www.nanohub.org
[9] http://www.tupeloproject.org
[10] http://mulgara.org/
[11] http://www.openrdf.org/

data volume and data rate requirements in sciences. Among the available scientific workflows, one could list Cyberintegrator [15], Kepler [18, 19], D2K [20], D2KSL, OGRE, Ensemble Broker [21], ArcGIS ModelBuilder (ESRI ArcGIS), SciFlo, DAGMan, CCA[12] or Taverna[13]. Nonetheless, only a few of these systems would accommodate the requirements dictated by the community scale and exploratory science needs of end-to-end solutions.

Security layers are necessary for scientific community for multiple reasons. First, it is the nature of distributed resources on the Internet, where clients are connecting and browsing multiple remote hosts, as well as building channels for communications, data transfers and utilization of computational resources. Second, it is the nature of conducting a cutting edge research where scientific collaborations constrained to teams of various sizes. The results and ideas of research teams have to be protected to guarantee the recognition of individual researchers after the results and discoveries are shared. Without providing too much details on security technologies, any applications that requires communication on the Internet (e.g., web browsing, email, instant messaging, or any data transfer) should include simple authentication and security layer (SASL). Among the existing open source security technologies, one could use myProxy[14]or DIGEST-MD5[15]. NCSA MyProxy is software for managing X.509 Public Key Infrastructure (PKI) security credentials and supporting authentication mechanisms, including passphrase, certificate, Kerberos, Pubcookie, VOMS, PAM, LDAP, SASL and One Time Passwords (OTP). DIGEST -MD5 uses the Java Cryptography Extension (JCE) and is already included in Java SDK. It might be also of interest to assess the security risks of developed Cyberenvironment using tools from the Open Web Application Security Project[16] (OWASP).

There exists a multitude of data transfer protocols, some are written for specific application and are built on top of Transmission Control Protocol (TCP) other are more generic. One of the oldest protocols around is File Transfer Protocol (FTP, also denoted as RFC 965] which was ratified as an Internet Request For Comment (RFC) in 1980. This allows sharing of data between two hosts, the first host is the server and puts the data up for download, and the second host can download the data from the server. Although this protocol has been around for a long time, due to its simplicity it is still used by many people to exchange data. Around the year 1985, the WWW became popular and people started to use WWW to transfer data. The WWW-based file transfers are similar to FTP-based transfers, one host is the server and the other host is the client that wants to download the data from the server. Most sites these days will have at least one HTTP (Hypertext Transfer Protocol) server running, allowing people to share data. Not all sites might have an FTP server anymore since the HTTP-based trasfers make sharing of information easier than FTP-based transfers. Another advantage of HTTP over FTP is the ability for secure transfers which makes nearly impossible for third parties to intercept the data.

In the mid 90's grid computing was introduced to denote a distributed computing infrastructure for science and engineering [9]. The grid adds the capability to both share data with other people (using GridFTP) and compute power. GridFTP is based on FTP but allows

[12] http://www.cca-forum.org/

[13] http://taverna.sourceforge.net/

[14] http://grid.ncsa.uiuc.edu/myproxy/

[15] http://java.sun.com/products/jndi/tutorial/ldap/security/digest.html

[16] http://www.owasp.org/

people to subsections of the data and the data can be retrieved from different servers, allowing the client host to select the server closest to them. The Globus Toolkit has been used by many people to create grid enabled applications, giving them an easy way how to share data While some people worked on the infrastructure for grid computing, others worked on building a protocol on top of HTTP to share computing resources and to access the data. The protocol was built as a remote procedure call using the web, which led to the introduction of web services. The web services are described in a description language that is independent of any programming language allowing easier integration with applications written in any programming language. Although web services do not directly focus on some of the issues with large and distributed data sets, they are still used to serve data because of their wide adoption.

Example Technologies Developed at NCSA

Example solutions of Cyberenvironments at NCSA have been developed in close collaborations with the communities of hydrologists, environmental engineers and earthquake engineers. Software requirements have been gathered based on the inputs from community workshops and surveys [11] and by forming inter-disciplinary teams (e.g., Environmental Cyberinfrastructure Demonstrator (ECID), GeoLearn, SP2Learn or MAEVis projects at NCSA/UIUC). As the outcome of multiple efforts, there are example solutions at NCSA that are available for deployment when science gateways, process management (workflow) environments, security layers, or targeted solutions following domain methodologies of interest. The list of these example solutions includes Liferay based CyberCollaboratory portal, Eclipse RCP based Cyberintegrator as a generic workflow environment, Java based GeoLearn and SP2learn informatics driven linear workflows, and Eclipse RCP based MAEViz informatics solution.

The CyberCollaboratory [22] promotes the role of contexts (social context, geospatial context, provenance, etc.) and the use of the Resource Description Framework (RDF) [23] to enable flexible and lightweight means to produce and share metadata. Like MyExperiment.org and Nanohub.org, the CyberCollaboratory embraces design strategies of the Web 2.0, such as user generated content [36]. Traditional science gateways such as the TeraGrid User Portal[17] focus on providing access to data and computational resources and "usually do not provide extensive social networking interaction or social context." [22]. By providing better tools to support participatory science, this new wave of scientific portals could truly revolutionize scientific gateways for the end users.

Demonstrations at the SC06 (International Conference for High Performance Computing, Networking, Storage and Analysis, 2006) and at the American Geophysical Union Fall 2006 meeting [37-39], showed how users of the CyberCollaboratory could start new communities, write blogs, wikis, share documents and share workflow representations. Furthermore, users could change workflow parameters on-the-fly and publish and execute the new workflow remotely to dynamically create new data streams, discuss the changes to the parameters online with collaborators and make them available to the rest of the community. This provided an example of how user generated content can go beyond text, pictures and videos,

[17] http://portal.teragrid.org/

to embrace more complex resources, such as workflow representations, and associated operations on these resources, such as workflow executions, workflow modifications and dynamic generation of derived data sets.

The Cyberintegrator [15] scientific workflow management system attempts to broaden he feature set of traditional workflow systems by focusing on the community and collaboration aspect and by enabling creations of new workflows in an experimental way. It allows users to share annotations and tags about tools, workflow steps, data sets and workflow representations from inside the editor. Information is stored using RDF (figure 1) and distributed repositories using the Tupelo semantic content repository. This enables a level of sharing of resources that is built into the system from the ground up.

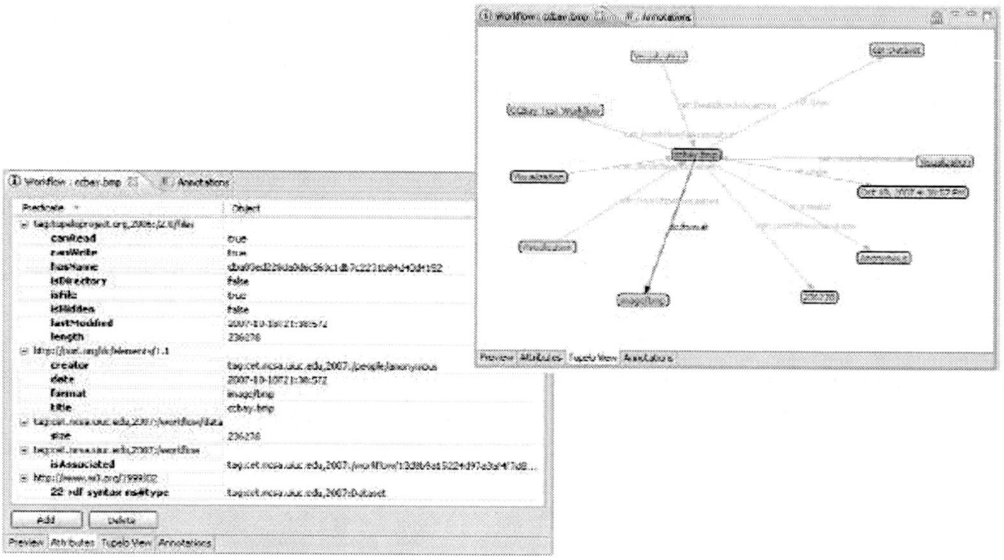

Figure 1. Metadata views in Cyberintegrator.

Tupelo provides a library of utilities for managing large RDF graphs. It abstracts the underlying RDF database implementations, so that application does not have to worry about the different stores and the stores can be swapped based on requirements. For example, Cyberintegrator currently supports the ability to store the data on the local file system, in a MySQL database (DB), or in a remote store, such as Sesame or WebDAV[18] instances. MySQL DB stores the RDF statements. Sesame and WebDAV store the binary data ingested and produced by the system. This gives the end user the ability to connect to multiple remote repositories that could be shared across groups of researchers. The Cyberintegrator takes advantage of shared repositories by providing ways of adding shared annotations and tags to data sets, tools and workflows.

Exploratory workflow creation is enabled by providing an alternative to the conventional graph-based visual programming metaphor. Users interact with a pool of available datasets and related tools that can be executed on specific data sets. As the user executes actions implemented by the tools, more data sets are created and the overall process is developed behind the scenes. The user experiments with different options based on the current data sets available (figure 2). To make the system extensible, the Cyberintegrator framework is built on top of the Eclipse Rich Client Platform (RCP), a plug-in based architecture with strong support from both the industry and the open source world.

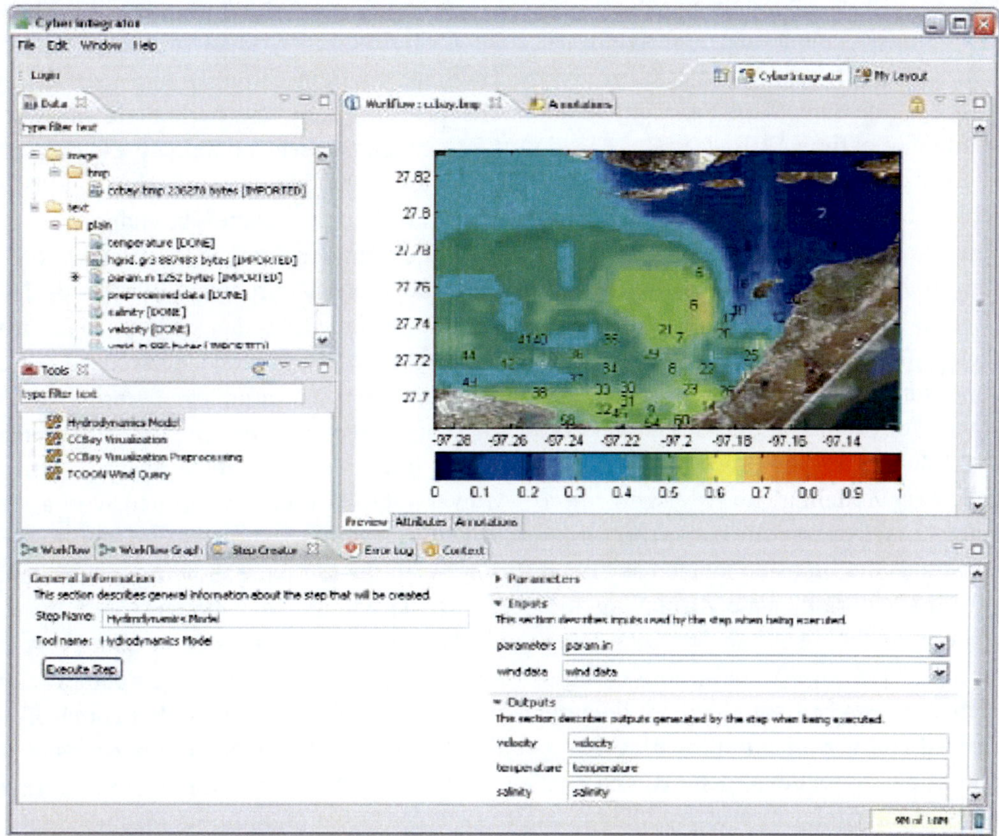

Figure 2. Cyberintegrator editor.

Example Applications Supported by NCSA Technologies

One of the applications is the study and prediction of hypoxia in the Corpus Christi Bay, TX . As part of the Corpus Christi Bay WATERS Test Bed Site, the Cyberintegrator has been used for connecting a hydrodynamics model based on ELCIRC [32] to sensor feeds from the Texas Coastal Ocean Observation Network and from the Shoreline Environmental Research Facility via the CUAHSI HIS web services[19].

Another application is the optimal redesign of a sewage network in the city of Cali, Columbia. Torres [31] developed a model for urban drainage systems based on NSGA-II [33] and SWMM 5.0 [34] software packages linked into a workflow in Cyberintegrator. Instead of manually configuring the execution of models every time, the sewage network planners can

[18] http://www.webdav.org/
[19] http://www.cuahsi.org/his.html

execute an optimization workflow that embeds how the different steps link together and exposes only the parameters necessary to drive a particular execution.

GeoLearn [24-26] was developed to better understand scientific questions raised by the hydroclimatology and terrestrial hydrology communities. Specifically the scientific questions related to causes and consequences of global changes of hydrologic variables through phenomenology, modeling, and synthesis. Like Cyberintegrator, GeoLearn is a workflow system. However, unlike Cyberintegrator the GeoLearn system can only run a set of linear workflows constrained by the steps and options predefined by the particulate informatics flow. In GeoLearn workflow, one can load multiple image datasets, combine these datasets to have consistent spatial and temporal resolution as well as geographic projection, and extract variables over user driven masks derived using additional image, boundary or point information. These steps could be also viewed as the common methodology steps that are present in the majority of scientific analyses in hydrology and incorporated into the design of GeoLearn. The rest of GeoLearn workflow is custom-designed for data mining (data-driven predictive modeling) and geo-spatial visualization of the resulting models, such as predicted and measured values, spatially distributed errors and ranked variables in terms of their relevance for predicting the output variables. The major steps of GeoLearn are illustrated in figure 3 with the tabs labeled as "Load Raster", "Integration", "Create Mask", Attribute Selection", "Modeling" and "Visualization." At every step, a user is presented with a set of parameters and operations to perform as well as with the visualization of the data. Figure 3 demonstrates the first step labeled as "Load Raster" with the left pane showing the raster files already loaded, right pane displaying the selected file (digital elevation map of Illinois), parameters of the display below athe image and the operations/buttons for loading ("Add"), removing ("Remove") and mosaicking ("Mosaic") raster files. The computer science problems and several solutions are documented in [7]. The problems include (a) out-of-core processing and visualization for files that would not fit to RAM, (b) integration of raster files with multiple projections, datums, and spatial resolutions, and (c) ranking of input variables in terms of their relevance to predicting output variables based on multiple machine learning techniques. The GeoLearn system has been used for analyzing the spatial and temporal dependencies of vegetation indices on 29 terrain, climate and land use variables at the US continental scale with 1KM spatial resolution and one month temporal resolution.

Like GeoLearn the SP2Learn [27, 28] system encapsulates a linear workflow assisting with modeling groundwater recharge and discharge rates. In this case, the design of SP2Learn workflow supports a portion of the informatics methodology for taking multiple clouds of geospatial point measurements, interpolating them into raster sets (spatial grids), predicting groundwater recharge and discharge (R/D) rates using physics-based models, introducing boundary conditions to R/D rate predictions by spatial image filtering, integrating auxiliary variables with the filtered R/D rate predictions, deriving rules for relating auxiliary variables with R/D rate predictions and applying the rules to the R/D rate maps to obtain the best accuracy and spatial consistency of the results. The phenomena related to groundwater recharge and discharge result from a set of complex, uncertain processes and are generally difficult to study without visual explorations and simulations. SP2Learn is an example of a framework supporting such exploratorations and simulations. By combining multiple pattern

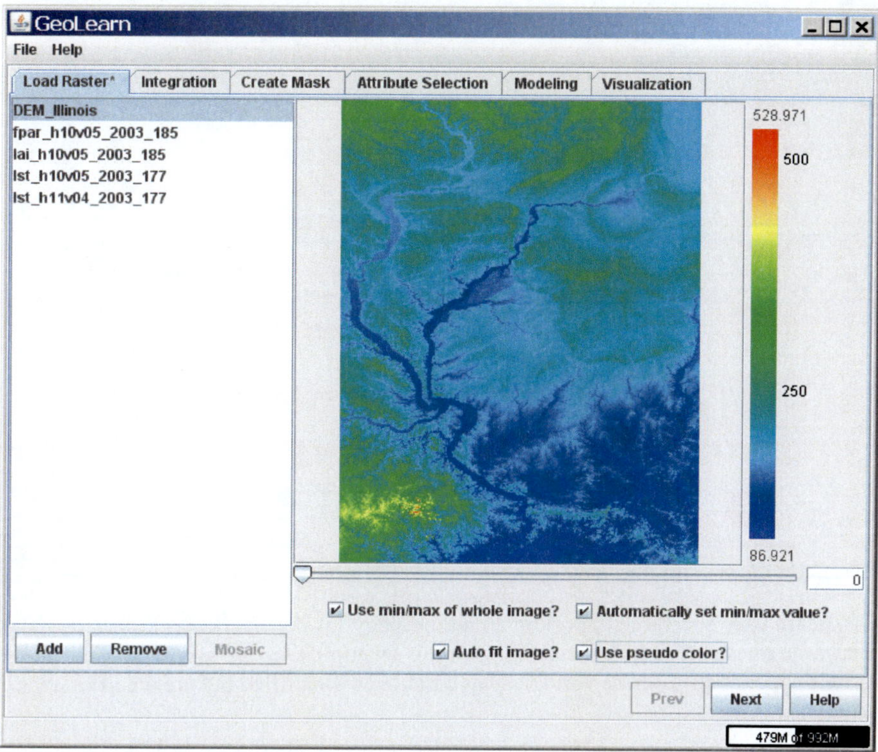

Figure 3. GeoLearn user interface supporting a methodology for data-driven exploratory analyses of hydrologic variables and their relationships.

recognition and data mining methods into a linear workflow, scientists can explore and discover relationships between recharge and discharge rates and other variables. In addition, the SP2Learn workflow can be used for optimization of all parameters of this process including interpolation, filtering, integration and machine learning algorithms. Figure 4 illustrates the main workflow steps as the tabs labeled "Load Raster", "Registration", Create Mask", Attribute Selection", "Rules" and "Apply Rules." These steps of the methodology assume that interpolation, physics-based prediction and spatial image filtering have been completed and the focus is on the auxiliary variables and their relationships to the R/D rate (or any arbitrary variable of interest). SP2Learn have been used in the past for exploring the relationships between R/D rate and slope, soil type and proximity to water bodies, and for optimization of the parameters of the process.

The earthquake hazard risk management called MAEViz [29] is another application of NCSA technologies. Based on the Mid-America Earthquake (MAE) Center research in Consequence-based Risk Management (CRM) [35], it constitutes an example of a domain specific solution that contains several elements described so far. Like the Cyberintegrator, the framework is built on top of the Eclipse RCP, to make the system modular and easily extensible to new analysis and data formats. A user can create scenarios, access local and remote repositories, and execute decision support models [30]. As new algorithms are developed, they can easily be added to the framework. One specific use case developed in the MAEViz framework investigated the earthquake risk to buildings in the Zeytinburnu district of Istanbul, Turkey [30]. Simulations of physical damage from ground shaking to buildings

Figure 4. SP2Learn user interface supporting a methodology for discovering rules between recharge/discharge rates and auxiliary variables (in this example they are slope, proximity to water or soil type). The relationships of these variables have not been quantified but are a part of experts' tacit knowledge.

were run. Based on the results, critical buildings that should be considered for retrofit were identified. Users can apply advanced filtering capabilities to visualize the most critical buildings and start exploring retrofit options (figure 5). Economic impacts such as property tax or sales tax changes from the loss of buildings can also be assessed from this interface. Following the CRM paradigm, retrofits can be applied to the critical buildings in the simulation and new simulations can be executed.

Conclusions and Future Directions

The experience of working on cyberinfrastructure research and development can be described as a highly inter-disciplinary effort involving cross-disciplinary education and multi-disciplinary solutions. In contrary to many other efforts, the process involves community engagement to address the community scale science and has to support scientific explorations to enable discoveries. We presented community informatics driven requirements for building CI software to highlight key features and/or performance expectations. Based on the lessons learnt from our past development, we have also discussed some of the basic design principles for building new software for CI and listed several NCSA technologies available today.

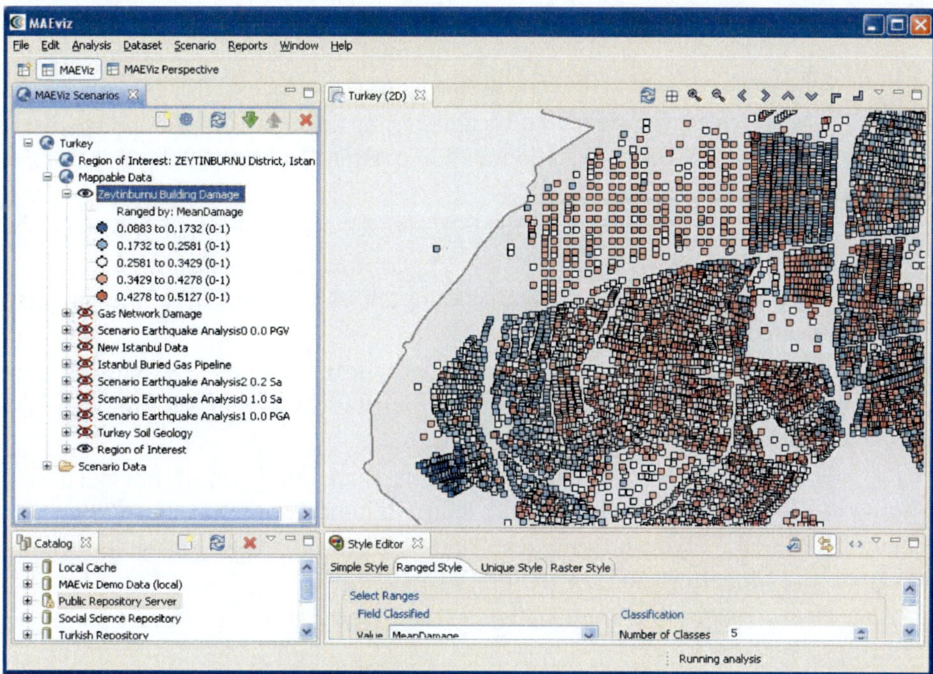

Figure 5. MAEViz interface for earthquake hazard risk management. Buildings in the right pane are colored bytheir mean damage, where blue refers to the lowest damage and red denotes the highest damage.

The future direction of working on cyberinfrastructure research and development is driven by the distributed nature of the resources (e.g., data access, access to services), by data-driven analyses in systems science, and by the capabilities needed for making discoveries and innovations in virtual observatory/digital watershed-type environments. Furthermore, modern digital publishing is changing the flow of information in fundamental ways. The amount of data and the diversity of data types increase, creators can (and often must) publish directly, dissemination costs are negligible, and users participate in editing and curating. Research is often hampered not by lack of data, but by the problem of finding them and interpreting them within context. Therefore the value of content repositories increasingly lies in the metadata and services that link core artifacts into a semantic web, which provides them meaning and facilitates their access. Domain scientists can and must contribute to the process of publication and curation. Without their active participation, the scientific discovery process becomes too onerous to curate large amounts of data, and information that is important to the community but not to the individual researcher will be lost (e.g., proper interpretation of a procedure).

References

[1] NSF, NSF's Cyberinfrastructure Vision for 21st Century Discovery, NSF Cyberinfrastructure Council. 2006.

[2] Maidment, D. R., ed. (2005), Hydrologic Information System Status Report, Version 1, Consortium of Universities for the Advancement of Hydrologic Science, Inc, 224 p, http://www.cuahsi.org/docs/HISStatusSept15.pdf (Last accessed on October 2nd, 2008).

[3] Atkins, Revolutionizing Science and Engineering through Cyberinfrastructure: Report of the National Science Foundation Blue-Ribbon Advisory Panel on Cyberinfrastructure. 2003.

[4] Finholt, T., and Van Briesen, J. (2007). WATERS Network Cyberinfrastructure Plan: the WATERS, Network Project Office Cyberinfrastructure Committee. Unpublished TechnicalReport., http://www.watersnet.org/docs/CyberinfrastructurePlan.pdf. Last accessed on October 2nd, 2008)

[5] Bajcsy, P., et al., A Meta-Workflow Cyber-infrastructure System Designed for Environmental Observatories. Technical Report: NCSA Cyber-environments Division, ISDA01-2005, December 30, 2005.

[6] Bajcsy, P., et al., Survey of Bio-Data Analysis from Data Mining Perspective, in Data Mining in Bioinformatics, J.T.L. Wang, et al., Editors. 2004, Springer Verlag. p. 9-39.

[7] Kumar, P., et al., Hydroinformatics: Data Integrative Approaches in Computation, Analysis, and Modeling. 2005: CRC Press LLC.

[8] Gil, Y., et al., Examining the Challenges of Scientific Workflows. *Computer*, 2007. 40(12): p. 24-32.

[9] Foster, I. and C. Kesselman, Computational Grids, in *The Grid: Blueprint for a New Computing Infrastructure*. 1999, Morgan-Kaufman.

[10] Karo, M., et al., Applying Grid technologies to bioinformatics, in *10th IEEE International Symposium on High Performance Distributed Computing*. 2001. p. 441-442.

[11] Spencer, B. F. Jr., et al. (2006). Cyberenvironment project management: lessons learned., 26 p, September 5, 2006, URL : http://www.nsf.gov/od/oci/CPMLL.pdf (Last accessed on October 2nd, 2008).

[12] Myers, J.D., et al., Re-integrating the research record. Computing in Science and Engineering, 2003.

[13] Dasu, T. and T. Johnson, Exploratory data mining and data cleaning. 2003, New York Wiley-Interscience.

[14] Pyle, D., Data preparation for data mining. 1999, San Francisco: Morgan Kaufmann Publishers.

[15] Marini, L., et al., Supporting exploration and collaboration in scientific workflow systems, in AGU, Fall Meet. Suppl., Abstract IN31C-07. 2007: San Francisco, CA.

[16] Daigle, S.L. and P.M. Cuocco, Portal Technology Opportunities, Obstacles, and Options: A View from the California State University, in *Web Portals and Higher Education Technologies to Make IT Personal*, Richard N. Katz and Associates, Editor. 2002, Jossey-Bass, A Wiley Company.

[17] IBM Global Education Industry, Higher Education Portals: Presenting Your Institution to the World. 2000. Available as Sue Hoffman, IBM brings strategic business intelligence technology to the college campus, IBM Press Room, NASHVILLE, TN - 10 Oct 2000: URL: http://www-03.ibm.com/press/us/en/pressrelease/1523.wss, Last accessed on October 2nd, 2008

[18] Altintas, I., et al., A Framework for the Design and Reuse of Grid Workflows, in SAG 2004. 2005, Springer-Verlag Berlin, Heidelberg. p. 120-133.

[19] Ludäscher, B., et al., Scientific Workflow Management and the KEPLER System. *Concurrence and computation: Practice and Experience, Special Issue on Scientific Workflows*.

[20] Welge, M., et al., Data to Knowledge (D2K): A Rapid Application Development Environment for Knowledge Discovery in Database. 1999, National Center for Supercomputing Applications, University of Illinois at Urbana-Champaign, Champaign.

[21] Alameda, J., et al., Ensemble Broker Service Oriented Architecture for LEAD, in the *22nd International Conference on Interactive Information Processing Systems for Meteorology, Oceanography, and Hydrology*. 2006.

[22] Liu, Y., et al., Towards A Rich-Context Participatory Cyberenvironment, in *International Workshop on Grid Computing Environments 2007, Supercomputing Conference 2007 (SC07)*. 2007: Reno, NV.

[23] Beckett, D., RDF/XML Syntax Specification (Revised). 2004, W3C.

[24] Bajcsy, P., et al., GeoLearn: An Exploratory Framework for Extracting Information and Knowledge from Remote Sensing Imagery, in *32nd International Symposium on Remote Sensing of Environment Sustainable Development Through Global Earth Observations*. 2007: San Jose, Costa Rica.

[25] Bajcsy, P., et al., GeoLearn: Prediction Modeling Using Large Size Geospatial Raster and Vector Data, in EOS Trans. AGU, 87(52), Fall Meet. Suppl., Abstract IN41C-06. 2006: San Francisco, CA.

[26] Kumar, P., et al. Data Driven Discovery from Satellite Remote Sensing: System Development and Analysis of Vegetation Indices, in ESTO 2006. 2006.

[27] Lin, Y.-F., et al., Evaluation of Alternative Conceptual Models Using Interdisciplinary Information: An Application in Shallow Groundwater Recharge and Discharge, in AGU, Fall Meet. Suppl., Abstract H31G-0738. 2007: San Francisco, CA.

[28] Lin, Y.-F., et al., Development of Point-To-Zone Pattern Recognition And Learning Utilities For Groundwater Recharge And Discharge Estimation, in The geological society of America 2006 Annual Meeting. 2006: Philadelphia.

[29] Elnashai, A., et al., MAEviz Architecture, in HAZTURK Workshop 2007. 2007: Istanbul, Turkey.

[30] Elnashai, A., et al., Overview of Applications of MAEviz-Istanbul, in HAZTURK Workshop 2007. 2007: Istanbul, Turkey.

[31] Sanchez A., Towards a demonstrator of an urban drainage decision support system, MSc thesis, 2007, UNESCO-IHE Institute for Water Education, Delft, the Netherlands.

[32] Zhang, Y.-L., Baptista, A.M. and Myers, E.P. (2004) "A cross-scale model for 3D baroclinic circulation in estuary-plume-shelf systems: I. Formulation and skill assessment". Cont. Shelf Res. 24: 2187-2214

[33] Deb, K., Pratap, A., Agrawal, S. and Meyarivan, T. (2000). A fast and elitist multiobjective genetic algorithm: NSGA-II. Technical Report No. 2000001. Kanpur: Indian Institute of Technology Kanpur, India.

[34] Rosman, L. A. (2005) Storm water management model user's manual Version5, U.S Environmental Protection Agency.Cincinnati, Ohio. USA.

[35] Elnashai, A., Hajjar, J. (2006). “Mid-America Earthquake Center Program in Consequence-Based Seismic Risk Management,” *Proc. of the 100th Anniversary*

Earthquake Conference: Managing Risk in Earthquake Country, San Francisco, California, Abstract/Paper Number: 8NCEE-002047

[36] Liu, Y., Myers, J., Minsker, B., Futrelle ,J. (2007), Leveraging Web 2.0 technologies in a Cyberenvironment for Observatory-centric Environmental Research, OGF-19 Semantic Web 2.0 and Grid Workshop, Jan. 29, 2007, the 19th Open Grid Forum, Chapel Hill, North Carolina, USA.

[37] Minsker, B., J. Myers, M. Marikos, T. Wentling, S. Downey, Y, Liu, P. Bajcsy, R. Kooper, L. Marini, N. Contractor, H. Green, Y. Yao, J. Futrelle. Environmental CyberInfrastructure Demonstrator Project: Creating Cyberenvironments for Environmental Engineering and Hydrological Science Communities. Presented at Supercomputing Conference 2006 (SC06), Tampa, FL, November 13-17, 2006.

[38] Liu, Y., S. Downey, B. Minsker, J. Myers, T. Wentling, and L. Marini, Event-Driven Collaboration through Publish/Subscribe Messaging Services for Near-Real- Time Environmental Sensor Anomaly Detection and Management, Eos Trans. AGU 87(52), Fall Meet. Suppl. 2006.

[39] Marini, L., B. Minsker, R. Kooper, J. Myers, and P. Bajcsy, CyberIntegrator: A Highly Interactive Problem Solving Environment to Support Environmental Observatories, Eos Trans. AGU 87(52), Fall Meet. Suppl. 2006.

In: Cyberinfrastructure Technologies and Applications
Editor: Junwei Cao
ISBN: 978-1-60692-063-3

Chapter 13

CYBERINFRASTRUCTURE FOR BIOMEDICAL APPLICATIONS: METASCHEDULING AS AN ESSENTIAL COMPONENT FOR PERVASIVE COMPUTING

Zhaohui Ding[1], Xiaohui Wei[2], Osamu Tatebe[3], Peter W. Arzberger[4], Philip M. Papadopoulos[5] and Wilfred W. Li[6]

[1,2] College of Computer Science and Technology, Jilin University
Changchun, Jilin, 130012 P. R. China
[3] Department of Computer Science, Tsukuba University
Tsukuba, Ibaraki, 3058573, Japan
[4] National Biomedical Computation Resource, University of California
San Diego, CA 92093, United States
[5,6] National Biomedical Computation Resource,
San Diego Supercomputer Center
University of California, San Diego, CA 92093, United States

ABSTRACT

Biomedical, translational and clinical research through increasingly complex computational modeling and simulation generate enormous potential for personalized medicine and therapy, and an insatiable demand for advanced cyberinfrastructure. Metascheduling that provides integrated interfaces to computation, data, and workflow management in a scalable fashion is essential to advanced pervasive computing environment that enables mass participation and collaboration through virtual organizations (VOs). Avian Flu Grid (AFG) is a VO dedicated to members from the

[1] E-mail address: zhaohui.ding@email.jlu.edu.cn
[2] E-mail address: weixh@email.jlu.edu.cn
[3] E-mail address: tatebe@cs.tsukuba.ac.jp
[4] E-mail address: parzberg@ucsd.edu
[5] E-mail address: phil@sdsc.edu
[6] E-mail address: wilfred@sdsc.edu

international community to collaborate on antiviral drug discovery for potential pandemic influenza viruses. The complex and dynamic drug discovery workflow requirement in the AFG-VO is met through innovative service oriented architecture with metascheduling playing a key role. The community scheduler framework (CSF4) is a web service resource framework (WSRF)-compliant metascheduler with an extensible architecture for customized plugins that provide cross site scheduling, workflow management, and data-aware scheduling on the Grid Datafarm (Gfarm) global filesystem. The Opal web service toolkit enables existing scientific applications to be deployed as web services, accessible by various types of clients including advanced workflow management tools such as Vision and Kepler. Molecular dynamics and virtual screening applications exposed as Opal based services are metascheduled using CSF4 to access distributed resources such as the Pacific Rim Applications and Middleware Assembly (PRAMGA) grid and the TeraGrid. Emerging trends in multicore processors, virtualization and Web 2.0 continue to shape the pervasive computing environment in the years to come and pose interesting opportunities for metascheduling research and development.

INTRODUCTION

From Metacomputing to Cyberinfrastructure

About 20 years ago, the term “metacomputing” was coined by Larry Smarr to describe a connected network computing environment [1], which would eventually enable everyone to obtain information on demand “all from their own desktop workstations”. With the emergence of the globus toolkit [2], metacomputing has been popularized as grid computing, with the promise of computing resources as dynamically accessible as the electricity grid. Various countries have made major investments in the development of the grid middleware, software stacks that bridge the network and computing resources to the domain specific applications and users [3].

Since 1997, the US funding agencies such as National Science Foundation (NSF), Department of Energy (DOE), have funded the development of cyberinfrastructure through a series of initiatives including but not limited to the National Partnership for Advanced Computational Infrastructure (NPACI) [4], National Middleware Initiatives (NMI), Software Development for Cyberinfrastructure (SDCI), Science Discovery through Advanced Computing (SciDAC) for software; the OptIputer [5], the Global Ring Network for Advanced Application Development (GLORIAD) [6] for network; the TeraGrid [7], the Open Science Grid (OSG) [8], and the Petascale Computing Environment for high throughput and high performance computing. A number of international grid activities have also sprung up, such as UK e-Science Programme [9], The Enabling Grids for E-science (EGEE) [10] from the European Union, and the Pacific Rim Grid Applications and Middleware Assembly (PRAGMA) grid [11] supported by NSF and member institutes in the Asian Pacific region. In the public health and biomedicine sector, the National Center of Research Resources (NCRR), National Institutes of Health (NIH) has funded the development of the Biomedical Informatics Research Network (BIRN) [12] for data sharing; the National Cancer Institute (NCI) has supported the development of the Cancer Biomedical Informatics Grid (caBIG) [13] for grid services for biomedical data models and integration. With the release of the Atkins report in 2003, cyberinfrastructure is generally used to refer to the national and international network of computers, storage, software and human resources dedicated to support the advancement of science, engineering and medicine [14, 15].

Today, as Moore's law continues to hold at doubling the amount of transistors in an integrated circuit every 18 to 24 months [16], several new trends are developing. Electricity usage and cooling requirements have limited the clock speed of the processors, and led the chip designers to the production of multi-core processors, with multiple processors in the same silicon chip [17]. Soon, workstations may be configured with 16 or 32 CPUs, each of which with 8 or 16 cores, enunciating the dawn of personal supercomputers. More specialized accelerators such as multi-core graphical processing units (GPUs) [18], and field programmable gate array (FPGAs) [19] make up an even more heterogeneous computing environment even within the same machine, posing additional challenges to the traditional programming models. The "desktop workstations" in the vision of metacomputing are appearing as either multi-core personal supercomputers or as increasingly small form but powerful devices such as laptops, pocket PC's and smart phones. Projects that thrive upon "salvaging" spare cycles from the general consumers, such as SETI@home [20], Folding@home [21], or FightAids@home [22] on the World Community Grid, have taken the claim of "theoretical" or "peak" petaflop computing power even before the leadership-class "sustained" petascale supercomputers become assembled. High speed satellite, wireless and cellular networks are making information readily available anywhere on demand, with ultra-broadband networks emerging for scientific and medical applications.

With the commoditization of computing resources and the prevalence of high speed network, optical [23] or wireless, a next generation grid computing paradigm, termed "Cloud Computing", has emerged. Computing clusters and data storage may be provisioned on demand transparently from pools of highly automated and massively parallel grids of data and compute resources, often serviced by third party providers [24, 25]. Cloud computing, shifting the burden of computer, network, software and content management away from end users, could eventually become the backbone for the popular online social network, gaming, information sharing, education, research and public health, and an integral part of the cyberinfrastructure for the 21st century.

While cloud computing makes utility computing [26] possible in a large scale, the way the applications are deployed and accessed are best described using virtualization [27] and Web 2.0 [28]. These two paradigms significantly reduce the application deployment overhead, and increase the ease of user participation and contribution through the service oriented architecture. Virtualization describes the portable, replicable and scalable computing environment provisioned through virtual machines within cloud or grid computing facilities; whereas Web 2.0 describes the participatory nature of a web-based platform for users to compose applications, and complex workflows with easy to use tools. Together these new trends influence how the cyberinfrastructure for biomedical computing evolves in the years to come.

New Collaborative e-Science through Virtual Organizations

The impact of cyberinfrastructure on e-science is most felt through the collaborative research it enables on a scale never imagined before. One of the key components is the ability to form virtual organizations, where researchers and professionals alike are able to share data, applications, and visualization environment transparently and seamlessly with the necessary authentication and authorization [29]. For example, the BIRN project enables the biomedical researchers to share neuroscience imaging datasets, analysis tools, as well as access to TeraGrid resources through its portal environment [12]. The OptIportals allow researchers to visualize large datasets over high speed optical networks. [5]. Similarly, the OSG supports high energy physics research, as well as computational biology virtual organizations [8, 30].

Virtual Organizations

Virtual Organizations provide applications and resources meeting specific user requirements through unified user, job, and resource management schemes. User management includes the accounting of allocations for individual grid users or shared community users, authentication and authorization of independent or federated identities [31]. Job management includes the classification of jobs based on priority or type (array job, serial or parallel), the monitoring of job status, information cache, fault-tolerance and job migration. Resource Management includes resource classification (compute, data and application resources), availability, reservation and virtualization [32]. Some comprehensive software stacks for developing VO-based grid computing environment have emerged over time, e.g., the globus toolkit, Condor [33], VDT (Virtual Data Toolkit) [34], gLite [35], OMII [36], and NAREGI [37].

The Avian Flu Grid (AFG) is a virtual organization in the PRAGMA grid designed to provide a computational data grid environment, with databases and shared information repositories in the fight against the pandemic threat of avian influenza viruses. Figure 1 illustrates the AFG VO in terms of the specific requirements mentioned above. It relies on the PRAGMA certificate authority (CA), as well as individual member organization CA's to issue user certificates; a VOMS server is coupled with GAMA to help manage user credentials. Job management is handled through the metascheduler CSF4 [38], and data management is achieved using Gfarm [39]. The molecular dynamics (MD) simulations are stored in the M*Grid [40]. It takes advantage of the distributed resources in the PRAGMA grid for virtual screening experiments, as well as the high performance computing resources for MD studies. Such virtual organizations greatly increase the productivity of researchers by providing defined services desired by groups of users on top of the basic security, resource, and most importantly, very specific user requirements.

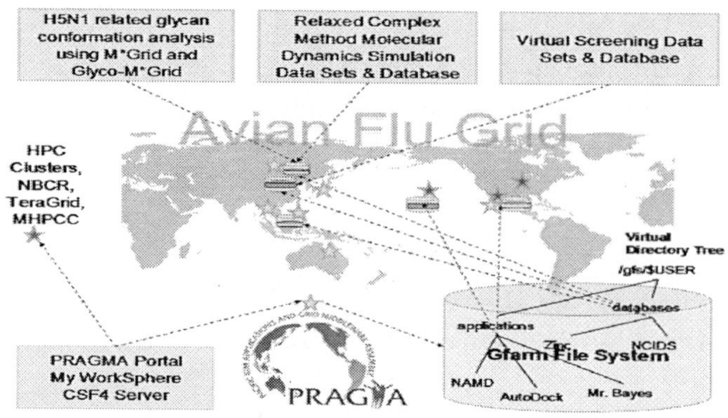

Figure 1. Avian Flu Grid Overview of Activities.

VO User, Computation and Data Management

In general, the basic security mechanism for user management involves the use of Grid Security Infrastructure (GSI) [41], implemented in the Globus toolkit (GT) [42]. X509 based certificates are created and signed by a trusted CA (certificate authority) for user identification, and the associated user proxies are used as user credentials for authentication and authorization in different VO's, often managed through a service such as MyProxy [43], GAMA (Grid Account Management Architecture) [44], or VOMS (Virtual Organization Management System) [45]. VOMS provides a centralized certificate management system with extended attributes which enables role based authentication and authorization with support for multiple VO's. A user may use a command-line tool (voms-proxy-init) to generate a local proxy credential based on the contents of the VOMS database and VOMS-aware applications can use the VOMS data, along with other services such as GUMS to support various types of usage scenarios and proper accounting practices [46].

The computational infrastructure often comprises grids of grids, grids of clusters or just clusters built upon proprietary or open source operating systems. For example, Rocks cluster environment is a popular distribution for building Linux based clusters and grids [47], with the appropriate grid rolls [48]. The BIRN infrastructure is replicable and expandable using the BIRN Racks, a minimum set of hardware requirement, with a customized biomedical software stack encapsulated using Rocks rolls. Condor, VDT, GT4, e.g., may be installed on Rocks clusters to build a grid of clusters. More than half of clusters in the PRAGMA grid use the Rocks cluster environment [39], with the entire grid integrated through the globus toolkit.

Data services, which include basic data handling, metadata management, automatic data staging, replication and backup, are often provided using gridFTP [49], RFT [50], Gfarm [51], Storage Resource Broker [52], dCache [53] or similar services. Resources such as the TeraGrid may also employ high performance file systems such as GPFS (General Parallel File System) [54] or Lustre file system [55] for large I/O requirements.

Metaschedulers Reduces Complexity for Users and Applications Developers

Each of the major middleware packages includes some form of resource brokers or metaschedulers, which ensures that the data and computational requirements are met across sites or clusters. Metascheduler services range from resource discovery, application staging, data I/O management, and on-demand computing resources [56]. However, they are often tied to a particular unified software stack in grid environments such as EGEE or OSG. Quite often, grids comprise heterogeneous resources distributed in multiple VOs which may have different local policies. Users and application developers are faced with many different choices of local schedulers, grid protocols, and resource brokers. It is desirable to have a lightweight metascheduler that may enable a biomedical researcher to access any grid through standardized protocols. Different approaches have been taken to achieve this goal. APST is an agent based approach for parameter sweep applications [57, 58]. Nimrod/G provides an economy based scheduling algorithm when choosing computational resources [59]. The Globus GRAM (Grid Resource Allocation and Management) provides a standardized interface to access different local schedulers [60]. Condor-G offers users of condor pools to also take advantage of Globus-based grids [61]. Gridway supports the DRMAA (Distributed Resource Management Application API) specification [62], as well as the Unicore, and EGEE infrastructure [63]. CSF4 leverages the Globus GRAM, and provides an advanced plugin mechanism for cross domain parallel applications, as well as workflow management [56].

Since a metascheduler is able to provide a virtualized resource access interface to end users, and enforce global policies for both resource providers and consumers as well, it plays an increasingly important role in efficient management of computational and data grids. Metascheduling is essential to workflow management and seamless access to distributed resources.

Emerging Technology for Pervasive Computing

Virtualization and Cloud Computing

In the age of cloud computing, virtualization of physical resources have added a dynamic twist to how we think about providing a customized computing environment. In order to cater to various user requirements for applications, memory, storage, and network infrastructure, it is desirable to provision such customized environments through the use of virtual machines [27]. These virtual machines encode the optimized requirements of the included applications, and may be scheduled on demand, within a compute cloud of hundreds and thousands of processing cores. Cloud computing facilities such as the Amazon EC2 (Elastic Compute Cloud), or the IBM Blue Cloud may be economical enough for occasional users to deploy their applications and have their problems solved without investing in the computational infrastructure. At the campus level, the large compute and data centers such as the Tsubame supercluster [64], and the progress in developing virtual clusters [65] define a new outlook for grid computing, where the grids are completely virtualized, within a single compute cloud, or across sites. Such virtual organizations are distinct in the sense that the applications within the VO could come with the virtual machine images completely. This enables the virtual deployment of "everyone's supercomputers" since the application environment are virtual images of one's familiar desktop, workstation or cluster environment, optimized and with extreme scalability.

Web 2.0

While virtualization, virtual organizations are buzz words for the computer scientists, the biomedical researchers are excited by different tunes. Applications in systematic modeling of biological processes across scales of time and length demand more and more sophisticated algorithms and larger and longer simulations. The grid computing technologies are enabling the creation of virtual organizations and enterprises for sharing distributed resources to solve large-scale problems in many research fields. One big challenge for building cyberinfrastructure is to enable Web 2.0 for scientific users, based on user requirements and the characteristic of grid resources, as exemplified by the massive collaborations that brought into fruition such public resources as Wikipedia [66]. Scientific users need the ability to use distributed resources for the scalability without knowing how it's provided.

The Web 2.0 paradigm provides several lessons that can be leveraged by the scientific and grid communities. In "What is Web 2.0" [28], O'Reilly lists some design patterns and general recommendations that are a large part of the Web 2.0 paradigm: (1) use the Web as a platform for development, (2) harness the collective intelligence of the community, (3) focus on services, and not prepackaged software, with cost-effective scalability, (4) enable light-weight programming models, and (5) provide a rich user experience. In principle, the Web 2.0 paradigm allows users to contribute their applications, share data, and participate in collaborative activities with ease and keep their focus on the process of creation, instead of maintaining and learning new tools and interfaces. The virtualization trend in computational resources is going to play an important role in enabling Web 2.0 for the masses.

Opal Toolkit for Virtualized Resources

The disparity between the exponential growth of capacity for computing, optical fiber and data storage [23] and the ability of researchers to take advantage of the computing power, and visualizing the information has been growing. In order to allow user to truly enjoy the benefits of pervasive computing, software design must observe human social behaviors, lay people or scientists alike. The popularity of social networks such as MySpace [67], FaceBook [68], even among educated researchers, suggests that Web 2.0 or easily accessible participatory, pervasive computing, is for everyone, and the next generation of researchers certainly expects it.

The Opal toolkit models applications as resources while exposing them as Web services [69]. The ability to deploy any application within any compute environment and exposes it as a web service with an automatically generated user interface accessible from any type of client significantly reduces cost of developing reusable and scalable applications. It also reduces the cost of service mashups because legacy applications may now be made accessible a standardized interface, with scalable computing and data backend support. Lastly, the automatic interface generation feature enables the user with to receive the latest updates automatically, and with minimal changes from what he's already familiar with. The purpose of Opal, therefore, is to shield the scientific application developers from the fast changing grid computing environment, and make the underlying grid middleware, currently using WSRF [70], compatible with standard web services. Opal is one of the tools that enable the Web 2.0 paradigm for scientific users and developers.

The base version of Opal by default submits jobs to a local scheduler. This places a limit to its scalability to a single cluster, physical or virtual. For greater scalability, it is necessary to leverage multiple clusters, physical or virtual, located at different locations or within the same compute cloud. With that in mind, we have extended the concept of an application as a requestable resource to the metascheduler CSF4 [7], the open source meta-scheduler released as an execution management component of the Globus Toolkit 4 [10]. CSF4 can coordinate heterogeneous local schedulers and provide an integrated mechanism to access diverse data and compute resources across the grid environment. More information about the Opal toolkit [71] and its usage is available online [72].

INTEGRATED COMPUTATION, DATA AND WORKFLOW MANAGEMENT THROUGH METASCHEDULING

Metaschedulers and Resource Brokers

As described earlier, many kinds of grid resource brokers, meta-schedulers and utilities have been proposed and developed for parallel computing and scheduling in the heterogeneous grid environment. They are quite different in resource management protocols, communication mode and applicable scenario, etc. A typical metascheduler handles the data stage in, credential delegation, job execution, status monitoring, and data stage out. Most existing metaschedulers only sends jobs to a single cluster, and transfer the data output to the user's machine or a central data server. Here we discuss some challenges in providing integrated computation, data and workflow management in biomedical applications which must be met through better and more intelligent metaschedulers and resource brokers.

Cross-site Job Scheduling

Running parallel jobs crossing sites in a grid environment is still a challenge. Parallel Virtual Machine (PVM) [73] and Message Passing Interface (MPI) [74] are usually used in the traditional single cluster environment. Grid enabled MPICH (MPICH-G2) is a Grid-enabled implementation of MPI, which is able to execute a parallel job across multiple domains [75] and has been used in many life science applications. Nonetheless, MPICH-G2 does not synchronize the resource allocation in multiple clusters, which would cause resource waste or even dead lock problems (see section 3.1). Many researchers have found that the MPICH-G2 jobs cannot execute successfully unless manual reservations are made in advance at destination clusters. Apparently, without an automated high-level coordinator's participation, these problems are hard to resolve.

The Moab grid scheduler (Silver) consists of an optimized and advanced reservation based grid scheduler and scheduling policies, which guarantee the synchronous startup of parallel jobs by using advanced reservation [76]. Silver also proposed the concept of synchronized resource reservation for cross-domain parallel jobs. GARA [32] splits the process of resource co-allocation into two phases: reservation and allocation. The reservation created in the first phase guarantees that the subsequent allocation request will succeed. However, GARA did not address synchronized resource allocation. Furthermore, GARA and Silver both require that local schedulers support resource reservation. Many other meta-schedulers, such as Gridway [63], have yet to make optimizations for cross-domain parallel jobs.

Grid Filesystems and Data Aware Scheduling

One significant bottleneck of the existing computational platforms, especially in the age of petascale computing, is the efficient data management, transfer, replication and on demand access to large datasets. Existing data management, high performance I/O, and data sharing solutions include but not limited to Gfarm, SRB, GPFS, and GSI-FTP. However, a number of these require the use of specific API's and places special burden on application developers and users. It has previously been shown that Gfarm-FUSE (File System in User Space) may enable legacy applications to leverage grid resources without modification [77], and data aware scheduling may provide efficient support for scheduling the computation to where the data resides [78].

When users have to access different compute resources available in different virtual organizations, file transfer between sites become a bottleneck due to network latency, especially when large amounts of small files are created. For a round trip time (RTT) of 0.3 ms, 10,000 files create 30 min of delay, 48,000 files mean a whole day of delay for a single user, and assuming only one exchange is required. This multiplies with the number of sites involved, and the number of services involved due to security requirements. The solution, until the ubiquity of dedicated high speed fiber optical networks is achieved, is to take the computation to the data. The success of Google lies in its technical infrastructure which enables localized data centers to provide the best possible service to the local population, with the necessary replication of data using dedicated connections. However, scientists always need to take some data with their own computer, send some data to their collaborators or share large datasets for data mining purposes, or visualization on specialized hardware, after the computation is done.

On the other hand, there is always need to transfer tens to hundreds of GB's of data from instruments such as electron microscopes (EM's), mass spectrometers, sequencing centers, and metagenomic or ecological sensory network, for storage and analysis, where the data come in streams and need to be processed on the fly or stored. In the case of EM or large scale simulations, there is the requirement of real time feedback or progress monitoring, this also places additional constraints on how fast the data may be analyzed and/or visualized. How to manage these data streams and make them available to legacy applications efficiently remains a challenge and is an area of active research [79].

Workflows in Translational Biomedical Research

Workflow is a group of correlative tasks, procedural steps, organizations or roles involved, required input and output information, and resource needed for each step in a business process, can be executed automatically or semi-automatically. Workflows may be described at different levels through process modeling, and more than 100 workflow patterns have been extracted [80]. XPDL (XML Process Definition Language) is often used to store and exchange the resulting process diagram [81], which may be mapped to an execution language such as BPEL (business process execution language), and executed in an execution engine such as ActiveBPEL [82]. However, workflows in business processes exhibit different characteristics from the workflows in scientific investigations. Grid Process Execution Language (GPEL) [83], an extension of WS-BPEL, and the corresponding grid workflow engines are being developed [84].

In biomedical research, grid based workflows require additional parameters such as statefulness, performance and reliability [85]. Some typical examples of workflows are illustrated in figure 2. For example, a user often needs to connect the input and output of several applications into a workflow, such as the use of a MEME-based position specific score matrix for a MAST-based database search [86]. Or in the case of electron microscopy, the use of MATLAB and a series of steps need to be done in series with large datasets to construct 3D volume reconstruction using 2D transmission EM tilt imaging series [87]. Other workflows involve sophisticated facial surgery simulation software through a series of image segmentation, mesh generation, manipulation and finite element simulation (see figure 3 and [88]).

The workflows often encompass web services with strongly type data [89] or legacy applications exposed as web services through a web service wrapper such as Opal [90]. The flexibility of web services interface and its accessibility by different clients have contributed

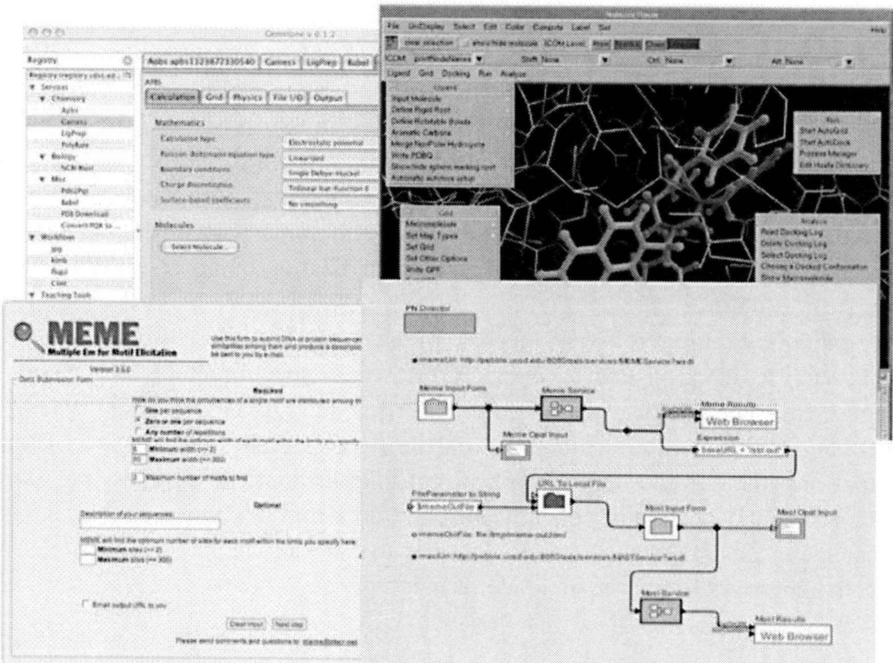

Figure 2. Sample workflows involving sequence analysis, visualization and electrostatics calculations.

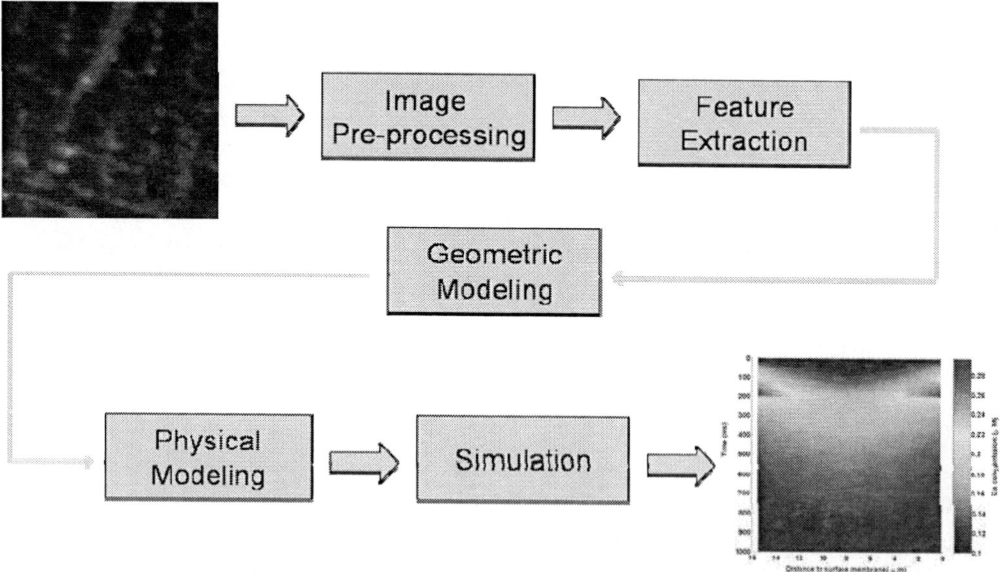

Figure 3. Image processing workflow: steps involved from light microscopy images to mesh generation, refinement to computational simulations. Courtesy of Zeyun Yu.

to its popularity as the main standard for developing a loosely coupled computing framework for different types of clients, and also make them suitable for relatively tightly coupled service mashups, in addition to the simple and popular REST (Representational State Transfer) protocol [91].

Quite often, in the field of genomic analysis, or virtual screening experiments, in a way that is very similar to parameter sweep studies, one often needs to rerun the same application with different parameters, e.g., input sequences, or different ligands. In such cases, the workflow or composite of workflows may be very tightly coupled applications or loosely couple services. Such scientific workflow requires special treatment since they require high throughput, fault tolerance, checkpointing and restarts, and pose particular challenges to metascheduling.

The main problems we encountered when running biomedical application on grids include huge amount of distributed data, distributed computing resource and sub jobs dependency in a large task. Although the global file system like Gfarm and the Meta-Scheduler like CSF4 are able to handle the distributed data and computing resources. Evaluation of job status, quality of service in terms of network bandwidth, queue length, and job turn-around time, are still difficult challenges to be met.

The job dependency issue can be handled by meta-scheduler easily, since meta-scheduler is able to queue and disassemble tasks and monitor the jobs status. Driven by this, we have developed a grid workflow plug-in in CSF4 meta-scheduler and the details will be described in next section.

Customized Scheduling Policy Plug-ins for Biomedical Applications

Many meta-schedulers are built on the top of specific local resource management system, such as Silver and Condor-G. They provide plenty of policies derived from local schedulers. However, these policies are implemented under specified local system protocols, so they are not available in a heterogeneous environment. The Nimrod/G resource broker introduces the concept of computational economy to metascheduling [59]. It leverages the Globus Monitoring and Discovering System (MDS) to aggregate resource information and enforces scheduling policies based on an auctioning mechanism [92]. To process resource negotiation, it is required that resource brokers use a common protocol like SNAP [93], and the negotiation result is difficult to predict. Gridway's scheduling system follows the "greedy approach", implemented by the round-robin algorithm.

We have developed additional plug-in modules for the Community Scheduler Framework (CSF4), e.g., an array-job plug-in to schedule AutoDock [94] or Blast [95] like applications, and demonstrate that many popular life science applications can take the advantage of CSF4 meta-scheduling plug-in model. Newly developed data-aware plugin and workflow plugin are also described.

CSF4 Framework for Plug-in Policies

The framework is completely scalable to the developers, and customizable to user requirements. Here we provide an overview of the kernel plug-in, array job plug-in, functional plug-ins such as workflow and data-aware plugin, and how the different plugins operate together.

Kernel Plug-in

The kernel plug-in is the default plug-in, which will be loaded and enforced by all the job queues. It only provides the most general functionalities and making match decisions according to FCFS round-robin policies. The purpose of the kernel plug-in is not making efficient match decisions but ensure the jobs can be complete, i.e. the resource requirements of the jobs must be satisfied.

In the past, using CSF4 was fairly complicated – if users have specialized resource requirements, they needed to know details of each of the clusters, such as deployment of applications, location of replicated data, dynamic state of clusters, etc. The resource management of CSF4 paid more attention to the integration of heterogeneous local schedulers [9].

The kernel plug-in use a virtualized resource based method, in which all the user requirements, such as cpu, memory, disk space, data, cluster type, and even application, are abstracted a kind of resource owned by clusters.

First, we divide grid resources into two parts - application resources and cluster resources. The description of an application resource includes necessary parameters: "name", "path", "clusters deployed on" and optional parameters "version", "compiler" and "dependent libraries". The status of the application resource does not change frequently once an application is deployed – it can be updated by the local administrator, and queried via MDS [92] or SCMSWeb [96], or even an RSS (Really Simple Syndication) feed. To submit a job to CSF4, a client just needs to specify the name of the application resource. After the scheduling decision has been made, as described below, the scheduler can use an appropriate version of the application, taking into account the location of binaries and dependencies from the resource information, thus obviating the need for an end-user to specify it each time. This is analogous to our model for the Web services where a user just deals with a service, and is not bothered with the internal details.

The description of cluster resource includes necessary parameters: "name", "infrastructure type", "scheduler type", "master hostname", "available CPUs" and optional parameters: "scheduler version", "scheduler port", "CPU architecture", "memory", "disk space". Some parameters of cluster resources, such as "available CPUs", are finite and change very frequently -- although some monitoring tools such as Ganglia [97] provide some semi-real-time cluster status information via MDS, this information has been found to be not very dependable since they are not gained from local schedulers and are frequently out of date. Hence, we also maintain an inner resource status list in CSF4, and change the resource status on every scheduling cycle by reports on the status of completed jobs. When a user submits a job to CSF4, it queries the application and cluster resources from a provider. In the scheduling cycle, CSF4 regards the user job as a set of resource requirements and matches the requirements with resources available via FCFS (first come first server) policies. CSF4 locks the allocated resource by changing the inner resource status list after a successful match. The locked resource will be released after the job finished. CSF4 also adjusts the value of "available CPUs" after the completion of submitted jobs. On one cluster, if most of the dispatched jobs are finished, CSF4 will give a higher weighted value for the "available CPUs" of this cluster. On the other hand, if few jobs are finished, which suggests that the cluster may be busy or that the local scheduler did not allocate all the CPUs for Grid jobs – in this case, the value "available CPUs" for the cluster will be weighted down. For transfer of data between nodes, CSF4 uses the GridFTP protocol [49], though RFT (reliable file transfer) is [50] through the WSRF framework is also. Figure 4 shows how the CSF kernel plug-in works.

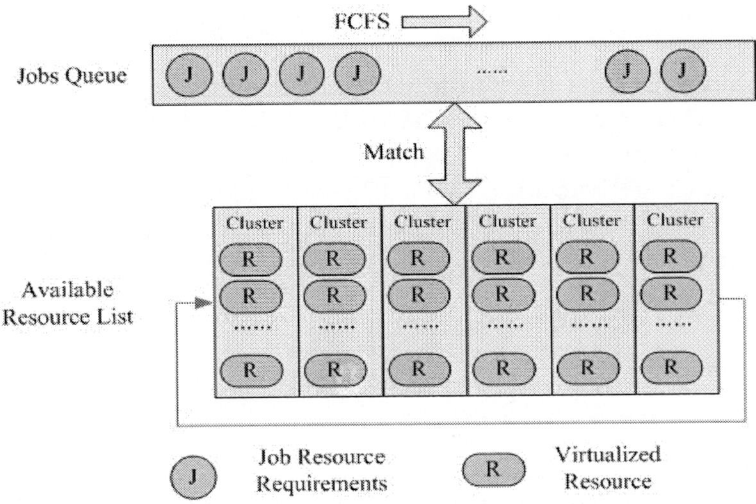

Figure 4. CSF4 kernel plug-in uses FCFS round-robin and virtualized resource based scheduling policies.

Array Job Plug-in

Quite frequently, life sciences applications are "pleasantly parallel", i.e., serial applications which may be used to handle many parallel data input. For example, AutoDock may be used to dock different ligands to a target protein structure, or Blast may be used with different input sequences to search for potentially related sequences within a target database. Here we show how a customized scheduling policy for these applications may be developed using the CSF4 plug-in model. A testbed of three clusters is setup with the Gfarm deployed, each with different host numbers and dynamic local work load. Normally AutoDock or Blast applications consist of a large number of subjobs. These subjobs execute same binary with different input/output files, so the plug-in is named array job plug-in, similar to what's available for some local schedulers. The metascheduling objective is to balance the load between clusters and complete the entire set of jobs as soon as possible. The array job plug-in call back functions implementation details are described in [56]. Briefly:

Initialize() sets up the maximum job load, 10, for example, for all local clusters. If the number of unfinished subjobs in a cluster exceeds this maximum job load, the metascheduler will not send any new job to it.

As the subjobs of AutoDock or Blast do not communicate to each other, and there is no dependency among them, the job execution order does not matter. Hence, Funjob–sort() is just a empty function in the plug-in.

As the input/output files are accessible in all the clusters through the Gfarm virtual file system, so any cluster can run AutoDock or Blast jobs as long as its local scheduler is up.

Therefore, FunCL−match() will simply select all clusters with local schedulers as matched clusters.

FunCL−sort() sorts the matched clusters according to their unfinished sub-job numbers. For example, in the beginning, the metascheduler dispatches 10 sub-jobs to both Cluster A and B. While doing the next scheduling, Cluster A finished 6 jobs and cluster B finished 4 jobs, then Cluster A would be preferred as the execution cluster for next job. If all the clusters have hit the maximum job load number set by Initialize(), no cluster is qualified as an execution cluster.

Dispatch() dispatches the first job in the job list to the execution cluster selected by FunCL−sort(), and increase the cluster's unfinished job number. Then, the metascheduler calls Funjob−sort(), FunCL−match(), FunCL−sort() and Dispatch() again for the next job until no execution cluster is returned by FunCL−sort(). In the case above, cluster A gets 6 new jobs, and cluster B gets 4. Although the local cluster's load and policies are not considered explicitly, the loads are balanced dynamically since the cluster which completes jobs faster will get more jobs.

Function Plug-in

CSF Workflow Plug-in

In the grid environment, a metascheduler has to accept and assemble grid tasks, match the jobs to qualified resources, and dispatch the jobs to distributed sites. Complicated grid tasks may be described and processed by using workflow techniques. Currently, there are a few popular workflow standards, such as XPDL [81], BPML and BPEL4WS. Since the XPDL focuses mostly on the distributed workflow modeling and supported by various workflow engines, it is convenient to use XPDL to describe the workflow in general.

Within the CSF scheduling plug-in framework, we have developed a plug-in to support the scheduling of grid workflow. As described before, the default kernel plug-in uses the FCFS round-robin policy to match jobs with appropriate resources, and is invoked by any job queue. However, the default plug-in cannot process complex jobs with intricate job dependencies. Hence, we have developed a separate plug-in named "grid workflow".

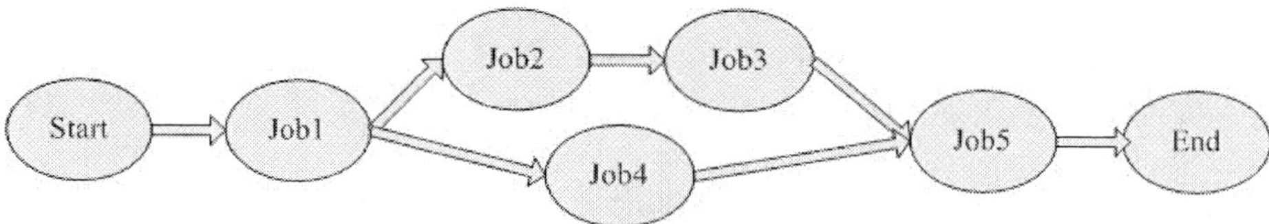

Figure 5. Example of a simple workflow for job dependency.

While XPDL is powerful enough to describe a grid task, there are many elements of XPDL, such as "visualized flow chart", which are not needed by a grid task. Therefore, we only implemented a subset of XPDL, but extended it with a few grid variables (such as 'Resource Specification Language', RSL for short).

There is only one <WorkflowProcesses> tag and one or more than one <WorkflowProcess> tags in a single grid workflow description. They present the main flow and the sub-flow. In a <WorkflowProcess> tag, there are two main tags, <Activities>, which includes all the atomic grid jobs (describe as an RSL file) in this sub-flow, and <Transitions>

presents the relationship of the grid jobs. For example, "<Transition Id="16" Name="", From="a" To="b"/>" represents that "a" is the precursory job of "b" and "b" is the subsequent job of "a". Then, any grid workflow can be described as a directed acyclic graph (DAG). A simple workflow is shown figure 5, and the corresponding workflow description is represented in XPDL in figure 6.

```
<?xml version="1.0" encoding="UTF-8" ?>
- <WorkflowProcesses xmlns="http://csf.jlu.edu.cn"
    xmlns:xsi="http://www.w3.org/2001/XMLSchema-instance"
    xsi:schemaLocation="http://csf.jlu.edu.cn xpdl.xsd">
  - <WorkflowProcess Id="1">
    - <Activities>
      - <Activity Id="start">
        - <Event>
            <StartEvent />
          </Event>
        </Activity>
      - <Activity Id="JOB1">
        - <Implementation>
          - <task>
              <rsl>/usr/examples/gram_job1.rsl</rsl>
            </task>
          </Implementation>
        </Activity>
      - <Activity Id="JOB2">
        - <Implementation>
          - <task>
              <rsl>/usr/examples/gram_job2.rsl</rsl>
            </task>
          </Implementation>
        </Activity>
      - <Activity Id="JOB3">
        - <Implementation>
          - <task>
              <rsl>/usr/examples/gram_job3.rsl</rsl>
            </task>
          </Implementation>
        </Activity>
      - <Activity Id="JOB4">
        - <Implementation>
          - <task>
              <rsl>/usr/examples/gram_job4.rsl</rsl>
            </task>
          </Implementation>
        </Activity>
      - <Activity Id="JOB5">
        - <Implementation>
          - <task>
              <rsl>/usr/examples/gram_job5.rsl</rsl>
            </task>
          </Implementation>
        </Activity>
      - <Activity Id="end">
        - <Event>
            <EndEvent />
          </Event>
        </Activity>
      </Activities>
    - <Transitions>
        <Transition Id="1" From="start" To="JOB1" />
        <Transition Id="2" From="JOB1" To="JOB2" />
        <Transition Id="3" From="JOB1" To="JOB4" />
        <Transition Id="4" From="JOB2" To="JOB3" />
        <Transition Id="5" From="JOB3" To="JOB5" />
        <Transition Id="6" From="JOB4" To="JOB5" />
        <Transition Id="7" From="JOB5" To="end" />
      </Transitions>
    </WorkflowProcess>
  </WorkflowProcesses>
```

Figure 6. The XPDL description of the workflow.

The main functionalities of workflow plug-in are analyzing the grid workflow, obtaining the RSL and the dependencies of all the jobs, and making the dispatch decisions. They were mainly implemented in the Funjob–sort() function. The input of Funjob-sort() is the tasks queue, the tasks can be single job, parallel job, array job or a grid workflow.

The Funjob-sort() will be invoked periodically, during the every invocation, each grid workflow will be decomposed into multiple single sorted jobs based on the job dependencies described in the workflow. If the job has an unfinished precursory job, its status will be set to "PENDING", if not, its status will be set to "SCHEDULE". The priority of a job is decided by the number of its subsequent jobs, a job has more subsequent jobs will get a higher priority. The output of the function is a single and sorted jobs queue. See the figure below.

Data-aware Plug-in

Emerging classes of data-intensive applications that both access and generate large data sets are drawing much more attention. High-performance data-intensive computing and networking technology has become a vital part of large-scale scientific research projects in areas such as high energy physics, astronomy, space exploration, human genome projects, and computational simulations in biomedical research. One example is the Large Hadron Collider (LHC) project at CERN. The so-called Data Grids provide essential infrastructure for such applications. Grid Datafarm (Gfarm), for example, is one of them.

Gfarm architecture is designed for global petascale data-intensive computing. It provides a global parallel file system with online petascale storage, scalable I/O bandwidth, and scalable parallel processing, and it can exploit local I/O in a grid of clusters with tens of thousands of nodes. Gfarm parallel I/O APIs and commands provide a single file system image and manipulate file system metadata consistently. If a huge amount of data I/O is involved, a network system's performance will be degraded by network congestions without proper data management and job scheduling. In Gfarm, gfrun and gfmpirun commands are able to allocate the file-affinity hosts for optimum execution of applications based on available metadata. However, the manual method is not scalable in a production environment with a large number of users running jobs concurrently. It is imperative to have an automated job scheduling and data management mechanism.

Utilizing the CSF4 extensible scheduling plug-in framework, we developed a scheduling plug-in for data intensive applications, called data-aware plug-in [78]. It works with Gfarm to manage the Gfarm file replicas, optimize the job distribution and reduce the use of I/O bandwidth to improve the system performance (figure 7).

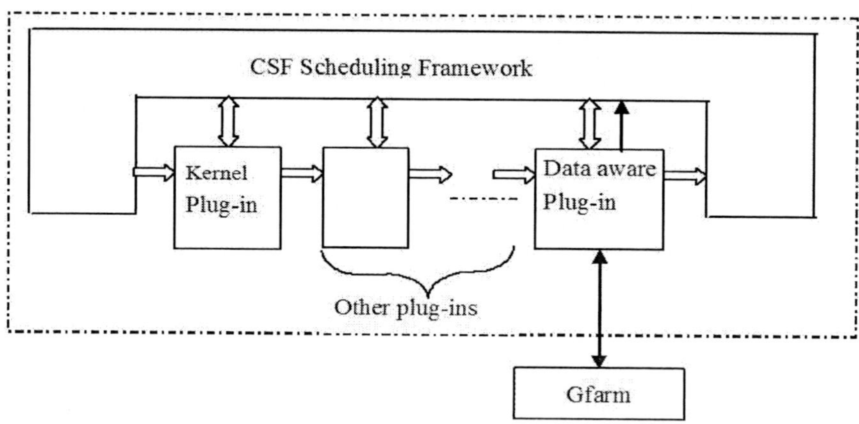

Figure 7. Data aware plugin of CSF4 for Gfarm.

The purpose of data aware algorithm are obtaining the replicas location of the required files from data grid, and making jobs/clusters match decisions based on the file replicas information. Since the scheduling targets of the meta-scheduler are not single machines but clusters or local resource managers, it is impossible to for a meta-scheduler to specify machines for jobs, so we just assume the local scheduler can dispatch the jobs to the machines with the necessary file replicas accessible through a local network file system or Gfarm file system. Then, the purpose of data aware plug-in is to dispatch the jobs to the clusters that have the required file replicas.

The functionalities of data-aware algorithm are implemented in Funjob-match(). The input of the function is a tasks queue, and the output is the tasks queue with dispatch decisions. Briefly, it entails getting the required files list from job resource description (rsl file); getting the metadata of the files (include file size, file replicas location) by using Gfarm API; getting the satisfied clusters list based on file replicas location; and making the jobs/clusters selection decisions. If the required file replicas are distributed at multiple clusters, the cluster that has most files will be selected to minimize the network transfer.

The data aware module can also work with other scheduling policies, such as array job plug-in and grid workflow plug-in to provide more flexibility as a standard CSF4 scheduling plug-in.

Parallel Job Scheduling on the Grid

Most shared grid resources, despite the collective computational power available, are often queued up with many jobs waiting to be executed. There are several ways to tackle this problem .One is through advanced reservation, where different resources are made available at the same time in order for a particular large experiment to run within allocated time period. A number of such experiments are used as demonstrations of the computing power in cross-site simulations on the TeraGrid, as well as federated grids [98, 99]. However, such experiments are still fraught with difficulties in network latency, data accessibility and resource co-allocation. To support cross-domain parallel jobs in different grids, we need to resolve these three issues. First, the input or output files should be accessible in all the candidate clusters/grids where the job executes. Second, the efficient inter-process communications of an application should be guaranteed between hosts in different domains. Last, the resource allocations in these clusters should be synchronized. Here we describe our experience on cross-site computations using commodity technologies. In particular, we use data grid technologies, like Gfarm [13], to provide the global data availability, and MPICH-G2, a grid enabled MPI implementation, to enable a user to run MPI programs across multiple sites [75]. We have also developed the Virtual Job Model (VJM) [100] for CSF4 to realize synchronized resource co-allocation for parallel jobs.

Synchronized Resource Co-allocation benefits from VJM

Since dynamic grid resources are distributed in different administrative domains, and there is not a global controller in grid system, synchronized resource co-allocation to running parallel jobs is a very complicated issue. When user submits a MPICH-G2 parallel job by mpirun (MPICH program starter), mpirun parses the job request first. Based on pre-defined information in a configuration file or MDS, mpirun decides the job execution clusters using round-robin policy and generates a RSL (Resource Specification Language) script. Then globusrun is used to distribute the sub-jobs to the clusters specified in the RSL file.

There are several difficulties for MPICH-G2 applications to succeed without manual intervention. First, since there is no cluster availability check, MPICH-G2 could submit sub-jobs to a cluster which is not even online. Second, when the sub jobs of a parallel job are submitted to local clusters, they are scheduled by local schedulers based on their local policies to compete for resources with local jobs instead of running immediately. If there is not a practical mechanism to support synchronized resource co-allocation, the jobs starting earlier have to wait for the other jobs that are still waiting for resource allocation, which will result in resource wasting. Lastly, when multiple parallel jobs were submitted, the resource allocation deadlock could occur due to the resource competition and the differences of various local scheduling policies.

VJM is a new meta-scheduling model designed for CSF4 to resolve the above problems in synchronized resource co-allocation. Before starting the actual job, VJM dispatches a virtual job to the candidate clusters to acquire the resources for real parallel jobs. VJM is designed to require only the minimal common features supported by all local resource managers, (no resource reservation), and to require no extra components or changes at any local site.

Scheduling Stages of VJM

VJM consists of three stages, resource availability check stage, resource allocation stage, and job startup stage (figure 8). First the meta-scheduler performs resource availability check (pre-check mechanism in CSF4) before sending resource allocation requests to local schedulers. The pre-check first checks if a cluster is online, then it checks the resource requirements of the job and the resource availability of the cluster. Based on the pre-check results, the meta-scheduler decides which clusters are qualified as execution clusters. With the pre-check, the jobs are never sent to the offline clusters, and a job with resource requirements that exceed the total resource capability of all clusters is rejected with a warning.

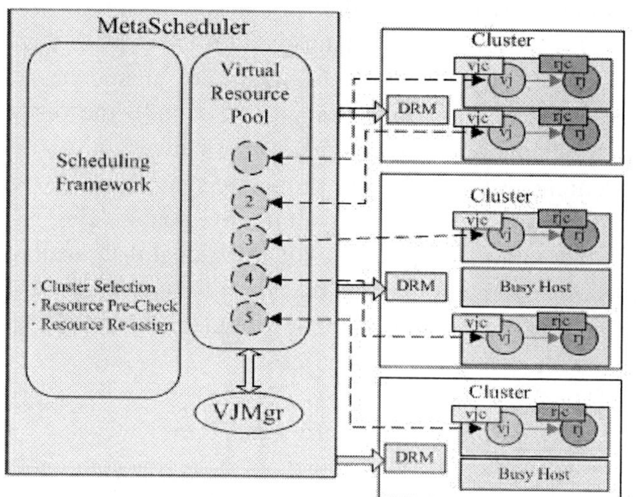

Figure 8. Architecture of Virtual Job Model (VJM).

After the pre-check, the meta-scheduler makes a temporary decision about on which clusters to execute the parallel job and how to distribute the sub-jobs among them according its meta-scheduling policies. Since the meta-scheduler is not the owner of local cluster resources, it cannot allocate resources directly. Hence, a virtual job mechanism is introduced to co-allocate the resource for a parallel job.

We design a virtual job manager (VJmgr) for dispatching virtual jobs, collecting the resource allocation information, managing a virtual resource pool and virtual job credential. VJmgr generate a temporary proxy credential, whose lifetime can be configured and the default value is 12 hours. As the proxy credential expired, all the existent virtual jobs will be cancelled. VJmgr will generate a new proxy credential and resend these virtual jobs automatically.

During the resource allocation stage, the virtual jobs with the virtual job proxy credential are dispatched to the clusters instead of the real jobs. In consideration of the resource accidental breakdown or heavy load, the number of the virtual job is generally more than resources required by the real job to obtain redundant availability. Virtual job's responsibility is to obtain the guaranteed resources for the real job and report the resource allocation status in each cluster to the meta-scheduler.

The virtual job will be scheduled by local scheduler like a normal serial job. By the proxy credential, virtual job can authenticate with VJmgr. As a virtual job gets the resource and starts up, it will send a "READY" notification to VJmgr, register the resource information (such as IP address) to virtual resource pool and wait for instructions from the VJmgr.

During the startup stage, after the VJmgr has received sufficient "READY" notifications from the virtual jobs, within a SSL channel, a "STARTUP" instruction is sent to every virtual job that has registered in resource pool. The "STARTUP" instruction contains the real job description and the proxy credential of real job owner. The subsequent redundant resource will be abandoned or serve other waiting jobs if needed. Since the virtual job can guarantee the resource availability, virtual job model can make sure that all the sub-jobs of a parallel job are synchronized.

The virtual job is not distinct from normal batch job, so VJM neither depends on advance reservation nor need to install extra resource manager on grid sites. Moreover, during the resource allocation stage, VJmgr is able to detect the potential dead lock and the cluster runtime error so that the meta-scheduler can adjust its scheduling decision accordingly. At the same time, VJmgr is able to backfill smaller jobs to the partially allocated resource of a larger parallel job to improve the resource usage of the system.

Deadlock Prevention

When the jobs were forwarded to local resource sites by meta-scheduler, they will not start up immediately but wait to be scheduled in the local queues. Since the resource availability is dynamic and the local scheduling policies are unknown, the waiting time is hard to predict. Hence, the deadlock prevention method based on a lexicographical order induced by host's IP address is quite inefficient in such scene. Motivated by this, we divide the deadlock prevention into two-phase: cluster selection phase and host selection phase.

Firstly, we give out some definitions. Since the master host's IP address, the available hosts, and the number of clusters are generally knowable. Let *Ci* (i=1,...,n) be a available cluster, *Ni*(i=1,...,n) be the available hosts number of the cluster *Ci*, and *Hij*(j=1,...Ni) be a host in the cluster *Ci*, *C1-Cn* are ordered by cluster's master host's IP address and *Hi1-Him* are ordered by host's IP address in the Cluster *Ci*.

In cluster selection phase deadlock prevention, we assume that a cluster represents a resource type and take the cluster's master host's IP address as unique resource type identifier, and the hosts in the same cluster represent multiple resource instances with identical resource type. Then we can avoid the resource allocation deadlock among clusters by using deadlock prevention method based on a global resource type order.

As presented in previous section, the virtual job can obtain host's information (such as IP address) and report it back to meta-scheduler after startup. We can prevent deadlock among hosts via resource order obtain from virtual job if we assume that a host represents a resource by taking the host's IP address as its unique local resource identifier. However, since the local scheduling decisions are unpredictable, we cannot know which hosts will be used before the virtual jobs startup, the method is still inefficient. For example, in cluster *Ci*, to prevent resource deadlock, the resource request cannot continue unless host Hi1 is held, even if *Ci*'s available resource are abundant. Hi1 becomes a bottleneck.

Therefore, we introduce an optimized algorithm to improve resource allocation in the host selection phase deadlock prevention. To describe the algorithm, we further define Ri(i=1,...,n) is the request resource number on cluster Ci. The pseudo code of optimized deadlock prevention algorithm is summarized in figure 9. When a virtual job obtains a resource Hij and notifies back to meta-scheduler, the algorithm is executed.

```
WHILE (Pooled Resource Total < ∑(i=1..N) Ni)
  WAITING For The Virtual Job Registration
  IF (Ni-j) > Ri
      Terminate the virtual job; // To Prevent Deadlock
  ELSE
      Pool The Reource;
      Ri:=Ri-1;
      Ni:=Ni-1;
  ENDIF
ENDWHILE
```

Figure 9. Pseudocode for deadlock resolution in VJM.

The (Ni-j) in the pseudo presents the number of the resource whose order is less than current registering resource. If it is larger than required resource on Cluster Ci, deadlock will not occur, otherwise, the resource should be abandoned to prevent the potential deadlock. Hence, the resource in smaller order will not be bottleneck.

As total resource number and their IP addresses are knowable, by using order-based deadlock prevention method, resource co-allocation will free from deadlock even if multiple meta-schedulers involve in resource competing. Moreover, the two-phase deadlock prevention improves the system performance and resource usage.

Besides the resource dead lock, the run-time cluster unavailability (such as server down or very heavy workload) can also be detected by VJmgr. VJmgr regards a cluster as invalid for a virtual job when it cannot startup within a predefined time. Then the meta-scheduler will re-dispatch the sub-jobs to another cluster.

Backfilling

In a distributed system, backfilling is widely used while scheduling parallel jobs to optimize resource usage. VJmgr is also able to backfill smaller jobs to the partially allocated resource of a larger parallel job to alleviate resource wasting. Such smaller jobs may be effectively jobs that may finish within a fixed amount of wall clock time, and their resource requirements are met with the resources currently allocated. In a single cluster, the local scheduler has full control of its jobs and resources, so it can backfill any job to the resources allocated for another job. In the grid environment, however, access control poses problems to the backfilling process. In VJM, the jobs are started by virtual job, instead of the scheduler, which only have the access permission of the desired job owner. Therefore, only the jobs from the same user can be backfilled safely. Otherwise, the backfilled job may fail to access the system resources at run time. After the investigation of grid applications, we found that most permission problems are caused of file accesses. Thus, we consider that if applications can access their desired files smoothly, the small job backfilling is enabled.

Data grid techniques like Gfarm which uses user proxies instead of user id or user/password to authenticate an access request. This provides more flexibility to backfilling when cooperating with VJM. In the implementation of VJM in CSF4, the meta-scheduler is able to delegate the user proxy of the backfilled job to the virtual job. Then, the virtual job can setup the correct credential context for those jobs to make sure they can access the Gfarm file systems successfully at runtime. After the backfilled jobs finished, virtual job can also clean the credential context to avoid the abuses of authority.

In summary, the VJM resolves the problems in synchronized resource co-allocation mentioned above, helps the deadlock prevention algorithm to adapt to grid computing and improve the performance and resource usage, and ensures proper backfilling of jobs when used with Gfarm.

USER INTERFACES TO METASCHEDULERS

Metascheduling is essential to pervasive computing in how it provides transparent access, ensures workflows and simplifies the users' interaction with the computing infrastructure. Even in the age of petascale or higher computing, the appearance of virtual machines or virtual clusters, with the rise of the virtual organizations, metascheduling is indispensible.

CSF Client Command-line

Since the CSF portlet is for grid-unaware users, we try to make the techniques details transparent to user and the functionalities were limited. For the advanced user, we provided a plenty of command-line tools to access CSF Server. A separate client package is also available for the client machine (such as a Linux server or a Mac laptop) without GT4 installed (figure 10).

Interface to Opal Based Web Services

As more and more clusters as well as and grids of clusters or virtual organizations (VO) spring up, the ability to simplify the scheduling of tasks across VO becomes critical. The difficulty in bridging the gap between the grid services and the target user communities remains significant due to the unfriendly interface of grid software. While the high energy physics community with computational prowess has taken a lead in the development of integrated systems for grid computing and data management, the biomedical research community requires additional effort to lower the barrier of entry by making the grid access transparent and easily accessible.

```
[21]: https://198.202.88.32:8443/wsrf/services/SecurityTestService
[22]: https://198.202.
[23]: https://198.202.
[24]: https://198.202.
[25]: https://198.202.
[26]: https://198.202.
[27]: https://198.202.
[28]: https://198.202
[29]: https://198.202
[30]: https://198.202
[31]: https://198.202
[32]: https://198.202
[33]: https://198.202
[34]: https://198.202
[35]: https://198.202
[36]: https://198.202.88.32:8443/wsrf/services/P
[37]: https://198.202.88.32:8443/wsrf/services/m
[38]: https://198.202.88.32:8443/wsrf/services/D
[39]: https://198.202.88.32:8443/wsrf/services/g
[40]: https://198.202.88.32:8443/wsrf/services/T
[41]: https://198.202.88.32:8443/wsrf/services/T
[42]: https://198.202.88.32:8443/wsrf/services/M
[43]: https://198.202.88.32:8443/wsrf/services/R
[44]: https://198.202.88.32:8443/wsrf/services/W
[45]: https://198.202.88.32:8443/wsrf/services/M
[46]: https://198.202.88.32:8443/wsrf/services/M
[47]: https://198.202.88.32:8443/wsrf/services/I
[48]: https://198.202.88.32:8443/wsrf/services/metascheduler/ResourceManagerLsfService
[49]: https://198.202.88.32:8443/wsrf/services/AuthzCalloutTestService
[50]: https://198.202.88.32:8443/wsrf/services/metascheduler/JobService
[51]: https://198.202.88.32:8443/wsrf/services/DelegationFactoryService
[52]: https://198.202.88.32:8443/wsrf/services/DefaultIndexService
```

```
[zding@rocks-32 ~]$ csf-resource-list
getRmlist() =>
NAME            HOST                  Type   RM PROTOCOL    CPU A   LOAD   QLENGTH
NBCR_Mirume     mirume.nbcr.net       fork   Pre-WS GRAM
NBCR_Puzzzle    puzzle.nbcr.net       fork   Pre-WS GRAM
```

```
[zding@rocks-32 ~]$ csf-job-create -rsl autodock.rsl -A 1-10 -stagein -stageout -nar
Service location: https://198.202.88.32:8443/wsrf/services/metascheduler/JobService
CreateJob Successfully ArrayJob : testautodock, jobcount : 10
[zding@rocks-32 ~]$ csf-jo
Array Job length:10
submit() => Successfully
```

```
[zding@rocks-32 ~]$ csf-job-list
getJoblist() =>
JOBNAME            RSL STYLE     STATUS    EXECUATION HOST
testautodock.10    Pre-WS GRAM   PDone     yuki.nbcr.net
testautodock.6     Pre-WS GRAM   PDone     yuki.nbcr.net
testautodock.4     Pre-WS GRAM   PDone     yuki.nbcr.net
testautodock.1     Pre-WS GRAM   PDone     mirume.nbcr.net
testautodock.7     Pre-WS GRAM   PDone     mirume.nbcr.net
testautodock.8     Pre-WS GRAM   PDone     yuki.nbcr.net
testautodock.9     Pre-WS GRAM   PDone     mirume.nbcr.net
testautodock.2     Pre-WS GRAM   PDone     yuki.nbcr.net
testautodock.5     Pre-WS GRAM   PDone     mirume.nbcr.net
testautodock.3     Pre-WS GRAM   PDone     mirume.nbcr.net
```

Figure 10. Screenshots of CSF4 command line client for remote job submission.

The Opal toolkit models applications as resources by exposing them as web services, however, the basic version can only submit jobs to a local scheduler. For greater scalability, it is necessary to leverage multiple clusters located at different geographic regions. However, our goal is to do this without compromising the ease of use of application deployment via Opal. As described in CSF4 kernel section, CSF4 virtualizes many kinds of resource include applications and is able to match the jobs to the clusters that have the special application. Hence, the only difference is user will not submit resource requirements to CSF4 but Opal web services. See figure 11.

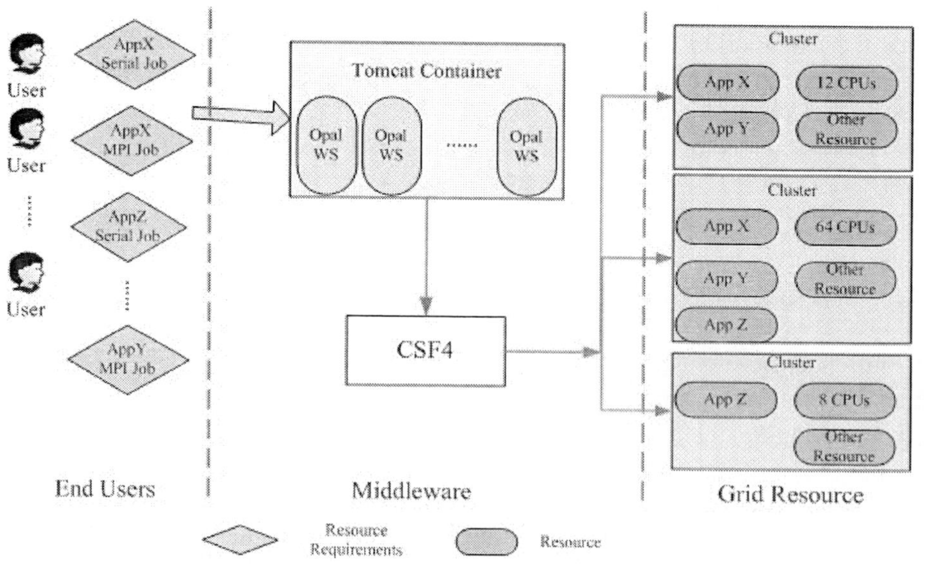

Figure 11. CSF4 Integrate with Opal based web services.

Portal Interface

My WorkSphere

Typically end users in life sciences access computational resources through a web based interface to a popular application such as Blast available at the National Center for Biotechnology Information (NCBI), or through desktop applications that may solve small problems locally. With Moore's law continue to hold true by increasing the number of computing cores per processor, desktops may become more and more powerful. However, the pervasive availability of distributed grid resources and petascale facilities will continue to entice end users as they begin to tackle problems at ever increasing scales. My WorkSphere is an integrative environment which leverages open source components to provide easy access to grid computing resources [101]. In this particular environment, a number of open source software packages are adopted and integrated, such as the GridSphere portal framework and Gridportlets, the GAMA server and portlet, the Opal web service toolkit, the community scheduler framework 4 (CSF4), Gfarm, Globus Toolkit and Commodity Grid Kits [102], and Rocks. The purpose of My WorkSphere is to prototype an environment where users can easily gain access to grid computation resources; run jobs without worrying about what resources they are using, and deploy applications easily for use on the grid. It is one of the TeraGrid Science Gateways for the biomedical community researchers under development by the National Biomedical Computation Resource, in collaboration with academic researchers and developers worldwide.

CSF4 Portlet

The CSF4 portlet v1.0 is developed through the collaboration between Jilin University and University of California, San Diego (UCSD) driven by the needs of My WorkSphere. It is a java based web application for dispatching jobs to a CSF4 metascheduler server through a web browser. The CSF4 portlet not only presents the functionality of CSF4 itself, but also provide a generic interface for users to submit jobs, view job specifications and job history, monitor job status, and get job output from remote sites using GridFTP. Additionally, the CSF4 portlet uses the gridportlets to provide methods for tracking Globus Security Service (GSS) credentials for a given user either from GAMA server or directly from a MyProxy server 28). Besides credential management, the APIs provided by GridSphere and the gridportlets are also used to implement account management services. The CSF4 portlet supports the integration of CSF4 and Gfarm file system (figure 12).

Vision Workflow Environment

Vision is a python based workflow program with support of a visual programming environment. The concept of the Vision's visual programming environment is very similar to the Web 2.0 "vision" of enabling the users to perform mashups of services [103]. An example is in (figure 13).

The ability of Vision or Vision-like visual programming environments allow users to compose workflows based on not just local software modules but remotely accessible services, whether data-centric or compute-centric, allow users to experience rich functionalities all from a user's desktop environment. Coupled with the ability of Opal

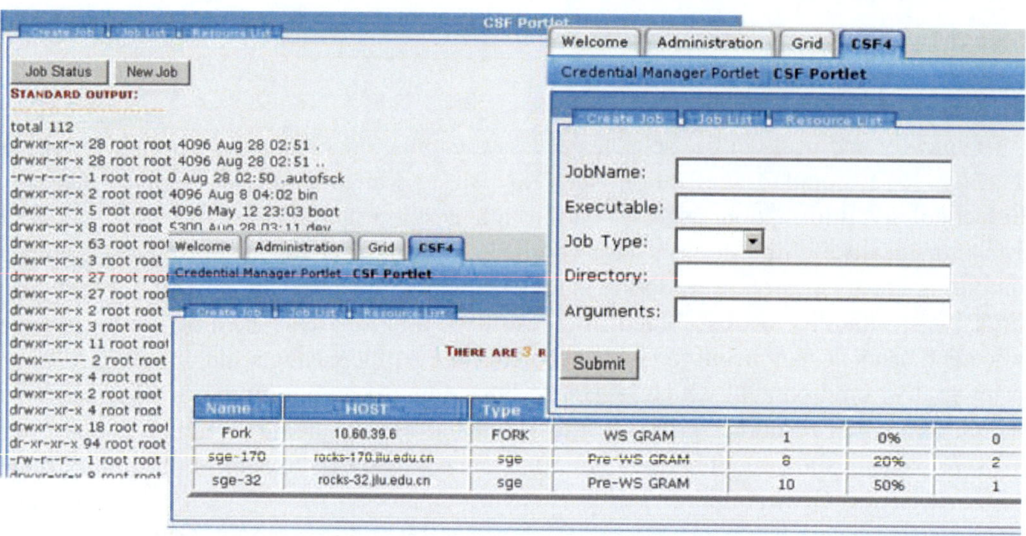

Figure 12. Screenshots of CSF4 portlet interface for GridSphere.

Toolkit's interface generation ability, using server provided configuration files, Vision is capable of providing user with up to date user interfaces and functionalities, relieving the users of the mundane tasks of maintaining and updating the software.

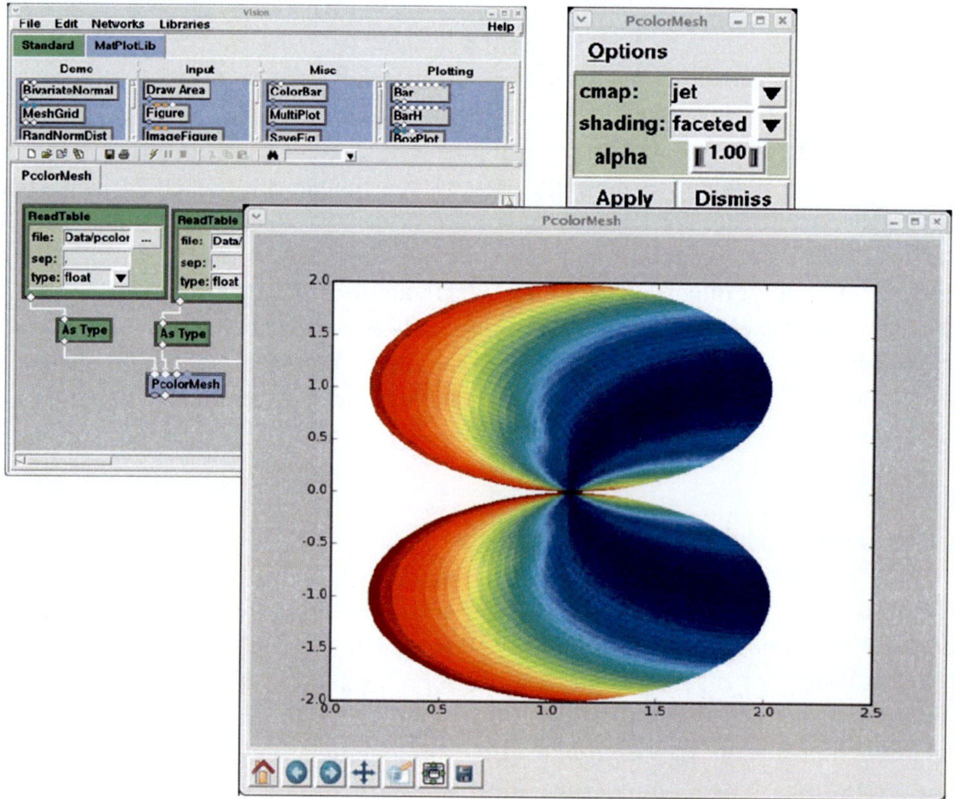

Figure 13. Vision based workflow for mesh rendering and manipulation. Courtesy of Michel Sanner.

CASE STUDIES AND PERFORMANCE EVALUATION

Avian Flu Grid is a VO that encompasses multiple grids and resources, including TeraGrid, PRAGMA Grid, Open Science Grid and NBCR compute resources. Figure 14 provides an illustration of the types of applications and workflows to be fully supported by the Avian Flu Grid project [104, 105], a virtual organization within the PRAGMA grid environment [11].

We have used AutoDock as a test case for the array job plug-in. The AutoDock case is docking 735 different ligands, each one needs around 2.5-3.5 hours running on a 2.4GHz CPU. The available resource involved two NBCR clusters and two clusters of PRAGMA grid test bed. The total jobs duration is 21 hours 30 minutes 44 seconds. We didn't make any advanced reservation, the job just utilizing the idle CPU hours of the distributed resource. We also ran the same case on our private productive cluster. We have to reserve 64 CPUs in advance, the total jobs duration is more than one and an half days.

The scheduling policies we configured include default plug-in and array job plug-in. The resource virtualized feature in default plug-in make the job description easier. The array job plug-in helps the case to achieve better job distribution and balanced workload. See (figure 15).

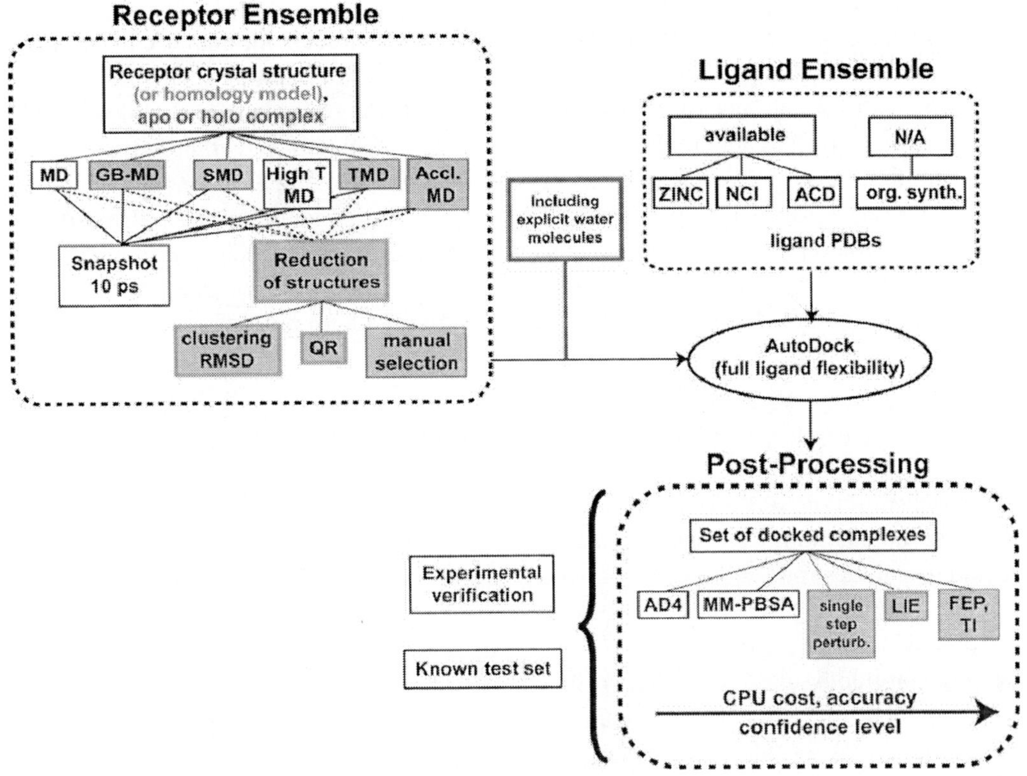

Figure 14. Overview of the applications supported in the Avian Flu Grid.

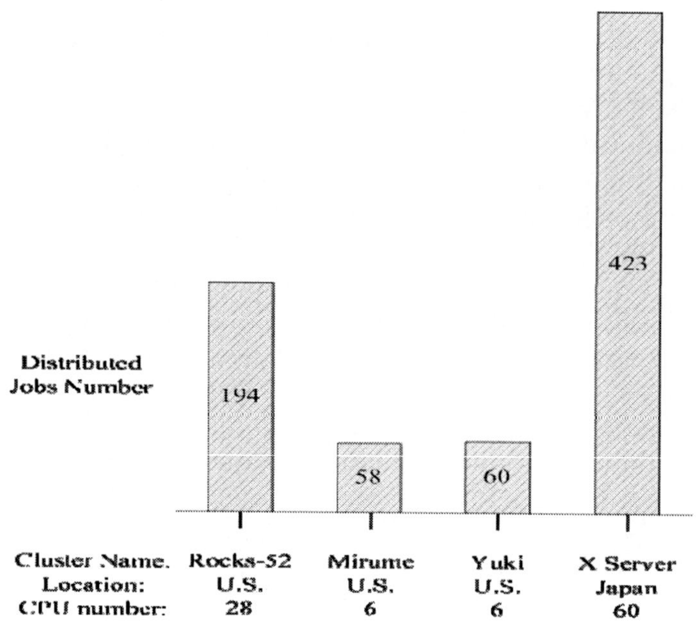

Figure 15. Distribution of AutoDock job in one virtual screening experiment.

We have also used NAMD to run jobs at different resources with higher CPU counts for NAMD [106], a parallel application for molecular dynamics simulations. This represents the class of applications that have a large CPU requirement, long running time, and a large data size for output, as well as input when the jobs have to be restarted. NAMD requires a number of pre-equilibrations and energy minimizations before the final extended free energy minimization run. It raises the possibility of using Gfarm to do automatic restarts on different clusters, automatic replication of large datasets as well as real time visualization of simulation results. The Avian Flu Grid has also provided a real life study case for developing advanced scheduling algorithm for QoS monitoring, fault detection and job migration.

Conclusion and Future Work

Metascheduling is critical in achieving the necessary transparency and scalability for biomedical applications and users. We've taken a bottom up approach by developing scheduling policies based on real scientific application requirements. The adoption of standardization effort at OGF on workflows, resource monitoring and reporting, and identify federation will further enhance the efficiency of metaschedulers or resource brokers, and thereby the productivity of biomedical researchers.

The number of workflow systems is increasing rapidly, along with the support of emerging standards for workflow models to be mapped into different execution environments. As better tools are built over time, our approach allows us to insulate scientific biomedical applications from the changing computational tools and infrastructure. As has been shown in the case of the Opal toolkit, by providing the functionalities that biomedical researchers can understand and use to meet the scalable requirement at hand, and the reporting requirements from their funding agencies, we can deliver virtualized compute resources without negatively impacting the productivity of multiscale research. In addition, by adding database support for logging and accounting, and job provenance, we have also provided scientific researchers with the ability to keep track of the history of their research, workflows.

Underlying these different advances, it is the ability to metaschedule that enables metacomputing, regardless of what buzz words we'd like to use for the next round of technology advances. Recently, software has been developed that adds the ability to schedule different web services [107]. This adds the higher level of support for ensuring the quality of service mashups, by providing QoS feedback based on the overall throughput of different web services. This complements the Opal-CSF4 based approach, which aims to optimize the performance of individual web services and the underlying resource usage.

Last but not the least, the ability to schedule jobs with different priorities and economic models based on the virtual organizations present an interesting new challenge for metascheduling, along with the ability to further optimize data aware scheduling, and the ability to metaschedule data transfers in concert with advanced reservations for computational requirements. The VJM for metascheduling also presents an interesting opportunity for co-scheduling Xen-based virtual machines, as well as MPICH-G2 based applications, which may be co-located within the same petascale resource, instead of disparate resources. The ability to handle large data streams and dynamically adjust the virtual cluster size presents another set of interesting possibilities for more research and development.

ACKNOWLEDGMENT

The authors would like to acknowledge support from the NIH P41 RR 08605 to NBCR; TATRC W81XWH-07-2-0014 to PWA and WWL; the NSF Grant No.INT-0314015 and OCI-0627026 for the PRAGMA project, the China NSF Grant No.60703024 for the CSF4 project and Jilin Department of Science and Technology Grant No.20070122 and 20060532.

REFERENCES

[1] NCSA. The Metacomputer: One from Many. Science for the Millennium 1995 [cited; Available from: http://archive.ncsa.uiuc.edu/Cyberia/MetaComp/MetaHome.html

[2] Foster, I. and C. Kesselman. Globus: a Metacomputing Infrastructure Toolkit. 1997 [cited; Available from: ftp://ftp.globus.org/pub/globus/papers/globus.pdf.

[3] Foster, I. and C. Kesselman, eds. *The Grid: Blueprint for a New Computing Infrastructure.* 1 ed. 1999, Morgan Kaufmann Publishers, Inc.: San Francisco.

[4] NPACI. National Partnership for Advanced Computational Infrastructure: Archives. 2002 [cited 2008; Available from: http://www.npaci.edu/About_NPACI/index.html.

[5] OptIPuter. A Powerful Distributed Cyberinfrastructure to Support Data-Intensive Scientific Research and Collaboration. 2005 [cited; Available from: http://www.optiputer.net.

[6] GLORIAD. Global Ring Network for Advanced Applications Development. 2007 [cited 2008; Available from: http://www.gloriad.org/gloriad/index.html.

[7] TeraGrid. TeraGrid. 2004 [cited; Available from: http://www.teragrid.org.
[8] OSG. Open Science Grid. 2006 [cited; Available from: http://www.opensciencegrid.org/.
[9] Hey, T. and A.E. Trefethen, Cyberinfrastructure for e-Science. Science, 2005. 308(5723): p. 817-21.
[10] EGEE. Enabling Grids for E-sciencE. 2007 [cited; Available from: http://public.eu-egee.org/.
[11] Arzberger, P.W. and P. Papadopoulos (2006) PRAGMA: Example of Grass-Roots Grid Promoting Collaborative e-Science Teams. CTWatch Quarterly Volume.
[12] Grethe, J.S., et al., Biomedical informatics research network: building a national collaboratory to hasten the derivation of new understanding and treatment of disease. Stud Health Technol Inform, 2005. 112: p. 100-9.
[13] Oster, S., et al., caGrid 1.0: An Enterprise Grid Infrastructure for Biomedical Research. J Am Med Inform Assoc, 2007.
[14] Atkins, D., et al., Revolutionizing Science and Engineering Through Cyberinfrastructure: Report of the National Science Foundation Blue-Ribbon Advisory Panel on Cyberinfrasturcuture. 2003, National Science Foundation: Arlington, VA.
[15] NSF. A brief history of NSF and the internet. 2008 [cited; Available from: http://www.nsf.gov/news/special_reports/cyber/internet.jsp.
[16] Moore, G.E., Cramming more components onto integrated circuits, in Electronics. 1965. p. 1-4.
[17] McCalpin, J., C. Moore, and P. Hester, The Role of Multicore Processors in the evolution of general-Purpose computing, in *The Promise and Perils of the Coming Multicore Revloution and Its Impact,* J. Dongarra, Editor. 2007. p. 18-30.
[18] GPGPU. General Purpose Graphics Processing Units. 2008 [cited; Available from: http://www.gpgpu.org/.
[19] OpenFPGA. Open FPGA. 2007 [cited; Available from: http://www.openfpga.org/.
[20] SETI@home. SETI at home. 2007 [cited; Available from: http://setiathome.berkeley.edu/.
[21] Folding@home. Folding@home distributed computing. 2007 [cited; Available from: http://folding.stanford.edu/.
[22] fightAIDS@home. fightAIDS@home. 2007 [cited; Available from: http://fightaidsathome.scripps.edu/.
[23] Stix, G., The Ultimate Optical Networks: The Triumph of the Light, in Sci Am. 2001. p. 80-86.
[24] Foster, I. There's Grid in them thar Clouds. 2008 [cited 2008; Available from: http://ianfoster.typepad.com/blog/2008/01/theres-grid-in.html.
[25] Hand, E., Head in the clouds. Nature, 2007. 449: p. 963.
[26] Ross, J.W. and G. Westerman, Preparing for utility computing: The role of IT architecture and relationship management. IBM Systems Journal, 2004. 43(1): p. 5-19.
[27] Xen. XenSource. 2008 [cited; Available from: http://xen.xensource.com/.
[28] O'Reilly, T. (2005) What is Web 2.0: Design patterns and business models for the next generation of software. O'Reilly.com Volume.
[29] Siebenlist, F., et al., Security for Virtual Organizations: Federating Trust and Policy Domains, in *The Grid: Blueprint for a New Computing Infrastructure*, I. Foster and C. Kesselman, Editors. 2004, Morgan Kaufmann: Amsterdam.

[30] Stevens, R.C., Trends in Cyberinfrastructure for Bioinformatics and Computational Biology, in CTWatch Quarterly, R.C. Stevens, Editor. 2006. p. 6-17.
[31] IGTF. International Grid Trust Federation. 2008 [cited; Available from: http://www.gridpma.org/.
[32] Foster, I., et al. A Distributed Resource Management Architecture that Supports Advance Reservations and Co-Allocation. in Intl. Workshop on Quality of Service. 1999. University College London, UK.
[33] Litzkow, M., M. Livny, and M. Mutka. Condor - a hunter of idle workstations. in Proceedings of the 8th International Conference of Distributed Computing Systems. 1988.
[34] VDT. VDT Documentation. 2007 [cited; Available from: http://vdt.cs.wisc.edu/documentation.html.
[35] gLite. Lightweight Middleware for Grid Computing. 2007 [cited; Available from: http://cern.ch/glite/.
[36] OMII. Open Middleware Infrastructure Institute. 2006 [cited; Available from: http://www.omii.ac.uk/.
[37] NAREGI. National Research Grid Initiative. 2006 [cited; Available from: http://www.naregi.org/index_e.html.
[38] Wei, X., et al. CSF4: A WSRF Compliant Meta-Scheduler. in International Conference 06' on Grid Computing and Applications. 2006. Las Vegas, USA.
[39] Zheng, C., et al. The PRAGMA Testbed: building a multi-application international grid. in CCGrid. 2006. Singapore.
[40] Choi, Y., et al. Glyco-MGrid: A Collaborative Molecular Simulation Grid for e-Glycomics. in 3rd IEEE International Conference on e-Science and Grid Computing. 2007. Banglore, India.
[41] Foster, I., et al. A Security Architecture for Computational Grids. in ACM Conference on Computer and Communications Security. 1998. San Francisco, CA.
[42] Globus. The Globus Alliance. 2004 [cited; Available from: http://www.globus.org.
[43] Novotny, J., S. Tuecke, and V. Welch. An Online Credential Repository for the Grid: MyProxy. in High Performance Distributed Computing (HPDC). 2001.
[44] Bhatia, K., S. Chandra, and K. Mueller. GAMA: Grid Account Management Architecture. in 1st IEEE International Conference on e- Science and Grid Computing. 2006. Melbourne, Australia.
[45] Alfieri, R., et al., From gridmap-file to VOMS: managing authorization in a grid environment. Future Generation of Computer Systems, 2005. 21: p. 549-558.
[46] GUMS. Grid User Management System. 2008 [cited; Available from: https://www.racf.bnl.gov/Facility/GUMS/1.2/index.html.
[47] ROCKS. Rocks Cluster Distribution. 2005 [cited; Available from: http://www.rocksclusters.org.
[48] Bruno, G., et al. Rolls: Modifying a Standard System Installer to Support User-Customizable Cluster Frontend Appliances. in IEEE International Conference on Cluster Computing. 2004. San Diego, USA.
[49] Allcock, W., et al. GridFTP: Protocol Extension to FTP for the Grid, Grid Forum Internet-Draft. 2001.
[50] Allcock, W.E., I. Foster, and R. Madduri. Reliable Data Transport: A Critical Service for the Grid. in Global Grid Forum 11. 2004. Honolulu, HI.

[51] Tatebe, O., et al. Gfarm v2: A Grid file system that supports high-performance distributed and parallel data computing. in 2004 Computing in High Energy and Nuclear Physics. 2004. Interlaken, Switzerland.

[52] SRB. The Storage Resource Broker. 2004 [cited; Available from: http://www.sdsc.edu/srb/.

[53] dCache. dCache.org. 2008 [cited; Available from: http://www.dcache.org/.

[54] IBM. General Parallel File System. 2007 [cited; Available from: http://www-03.ibm.com/systems/clusters/software/gpfs/index.html.

[55] CFS. Lustre Filesystem. 2007 [cited; Available from: http://www.lustre.org.

[56] Ding, Z., et al., Customized Plug-in Modules in Metascheduler CSF4 for Life Sciences Applications. *New Generation Computing*, 2007: p. In Press.

[57] Casanova, H. and F. Berman, Parameter sweeps on the Grid with APST, in *Grid Computing: Making the Global Infrastructure a Reality*, F. Berman, G.C. Fox, and A.J.G. Hey, Editors. 2003, Wiley Publishers, Inc.: West Sussex.

[58] Li, W.W., et al., The Encyclopedia of Life Project: Grid Software and Deployment. New Generation Computing, 2004. In Press.

[59] Abramson, D., J. Giddy, and L. Kotler. High performance parametric modeling with Nimrod/G: Killer application for the global grid? in IPDPS. 2000.

[60] Czajkowski, K., et al. A Resource Management Architecture for Metacomputing Systems. in Proceedings of IPPS/SPDP'98 Workshop on Job Scheduling Strategies for Parallel Processing. 1998.

[61] Frey, J., et al., Condor-G: A Computation Management Agent for Multi-Institutional Grids. Cluster Computing, 2002. 5(3): p. 237-246.

[62] DRMAA. Distributed Resource Management Application API. 2007 [cited; Available from: http://www.drmaa.net/w/.

[63] Huedo, E., R.S. Montero, and I.M. Llorente, A modular meta-scheduling architecture for interfacing with pre-WS and WS grid resourcement services. Future Generation of Computer Systems, 2007. 23: p. 252-261.

[64] Matsuoka, S., The TSUBAME Cluster Experience a year later, and onto petascale TSUBAME 2.0. Lecture Notes In Computer Science, 2007. 4757: p. 8-9.

[65] Nishimura, H., N. Maruyama, and S. Matsuoka. Virtual clusters on the fly -- fast, scalable and flexible installation. in Cluster Computing and the Grid, CCGrid 2007. 2007. Rio de Janeiro, Brazil.

[66] Tapscott, D. and A.D. Williams, Wikinomics: How Mass Collaboration Changes Everything. 2007: Portfolio Hardcover.

[67] MySpace. MySpace.com. 2008 [cited; Available from: http://www.myspace.com.

[68] Facebook. Facebook.com. 2008 [cited; Available from: http://www.facebook.com.

[69] Clementi, L., et al. Providing dynamic virtualized access to grid resources via the web 2.0 paradigm. in Grid Computing Environment 2007 (GCE 07). 2007. Reno, Nevada.

[70] WSRF. Web Services Resource Framework. 2004 [cited; Available from: http://www.oasis-open.org/committees/tc_home.php?wg_abbrev=wsrf.

[71] Krishnan, S., et al., Leveraging the Power of the Grid with Opal: A Guide to Biomedical Application Developers and Users, in *Handbook of Research on Computational Grid Technologies for Life Sciences, Biomedicine and Healthcare*, M. Cannataro, Editor. 2008: Milan. p. Invited.

[72] NBCR. National Biomedical Computation Resource. 2005 [cited; Available from: http://nbcr.net.

[73] Geist, A., et al., PVM:Parallel Virtual Machine—A User's Guide and Tutorial for Network Parallel Computing. 1994, Cambridge, MA: MIT Press.

[74] MPICH. MPICH-A Portable Implementation of MPI. 2006 [cited 2006; Available from: http://www-unix.mcs.anl.gov/mpi/mpich/.

[75] Karonis, N., B. Toonen, and I. Foster, MPICH-G2: A Grid-enabled Implementation of the Message Passing Interface. J. Parallel Distrib Comput, 2003. 63: p. 551-563.

[76] Silver. Moab grid scheduler. 2000 [cited; Available from: http://www.supercluster.org/projects/silver/index.html.

[77] Ding, Z., et al. My WorkSphere: Integrated and Transparent Acess to Gfarm Computational Data Grid through GridSphere Portal with Metascheduler CSF4. in 3rd International Life Sciences Grid Workshop. 2006. Yokohama, Japan.

[78] Wei, X., et al. Implementing data aware scheduling on Gfarm by using LSFTM scheduler Plugin. in International Symposium on Grid Computing and Applications. 2005. Las Vegas, NV.

[79] Zhang, W., et al., Grid Data Streaming, in *Grid Computing Research Progress*, J. Wong, Editor. 2008, Nova Science Publishers.

[80] WPI. Workflow Pattern Initiatives. 1999 [cited 2008; Available from: http://www.workflowpatterns.com.

[81] WFMC. XML Process Definition Language. 2007 [cited 3008; Available from: http://www.wfmc.org/standards/xpdl.htm.

[82] ActiveBPEL. ActiveBPEL Execution Engine. 2008 [cited 2008; Available from: http://www.active-endpoints.com/open-source-active-bpel-intro.htm.

[83] Wang, Y., C. Hu, and J. Huai. A New Grid Workflow Description Language. , . in *IEEE International Conference on Services Computing*. 2005. Los Alamitos: IEEE Computer Society Press.

[84] Zeng, J., et al. CROWN FlowEngine: a GPEL-Based Grid Workflow Engine. in *Third International Conference on High Performance Computing and Communications*. 2007. Houston, USA: Springer.

[85] gridworkflow.org. Grid Workflow. 2007 [cited 2008; Available from: http://www.gridworkflow.org/snips/gridworkflow/space/Grid+Workflow.

[86] Krishnan, S., et al. An end-to-end web services-based infrastructure for biomedical applications. in Grid 2005. 2005. Seattle, Washington.

[87] Lathers, A., et al. Enabling Parallel Scientific Applications with Workflow Tools. in Proceedings of the 6th International Workshop on Challenges of Large Applications in Distributed Environments. 2006.

[88] Cao, J., et al. Implementation of Grid-enabled Medical Simulation Applications Using Workflow Techniques. in 2nd Inter. Workshop on Grid and Cooperative Computing. 2003. Shanghai, China: LNCS.

[89] Baldridge, K., et al. GEMSTONE: Grid Enabled Molecular Science Through Online Networked Environments. in Life Sciences Grid Workshop. 2005. Singapore: World Scientific Press.

[90] Krishnan, S., et al. Opal: Simple Web Services Wrapers for Scientific Applications. in International Conference of Web Services. 2006. Chicago, USA.

[91] Fielding, R.T., Architectural Styles and the Design of Network-based Software Architectures, in Computer Science. 2000, University of California, Irvine: Irvine. p. 162.

[92] Zhang, X. and J. Schopf. Performance Analysis of the Globus Toolkit Monitoring and Discovery Service, MDS2. in International Workshop on Middleware Performance, co-located with the 23rd International Performance Computing and Communications Workshop. 2004.

[93] Czajkowski, K., et al., SNAP: A Protocol for negotiating service level agreements and coordinating resource management in distributed systems. Lecture Notes In Computer Science, 2002. 2537: p. 153-183.

[94] Goodsell, D.S., G.M. Morris, and A.J. Olson, Automated docking of flexible ligands: applications of AutoDock. J Mol Recognit, 1996. 9(1): p. 1-5.

[95] Altschul, S.F., et al., Gapped BLAST and PSI-BLAST: a new generation of protein database search programs. Nucleic Acids Res, 1997. 25(17): p. 3389-402.

[96] Uthayopas, P., J. Maneesilp, and P. Ingongnam. SCMS: An Integrated Cluster Management Tool for Beowulf Cluster System. in PDPTA. 2000. Las Vegas, Nevda, USA: IEEE.

[97] Massie, M., B. Chun, and D. Culler, The Ganglia Distributed Monitoring System: Design, Implementation and Experience. Parallel Computing, 2004. 30: p. 817-840.

[98] Boghosian, B., et al., NEKTAR, SPICE and Vortonics: using federated grids for large scale scientific applications. Cluster Computing, 2007. 10(3): p. 351-364.

[99] Dong, S., N.T. Karonis, and G.E. Karniadakis. Grid Soluations for Biological and Physical Cross-Site Simulations on the TeraGrid. in 20th International Parallel and Distributed Processing Symposium (IPDPS 2006). 2007. Rhodes Island, Greece.

[100] Ding, Z., X. Wei, and W.W. Li. VJM-A Deadlock Free Resource CO-allocation Model for Cross Domain Parallel Jobs in HPC Asia. 2007. Seoul, Korea.

[101] Ding, Z., et al. My WorkSphere: Integrative Work Environment for Grid-unaware Biomedical Researchers and Applications. in Supercomputing Conference 2006, SC06, 2nd Grid Computing Environment Workshop. 2006. Tampa, Florida.

[102] Von Laszewski, G., et al., Features of the Java Commodity Grid Kit. Concurrency and Computation: Practice and Experience, 2002. 14(13-15): p. 1045-1055.

[103] Sanner, M.F., A component-based software environment for visualizing large macromolecular assemblies. Structure (Camb), 2005. 13(3): p. 447-62.

[104] Amaro, R.E., et al. Avian Flu Grid: International Collaborative Environment for Team Science on Avian Influenza. in PRAGMA 13 Workshop. 2007. Urbana-Champaign, Illinois.

[105] Amaro, R.E., Baron, R., McCammon, J. A., An improved relaxed complex scheme for incorporating receptor flexibility in rational drug design. J Comput Aided Mol Des, 2008. in press.

[106] Phillips, J.C., et al., Scalable molecular dynamics with NAMD. J Comput Chem, 2005. 26(16): p. 1781-802.

[107] ichikawa, K., S. Date, and S. Shimojo. A Framework for Meta-Scheduling WSRF-based Services. in Grid Asia. 2007. Singapore.

In: Cyberinfrastructure Technologies and Applications
Editor: Junwei Cao
ISBN: 978-1-60692-063-3

Chapter 14

THE BRIDHING DOMAIN MULTISCALE METHOD AND ITS HIGH PERFORMANCE COMPUTING IMPLEMENTATION

Shaoping Xiao[1], Jun Ni[2] and Shaowen Wang[3]

[1] Department of Mechanical and Industrial Engineering,
Center for Computer-Aided Design, The University of Iowa, Iowa City, IA 52242
[2] Department of Mechanical and Industrial Engineering,
Department of Radiology; The University of Iowa, Iowa City, IA 52242
[3] CyberInfrastructure and Geospatial Information Laboratory (CIGI)
National Center for Supercomputing Application (NCSA)
University of Illinois at Urbana-Champaign, Urbana, IL 61801

ABSTRACT

This chapter presents a study on the feasibility of applying high performance computing (HPC) to the Bridging Domain Multiscale (BDM) method, so that featured scalable multiscale computations can be achieved. Wave propagation in a molecule chain through an entire computational domain is employed as an example to demonstrate its applicability and computing performance when multiscale-based simulations are conducted in a large-scale parallel computing environment. In addition, the conceptual idea and computing framework using Grid computing technologies is proposed to enhance future multiscale computations in nanotechnology.

INTRODUCTION

Nanotechnology is used to create innovative materials with new functional or multifunctional structures and novel devices on a nanometer (10^{-9}m) scale. Nanotechnology

[1] E-mail address: shaoping-xiao@uiowa.edu
[2] E-mail address: jun-ni@uiowa.edu
[3] E-mail address: shaowen@uiuc.edu

has emerged in multidisciplinary research fields. For instance, numerical methods in computational science play a crucial role in the research fields of nano- mechanics and materials science, especially in simulating and understanding the principal design or fabrication of novel nanoscale materials such as nanocomposites. Among these methods, molecular dynamics (MD) is one of the effective and efficient numerical methods that have been widely used in elucidating complex physical phenomena [1-5] on a nanoscale. Although MD has many advantages, it exhibits limitations with respect to both length and time scales. For example, a material with a cubic volume of 1 m^3 contains trillions of atoms, and a typical time-step in MD simulation is roughly one femtosecond (~10^{-15}s). Consequently, these characteristics limit the use of MD for simulation of many physical phenomena, such as material failure problems. At present, a complete MD modeling is unrealistic, especially for completely simulating a material system with heterogeneity, even with powerful high-end computers. Therefore, there exists an urgent need to develop a new and applicable methodology that can be used to efficiently simulate large nanosystems.

Recently, the development of efficient multiscale methods that are capable of addressing large length and time scales has been addressed in computational nanotechnology for the design of novel nanoscale materials and devices [6]. Multiscale methods can be divided into two classes: hierarchical [7-10] and concurrent multiscale methods [11-16]. In hierarchical methods, the continuum approximation is based on the properties of a subscale model, such as an MD model. The intrinsic properties of the material are determined at the atomic level and embedded in a continuum model using a homogenization procedure. Classical hierarchical multiscale methods include quasicontinuum methods [7] and discontinuous Galerkin (DG) methods within the framework of the Heterogeneous Multiscale Method (HMM) [8]. Most hierarchical methods assume that nanostructures are perfect molecular structures subject to zero temperature. Xiao and Yang [9, 10] proposed nanoscale meshfree particle methods with a temperature-related homogenization for nanoscale continuum modeling and simulation.

Concurrent multiscale methods employ an appropriate model in different subdomains to treat each length scale simultaneously. In a pioneering work, Abraham et al. [11, 12] developed a methodology called MAAD (Macro-Atomistic-Ab initio-Dynamics) in which a tight-binding quantum mechanical calculation is coupled with MD and, in turn, coupled with a finite element continuum model. Choly et al. [13] presented formalism for coupling a density-functional-theory-based quantum simulation within a classical simulation for the treatment of simple metallic systems. Recently, several concurrent multiscale techniques that couple continuum and molecular models in particular have been developed. Wagner and Liu [14] developed a multiscale method in which the molecular displacements are decomposed into fine scale (molecular) and coarse scale (continuum).

Belytschko and Xiao [15, 16] developed a bridging domain method for coupling continuum models with molecular models. In this method, the continuum and molecular domains are overlapped in a bridging subdomain, where the Hamiltonian is taken to be a linear combination of the continuum and molecular Hamiltonians. The compatibility in the bridging domain can be enforced by Lagrange multipliers or by the augmented Lagrangian method. An explicit algorithm for dynamic solutions was developed [16]. Results showed that this multiscale method could avoid spurious wave reflections at the molecular/continuum interface without any additional filtering procedures, even for problems with significant nonlinearities. The method was also shown to naturally handle the coupling of the continuum energy equation with the molecular subdomain. A multiple-time-step algorithm was also developed within this framework [16].

A concurrent multiscale method can be designed to span a range of physical domains of different length scales, from atomic to microscopic/mesoscopic to macroscopic scales. Unfortunately, most multiscale methods still require intensive computation for large nanoscale simulations, although such limitations are much smaller than those associated with full MD simulations. On the other hand, due to the highly intensive computation requirement, a single-processor computer is barely sufficient to handle simulations that typically involve trillions of atoms and up to several seconds. The limitation of computing capacity naturally motivates an alternative approach—to conduct concurrent multiscale computations based on high performance computing (HPC) or Grid-based distributed computing. To date, a few HPC-based multiscale simulations have been reported. For example, Yanovsky [17] utilized parallel computing technologies to study polymer composite properties. Ma et al. [18] implemented their continuum/atomic coupling algorithm in the Structural Adaptive Mesh Refinement Application Infrastructure (SAMRAI) using parallel processing to study two-dimensional crack propagation. However, most existing works do not focus on HPC algorithm development and efficiency.

In the U. S., the Cyberinfrastructure recently promoted by the National Science Foundation (NSF) provides a gateway to future science and engineering discovery. It enables large-scale resource sharing, especially distributed HPC systems with unprecedented computational capacity. Such capacity reaches several hundred teraflops, and upgrading to petaflops is just a few years away on the NSF TeraGrid—a Grid computing environment and key element of U.S. cyberinfrastructure. Grid computing technologies enable users to assemble large-scale distributed computational resources to create a secured virtual supercomputer that can be used to accomplish a specific purpose [19]. This assemblage of distributed resources is dynamically orchestrated using a set of protocols as well as specialized software referred to as Grid middleware. This coordinated sharing of resources takes place within formal or informal consortia of individuals and/or institutions often referred to as Virtual Organizations (VO) [20]. Grid computing technologies give scientists the ability to handle large-scale computations, especially in nanotechnology investigations.

In this chapter, we will develop a scalable, parallel bridge domain coupling algorithm for computations in nanotechnology applications. The bridging domain coupling method is described in Section 2 after the introduction. This coupling method is extended to a scalable parallel multiscale method in Section 3. Associated domain decomposition and communication algorithms are explained, and a one-dimensional example is studied for investigating computing performance. Section 4 offers a description of the feasibility of multiscale modeling and simulation using Grid computing, followed by a conclusion.

Bridging Doamin Coupling Method

Coupling Strategy

The Bridging Domain Coupling Method (BDCM) was proposed by Xiao and Belytschko [15, 16]. Here, we first provide a brief summary of this methodology, detailed in a one-dimensional molecule chain that includes a molecular dynamics domain, a finite element (continuum) domain, and a bridging domain where the molecular and continuum domains overlap, as shown in figure 1.

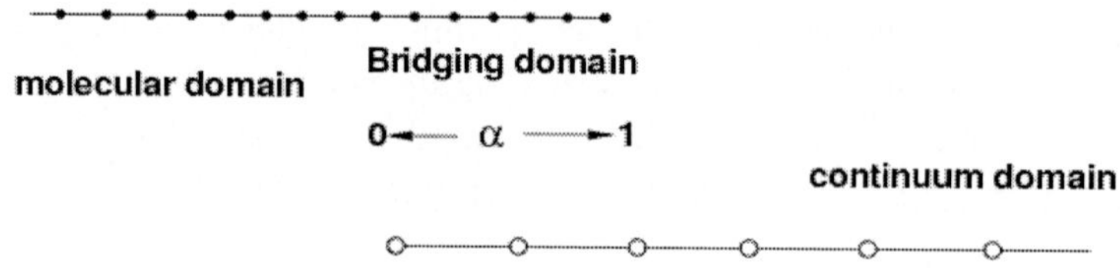

Figure 1. Bridging domain coupling model of a molecule chain.

Generally, a BDCM method serves as a linkage between molecular and continuum models through an overlapped domain. The continuum domain, Ω^C, modeled by a macroscopic continuum model, overlaps the molecular domain, Ω^M, modeled by a molecular model, through an interaction region, called a bridging domain, Ω^{int}. The superscripts "M" and "C" refer to the molecular and continuum domains, while the superscript "int" refers to the interaction domain or bridging domain where a bridging domain coupling technique is employed. In the BDCM method, the total energy is taken to be a linear combination of the molecular and continuum energies. Therefore, in the bridging domain, the molecular and continuum models co-exist. A linear switch scaling parameter α is introduced hereby in the overlapped bridging domain. For example, the parameter α can be proposed as

$$\alpha = \begin{cases} 0 & in \quad \Omega_0^M - \Omega_0^{int} \\ [0,1] & in \quad \Omega_0^{int} \\ 1 & in \quad \Omega_0^C - \Omega_0^{int} \end{cases} . \tag{1}$$

The Hamiltonian energy for the complete domain can be considered as a linear combination of the molecular and continuum counterparts. It can be expressed as

$$\begin{aligned} H = (1-\alpha)H^M + \alpha H^C &= \sum_I [1-\alpha(X_I)]\frac{m_I \dot{d}_I^2}{2} + (1-\alpha)W^M + \int_{\Omega_0^C} \frac{1}{2}\alpha\rho_0 v_I^2 d\Omega_0^C + \alpha W^C \\ &= \frac{1}{2}\sum_I [1-\alpha(X_I)]m_I \dot{d}_I^2 + (1-\alpha)W^M + \frac{1}{2}\int_{\Omega_0^C} \alpha\rho_0 v_I^2 d\Omega_0^C + \alpha W^C \end{aligned} \tag{2}$$

where m_I and d_I are the mass and displacement of the atom, ρ_0 is the initial density in the finite element domain, v_I is the velocity of node I, and W^C is the total strain energy in the continuum domain. In the molecular domain, $W^M(x)$ refers to the potential function that is the summation of all energies due to any force field (such as the pair-wise interaction of the atoms, three-body potentials, or others). Assume the potential is due only to a constant external force, $\bar{f}_I^{ext}$, such as electrostatic forces, and a pair-wise interatomic potential is denoted by $w_{IJ} = w_M(x_I, x_J)$, where x is the coordinate. The total potential can be expressed as

$$W^M = -W_M^{ext} + W_M^{int} = -\sum_I \bar{f}_I^{ext} d_I + \sum_{I,J>I} w_M(x_I, x_J). \tag{3}$$

The strain energy of the continuum model is defined by

$$W^C = \int_{\Omega_0} w_C d\Omega \tag{4}$$

where w_C is the potential energy per unit volume within the continuum domain. It depends on the elongations and angle changes of the atomic bonds that underlie the continuum model. The potential energy density can be calculated based on the homogenization techniques[7]. If temperature effects are considered, the free energy density, instead of the potential energy density, is employed in Eq. (4), and calculation can be performed based on the temperature-related Cauchy-Born rule [9, 10].

In the bridging domain, a Lagrange multiplier can be used to conjunct the molecular and the continuum domains, with the following constraints: $u(X_I,t) - d_I(t) = 0$, where $u(X_I,t) = \sum_J N_J(X_I)u_J$ is the finite element interpolated displacement of atom I. The total Hamiltonian energy of the system can then be expressed as

$$H = (1-\alpha)H^M + \alpha H^C + \sum_{I\in\Omega^{\text{int}}} \lambda(X_I)(u^C(X_I,t) - u_I^M(t)). \tag{5}$$

Based on Hamiltonian mechanics, the following discrete equations can be derived:

$$\begin{aligned} \overline{M}_I \ddot{u}_I &= f_I^{extC} - f_I^{intC} - f_I^{LC} \quad in \quad \Omega_0^C \\ \overline{m}_I \ddot{d}_I &= f_I^{ext} - f_I^{int} - f_I^{L} \quad in \quad \Omega_0^M \end{aligned} \tag{6}$$

where

$$\overline{M}_I = \alpha(X_I)M_I, \ \overline{m}_I = (1-\alpha(X_I))m_I$$

and M_I is the nodal mass associated with node I. The external nodal forces, including the scaling factor, are defined as

$$\begin{aligned} f_I^{extC} &= \int_{\Omega_0^C} \alpha(X)N_I \rho_0 b d\Omega_0^C + \int_{\Gamma_0^t} \alpha(X)N_I \bar{t} d\Gamma_0^t \\ f_I^{ext} &= (1-\alpha(X_I))\bar{f}_I^{ext} \end{aligned} \tag{7}$$

where b is the body force per unit mass and $\bar{t}$ is the traction applied on the boundary Γ_0 in the continuum domain. Similarly, the internal forces are

$$f_I^{intC} = \int_{\Omega_0^C} \alpha(X)\rho_0 \frac{\partial w_C(F)}{\partial u_I} d\Omega_0^C \tag{8}$$

$$f_I^{int} = (1-\alpha(X_I)) \sum_{I,J>I} \frac{\partial w_M(x_I, x_J)}{\partial d_I}. \tag{9}$$

The forces f_I^{LC} and f_I^{L} are due to the constraints enforced by the Lagrange multipliers, and they are

$$\begin{aligned} \mathrm{f}_\mathrm{I}^\mathrm{LC} &= \sum_\mathrm{J} \lambda_\mathrm{J} \frac{\partial \mathrm{g}_\mathrm{J}}{\partial \mathrm{u}_\mathrm{I}} = \sum_\mathrm{J} \lambda_\mathrm{J} \mathrm{N}_\mathrm{J}(\mathrm{X}_\mathrm{I}) \\ \mathrm{f}_\mathrm{I}^\mathrm{L} &= \sum_\mathrm{J} \lambda_\mathrm{J} \frac{\partial \mathrm{g}_\mathrm{J}}{\partial \mathrm{d}_\mathrm{I}} = \sum_\mathrm{J} (-\lambda_\mathrm{J} \delta_\mathrm{IJ}). \end{aligned} \tag{10}$$

An explicit algorithm to solve the above discrete equations is described below:
Initialize the domains, displacements, velocities, and accelerations
Calculate the trail displacements with constraints neglected:

$$\begin{aligned} u_{I(n+1)}^* &= u_{I(n)} + \dot{u}_{I(n)}\Delta t + \frac{1}{2}\ddot{u}_{I(n)}\Delta t^2 \quad in \ \ \Omega_0^C \\ d_{I(n+1)}^* &= d_{I(n)} + \dot{d}_{I(n)}\Delta t + \frac{1}{2}\ddot{d}_{I(n)}\Delta t^2 \quad in \ \ \Omega_0^M \end{aligned} \tag{11}$$

(The subscripts in parentheses are the time step numbers). In the equations above, the accelerations are obtained from Eq. (6) without considering the forces due to the constraints; therefore,

$$\begin{aligned} \ddot{u}_{I(n+1)}^* &= \frac{1}{M_I}\left[f_{I(n+1)}^{extC} - f_{I(n+1)}^{\mathrm{int}C}\right] \quad in \ \ \Omega_0^C \\ \ddot{u}_{I(n+1)}^* &= \frac{1}{m_I}\left[f_{I(n+1)}^{ext} - f_{I(n+1)}^{\mathrm{int}}\right] \quad in \ \ \Omega_0^M \end{aligned} \tag{12}$$

Calculate the trial velocities:

$$\begin{aligned} \dot{u}_{I(n+1)}^* &= \dot{u}_{I(n)} + \frac{1}{2}\left[\ddot{u}_{I(n)} + \ddot{u}_{I(n+1)}\right]\Delta t \quad in \ \ \Omega_0^C \\ \dot{d}_{I(n+1)}^* &= \dot{d}_{I(n)} + \frac{1}{2}\left[\ddot{d}_{I(n)} + \ddot{d}_{I(n+1)}\right]\Delta t \quad in \ \ \Omega_0^M \end{aligned} \tag{13}$$

Compute unknown Lagrange multipliers:

$$\sum_L A_{IL}\lambda_L = g_I^* = \sum_J N_J(X_I)\dot{u}_J^* - \dot{d}_I^* \tag{14}$$

where

$$A_{IL} = \Delta t \overline{M}_I^{-1} \sum_J N_J(X_I) N_L(X_J) - \Delta t \overline{m}_I^{-1} \delta_{LI}$$

Update the velocities:

$$\dot{u}_{I(n+1)} = \dot{u}_{I(n+1)}^* - \overline{M}_I^{-1} \Delta t \sum_J N_J(X_I)\lambda_J \tag{15}$$

$$\dot{d}_{I(n+1)} = \dot{d}_{I(n+1)}^* - \overline{m}_I^{-1} \Delta t \sum_J (-\delta_{IJ})\lambda_J \tag{16}$$

Repeat Step 2 through Step 5 until the end of simulation.

Example

In the first example, we simulate wave propagation in a one-dimensional molecule chain, which contains 431 atoms. The Lennard-Jones (LJ) 6-12 potential is employed to describe the inter-atomic interaction between two neighboring atoms. This potential is expressed as

$$W^M(r_{ij}) = 4\varepsilon\left[\left(\frac{\sigma}{r_{ij}}\right)^{12} - \left(\frac{\sigma}{r_{ij}}\right)^6\right] \tag{17}$$

where the parameters are $\varepsilon = 0.2J$ and $\sigma = 0.11nm$. The equilibrium bond length is given as $r_0 = 0.139nm$.

In the bridging domain coupling model of the molecule chain similar to that in figure 1, the molecular domain contains 211 atoms, and there are 40 elements of equal length in the continuum domain. Each element contains approximately 5 atoms. The bridging domain includes 6 finite elements. In the continuum model, all the bonds in a single element are assumed to be deformed uniformly. Consequently, the length of deformed bonds in this single element is $r = Fr_0$, where F is the deformation gradient in this element. In computing the nodal internal forces via Eq. (8), the strain energy density in an element is the potential energy density:

$$w_C(F) = \frac{4\varepsilon}{r_0}\left[\left(\frac{\sigma}{Fr_0}\right)^{12} - \left(\frac{\sigma}{Fr_0}\right)^6\right]. \tag{18}$$

During the simulation, the time step is 0.002 ps. There are initial displacements on the atoms in the left portion of the molecular domain. The initial displacements contain a combination of high-frequency and low-frequency modes. Once the simulation starts, there is a wave propagating from the molecular domain to the continuum domain. Since the length of the high-frequency wave is larger than the element length in the continuum domain, a nonphysical wave reflection phenomenon may occur in most multiscale simulations[16]. However, the bridging domain coupling technique can automatically eliminate such a phenomenon, as shown in figure 2.

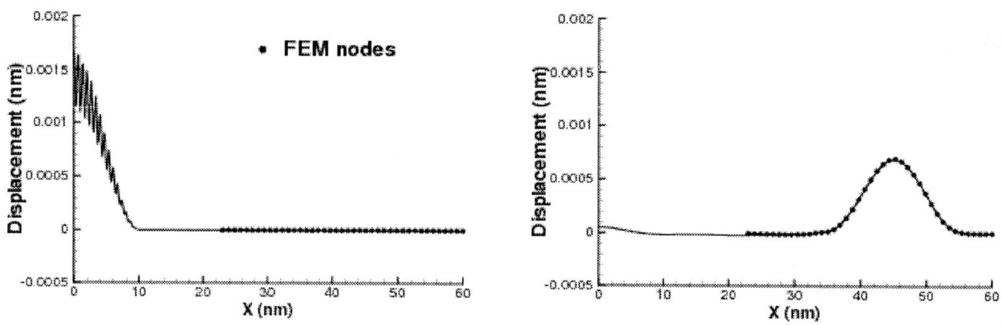

Figure 2. Wave propagation in 1D molecule chain: (a) Initial wave; (b) Wave propagation.

In this section, we employ the bridging domain coupling method to study material properties, especially Young's moduli, of carbon nanotube (CNT)-reinforced aluminum (Al) composites. It should be noted that we only consider pristine nanotubes since a defected tube plays a different role when embedding in composites; see details provided by Xiao and Hou [33]. We first consider single-walled nanotubes (SWNTs) and assume that long SWNTs are aligned unidirectionally and distributed homogeneously in the Al matrix. Therefore, we use a unit cell model, shown in figure 3, with periodic boundary conditions to investigate Young's moduli of nanocomposites.

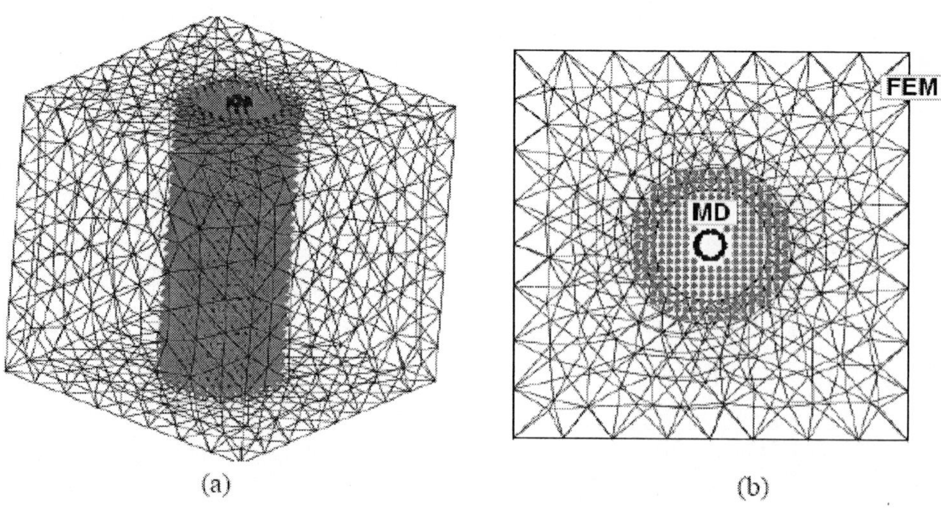

Figure 3. Bridging domain coupling model of SWNT/Al nanocompositesJ: (a) 3-D view; (b) top view.

The cell model in figure 3 contains an SWNT. In the coupling model, the SWNT is embedded in the center of the unit cell. The bridging domain is a circular band with the outer radius of r_o and the inner radius of r_i between the molecular and continuum domains. Therefore, the scaling parameter α is defined as a linear function of $(r-r_i)/(r_o-r_i)$, where r is the distance of the projection to the tube axis, as shown in figure 4. Except in the bridging domain, α is 0 in the molecular domain and 1 in the continuum domain. Obviously, the α is an axially-symmetric function.

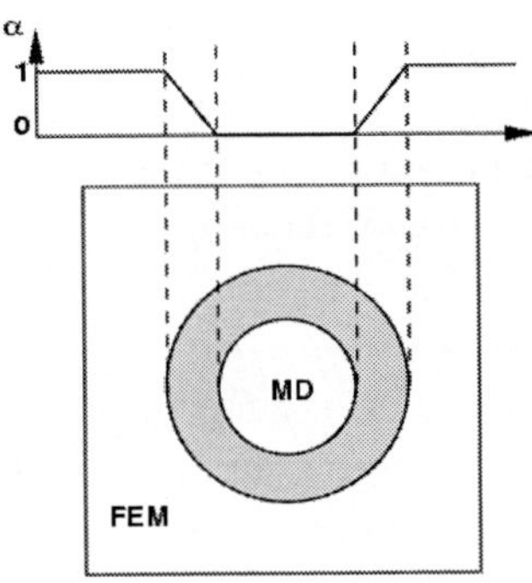

Figure 4. The scaling parameter in the multiscale model of nanocomposites.

We employ the finite element method with tetrahedral elements in the continuum model. The material properties at the room temperature of 300K are: Young's modulus of 78.02 GPa and Poisson's ratio of 0.325. In the molecular model, molecular dynamics with the Hoover thermostat is used. We use the modified Morse potential function [3] to describe interatomic interaction between bonded carbon atoms in SWNT. This potential can be written as:

$$
\begin{aligned}
E &= E_{stretch} + E_{angle} \\
E_{stretch} &= D_e\left\{\left[1-e^{-\beta(r-r_0)}\right]^2-1\right\} \\
E_{angle} &= \frac{1}{2}k_\theta(\theta-\theta_0)^2\left[1+k_s(\theta-\theta_0)^4\right]
\end{aligned}
\tag{19}
$$

where $E_{stretch}$ is the bond energy due to bond stretching or compressing, E_{angle} is the bond energy due to bond angle-bending, r is the current bond length, and θ is the angle of two adjacent bonds representing a standard deformation measure in molecular mechanics. The parameters are:

$$
\begin{aligned}
& r_0 = 1.42\times10^{-10}\,m,\ D_e = 6.03105\times10^{-19}\,Nm, \\
& \beta = 2.625\times10^{10}\,m^{-1},\ \theta_0 = 2.094 rad \\
& k_\theta = 1.13\times10^{-18}\,Nm/rad^2,\ k_s = 0.754 rad^{-4}
\end{aligned}
\tag{20}
$$

Since only weak CNT/Al interfaces were observed in the experimentation [34], the Lennard-Jones potential as follows is used to describe nonbonded interaction between the embedded carbon nanotube and the aluminum matrix.

$$E_{LJ} = 4\varepsilon\left[\left(\frac{\sigma}{r}\right)^{12} - \left(\frac{\sigma}{r}\right)^{6}\right] \tag{21}$$

The parameters for the interactions between carbon atoms and Al atoms are obtained from the Lorentz-Berelot combining rules: $\varepsilon = 0.038eV$ and $\sigma = 0.296nm$.

To calculate Young's moduli of nanocomposites, we investigate the stress-strain relations of nanocomposites first. The prescribed displacement is applied on the top of the unit cell, while the bottom of the unit cell is fixed. Since the long tube is considered and the periodic boundary condition is employed, the prescribed displacement is applied on both the nanotube and the Al matrix. We evaluate stress at every 0.1% strain during the simulation via summing the internal forces on the nodes/atoms on the top of the unit cell and then dividing it by the cross section area of the cell. The Young's modulus is then calculated as the slope of the stress-strain relation curve. In this section, we mainly investigate Young's moduli of nanocomposites with various volume fractions of embedded CNTs at the room temperature of 300K.

We consider two SWNTs, respectively, as inclusions in the Al matrix material. One is the (5,5) SWNT with a diameter of 0.68 nm, while the other is the (21,21) SWNT with a diameter of 2.85 nm. We vary the size of the unit cell so that various volume fractions of embedded tubes can be achieved. Table 1 illustrates the calculated Young's moduli of nanocomposites with various tube volume fractions at the room temperature. Apparently, a larger volume fraction of embedded SWNTs results in a larger Young's modulus of nanocomposites. It also can be seen that the Young's moduli of nanocomposites are similar no matter which SWNTs are embedded for a given tube volume fraction. In other words, nanotube size has no effect on the Young's modulus of nanocomposites. This is because size effects on nanotubes' stiffness are significant only for nanotubes with small diameters [35]. Therefore, there is no effect of tube size on Young's moduli of nanotube-reinforced composites.

Table 1. Young's moduli of SWNT-based Al nanocomposites with various tube volume fractions at the room temperature

SWNT volume fraction (%)	1.3	1.8	4.1	7.3	11.4	16.4
Young's moduli (GPa) of nanocomposites with (5,5) SWNTs	85.0	86.8	89.4	96.6	101.4	118.0
Young's moduli (GPa) of nanocomposites with (21,21) SWNTs	84.2	86.4	90.1	97.7	103.2	115.7

As utilized in the experiments [36], most fabricated CNTs are multi-walled nanotubes (MWNTs) or SWNT bundles. An MWNT contains a number of co-axial SWNTs, and the interlayer distance between SWNTs is 0.34 nm. In SWNT bundles, SWNTs are packed in two-dimensional triangular lattices. The nearest distance between two neighboring SWNTs is also 0.34 nm. The Lennard-Jones potential [37], similar to Eq. (21), is employed to describe nonbonded interaction between two carbon atoms that are located at different SWNTs in MWNTs or SWNT bundles. The parameters are: $\varepsilon = 0.0025eV$ and $\sigma = 0.34nm$.

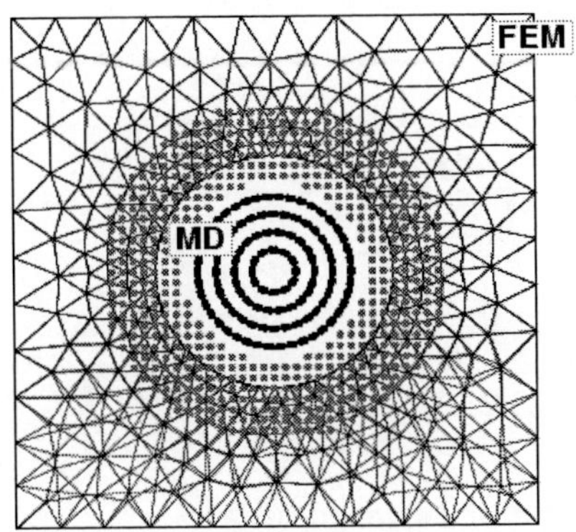

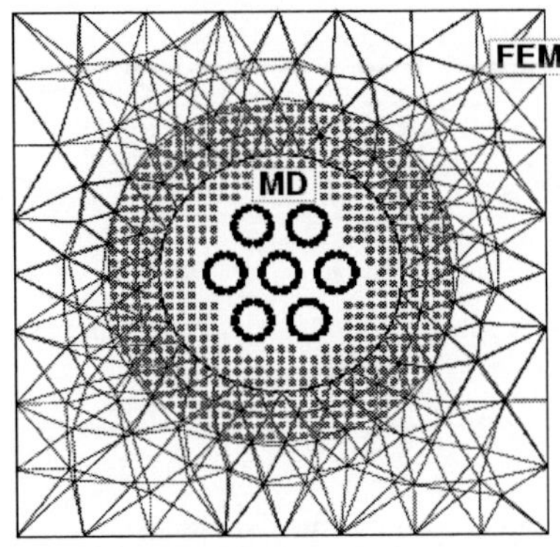

Figure 5. Multiscale models of nanocomposites (top views): (a) nanocomposite with an MWNT; (b) nanocomposite with an SWNT bundle.

In our multiscale simulations, we select a (21,21) MWNT, which contains (21,21), (16,16), (11,11), and (6,6) SWNTs, and a (5,5) SWNT bundle, which contains seven (5,5) SWNTs. The multiscale models are illustrated in figures 5(a) and 5(b), respectively. The scaling parameters in the bridging domains can be constructed similarly to the one in figure 4. It should be noted that all the (21,21) SWNT, (21,21) MWNT, and (5,5) SWNT bundles have the similar diameters. Figure 6 compares the roles of the above three nanotube inclusions in reinforcement of nanocomposites. It is evident that the MWNT results in the most significant reinforcement, followed by the SWNT bundle. The SWNT results in the least significant reinforcement compared to the other two inclusions. As an instance of 10% CNT volume fraction, the nanocomposite containing (21,21) SWNTs has the Young's modulus of 100 GPa. However, the nanocomposite containing (5,5) SWNT bundles can have the Young's modulus of 110 GPa, and the one containing (21,21) MWNTs can have the Young's modulus of as high as 135 GPa.

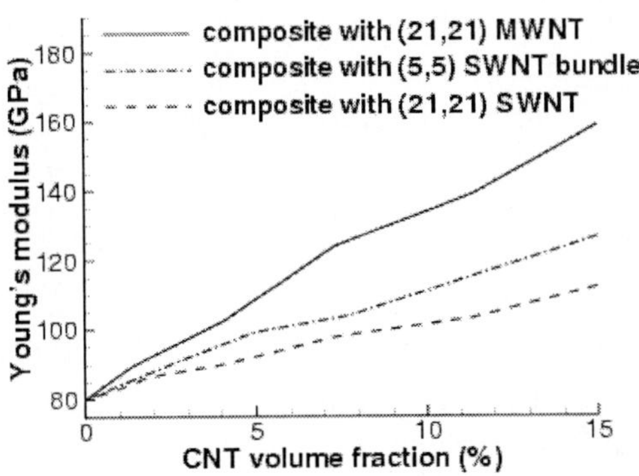

Figure 6. Young's moduli of nanocomposites at various CNT volume fractions.

Bridging Domain Multiscale Method with Parallel Computing

Bridging Domain Multiscale Method

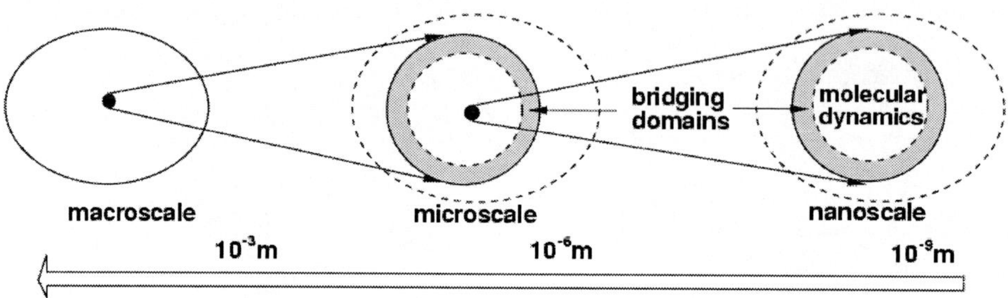

Figure 7. A bridging domain multiscale (BDM) method with different length scales.

The bridging domain coupling method can be extended for coupling the nanoscale to the microscale and macroscale for a given system, as shown in figure 7. A macroscale domain (~10^{-3} m) can be modeled as a linear elasticity domain using finite element methods [21]. The elastic properties are obtained using a Representative Volume Element (RVE) model from an MD simulation at a given temperature. A microscale domain (~10^{6}m) can be embedded in the macroscale domain, modeled using a nonlinear FE method [22]. The homogenization techniques, such as the TCB rule [9, 10], can be implemented in the nonlinear FE method to construct the temperature-dependent constitutive equations of the continuum. A subdomain in the microscale domain can be treated as a nanoscale (molecular) domain (~10^{-9} m) using MD if one is interested in studying physical phenomena in this subdomain. We can employ a Hoover thermostat [23] to maintain the nanoscale domain at a given temperature. It should be noted that a number of different length scales, such as the mesoscale domain, could be added as desired between the macroscale and microscale continuum domains. Furthermore, even the macroscale or the microscale can contain a number of different length scale subdomains. Different length scales are coupled via the bridging domain coupling technique [16].

Most concurrent multiscale methods employ multiple length scales but only a single time step. In coupling finite element methods and molecular dynamics, if the finite element mesh is graded down to the atomic spacing at the interface of the continuum and molecular domains [11], the time step must be restricted to the order of one femtosecond (10^{-15} s), due to the stability requirement in the molecular model. Consequently, significant computation time is wasted for large length scales in which large time steps can be used. In the bridging domain multiscale method, coupling different scales without such grading down of mesh sizes will be achieved by a straightforward implementation of different time steps via a multiple-time-step algorithm.

Since uniform meshes can be used in each length scale in the bridging domain multiscale method, it is possible to apply different time steps in different length scales, as shown in figure 8. A multiple-time-step algorithm is proposed for the bridging domain multiscale method. For example, to couple a molecular model at the nanoscale and a continuum model at

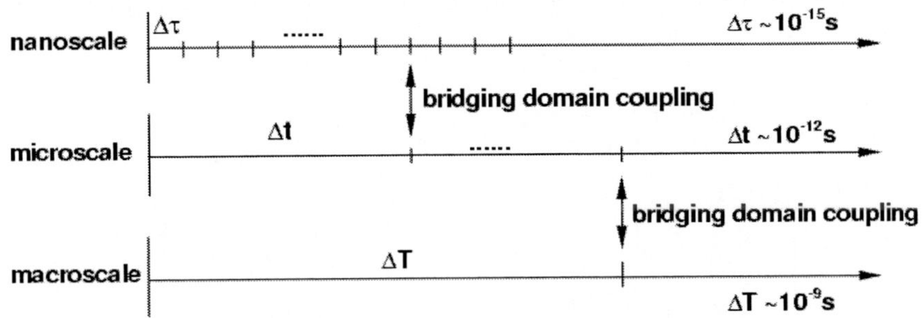

Figure 8. Multiple time steps used in the proposed multiscale method.

the microscale, a fine time step, $\Delta\tau$, is used in the molecular model and a coarse time step, $\Delta t = N\Delta\tau$, is used in the microscopic continuum model. Eq. (13) is then rewritten as:

$$\dot{u}^{*}_{I(n+1)} = \dot{u}_{I(n)} + \frac{1}{2}\left[\ddot{u}_{I(n)} + \ddot{u}_{I(n+1)}\right]\Delta t \qquad in \;\; \Omega_0^C$$
$$\dot{d}^{*}_{I(n+\frac{j+1}{N})} = \dot{d}_{I(n+\frac{j}{N})} + \frac{1}{2}\left[\ddot{d}_{I(n+\frac{j}{N})} + \ddot{d}_{I(n+\frac{j+1}{N})}\right]\Delta\tau \qquad in \;\; \Omega_0^M \tag{22}$$

The bridging domain then connects a fine length/time scale and a coarse length/time scale. Compatibility of different length scales will be enforced by means of a constraint imposed via the Lagrange multiplier method, i.e., Eq. (14), at each coarse time step, Δt, as shown in figure 8. Consequently, the equations of motion will be solved independently in different length scales with different time steps. At each coarse time step, velocities of nodes/atoms in the bridging domain will be corrected via the bridging domain coupling technique.

Domain Decomposition

To perform such multiscale simulations via high performance computing, efficient and effective domain decomposition strategies are necessary. In the bridging domain multiscale method, as illustrated in figure 9, a simulation domain can be easily hierarchically divided into subdomains, each of which can be processed in parallel.

The entire domain is divided into several sub-domains based on the scale difference. The primarily decomposed sub-domains are called the first-generation (FG) subdomains. The computational jobs on the subdomains can be distributed to different groups of HPC processors. For example, the computations in the molecular domain can be performed at Group 1 (a group of processors within a single HPC cluster), while those in the microscale/macroscopic domains can be performed at Group 2 (another group of processors within a single HPC cluster). The first-generation subdomain decomposition is parallel computing task decomposition. Each FG subdomain is further divided into a number of second-generation (SG) subdomains. Each SG subdomain is assigned to a single processor. In other words, as the secondary effort, the data within the subdomain are decomposed and transferred to each processor within a specified group for intensive computation. Obviously, it is a data-decomposition. This way, the task and data decomposition are combined.

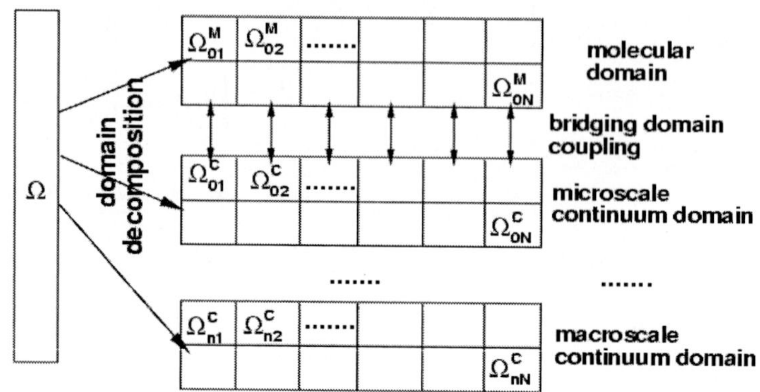

Figure 9. Domain decomposition for the BDM method.

A bridging subdomain is a special subdomain. It is shared by two SG subdomains, each of which belongs to two different length scales. Although one SG subdomain can overlap more than one other SG subdomain, the bridging subdomains do not overlap each other.

Inter-domain communications between two SG subdomains take place in each FG subdomain separately. Such inter-domain communications occur prior to solution of the equations of motion so that the motion of atoms or nodes at the SG subdomain boundaries is consistent.

The procedure for solving equations of motion in each FG subdomain is independent.

Once the equations of motion are solved at each time step (or coarse time step in the event that a multiple-time-step algorithm is used), the bridging domain communications take place between two processors. It should be noted that those two processors belong to two different groups of processors, shown in figure 9. The bridging coupling techniques are applied to correct the trial velocities of nodes or atoms in each bridging subdomain independently.

Inter-Domain and Bridging Domain Communications

Figure 10 illustrates the inter-domain communication in the BDM method. A simple BDM model contains two subdomains, Ω_A and Ω_B, in the molecular domain and two subdomains, Ω_C and Ω_D, in the continuum domain. Ω_A and Ω_B are allocated to two different processors in the same group of processors. Similarly, Ω_C and Ω_D are allocated to two processors in another group of processors. Inter-domain communications occur between Ω_A and Ω_B or between Ω_C and Ω_D prior to solution of the equations of motion. Such communications take place only inside each group of processors. In the molecular domain, Ω_A and Ω_B share atom E. Since the multi-body potential functions are generally utilized in MD simulations, slave domains Ω_B^{int} and Ω_A^{int}, associated with Ω_A and Ω_B, respectively, are introduced to support energy and force calculations. For example, the motion of atom H is updated in Ω_B. Such information is passed to the slave domain Ω_B^{int} to assist in the calculation of the interatomic force of atom E in Ω_A because the potential in the molecular domain includes the angle-bending potential of bonds GE and HE. Similarly, the motion of atom G is updated in Ω_A. Such information is passed to the slave domain Ω_A^{int} to assist in the calculation of the interatomic force of atom E in Ω_B. Consequently, after solving the equations of motion, the updated motions of atom E in either subdomain Ω_A or Ω_B are identical to avoid the occurrence of nonphysical phenomena. It should be noted that the size of a slave domain depends on the selected potential functions, in particular the cutoff distance for Van der Waals potential functions.

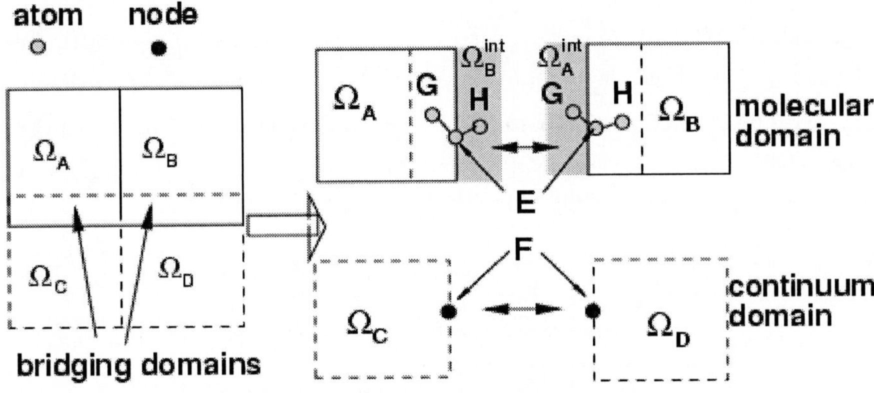

Figure 10. The inter-domain communication.

The inter-domain communication in the continuum domain has a different strategy from the one in the molecular domain. The continuum subdomains Ω_C and Ω_D share the boundary node F, as shown in figure 10. F^C and F^D are used to represent the same node F, but in different subdomains. Unlike inter-domain communication between neighboring molecular subdomains, inter-domain communication between neighboring continuum subdomains does not require slave domains, instead acting to aid the exchange of calculated internal forces of boundary nodes. For instance, the internal force of node F^C, calculated in subdomain Ω_C, is passed to subdomain Ω_D and is set as the negatively external force of node F^D. A similar procedure is performed to pass the internal force of node F^D, calculated in subdomain Ω_D, to subdomain Ω_C as the negatively external force of node F^C. Therefore, the motions of node F, updated by solving the equations of motion, Eq. (6), in both subdomains Ω_C and Ω_D, are identical.

Once the equations of motion are solved independently in each processor, the bridging-domain communication occurs separately in each bridging subdomain, e.g., Ω_{AC}^B and Ω_{BD}^B in figure 11. For example, trial velocities of atoms in Ω_{BD}^B of Ω_B are passed to Ω_D, while trial

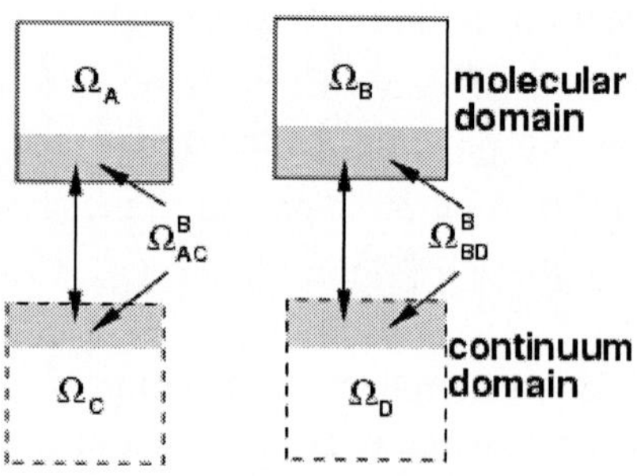

Figure 11. The bridging domain communication.

velocities of nodes in Ω^B_{BD} of Ω_D are passed to Ω_B. Then, the Lagrange multipliers in Ω^B_{BD} are solved via Eq. (14). Last of all, the trial velocities of atoms and nodes in the bridging domain Ω^B_{BD} will be corrected in Ω_B and Ω_D, independently through Eqs. (15) and (16).

Upon the above domain decomposition and data communication, the parallel computing processing can be implemented in the workflow illustrated in figure 12.

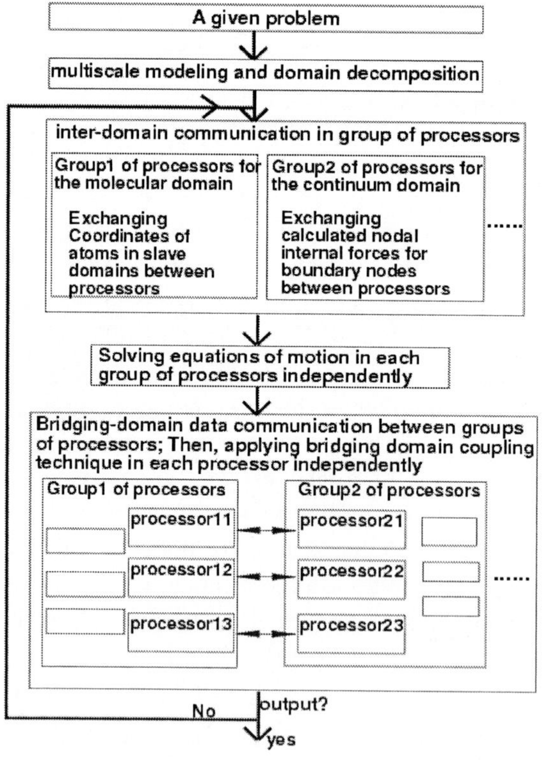

Figure 12. The workflow of the BDM method.

Complexity and Performance Evaluations

An important focus of the proposed method is the definition of a suitable computational intensity metric for sub-domains:

$$M(d) = f(D(d), \boldsymbol{O}, P) \tag{23}$$

where M is a function of the domain size D, i.e., the number of nodes or atoms having sub-domain d, the computing time complexity $\boldsymbol{O}$ of a particular component of a multiscale method, and parameters P that characterize the computing capacity of each specific HPC resource. M will be used to determine the size of sub-domains in domain decomposition. M is also an important parameter that can be used to guide decisions about when and where these sub-domains should be scheduled.

The computing time complexity of a continuum domain is $\boldsymbol{O}(n^2)$ while the complexity for a molecular domain is $\boldsymbol{O}(m^2)$, where m represents the number of atoms and n the number of finite element nodes. Although the complexity representation is the same for these two domains, m is often significantly larger than n. $\boldsymbol{O}(m^2)$ is approximately within the range $\boldsymbol{O}(n^3)$ to $\boldsymbol{O}(n^4)$, because m is approximately equivalent to $n^{3/2\sim2}$. The domain decomposition approach will address this complexity difference to produce sub-domains adaptively, the representations of which include the complexity information for tasking scheduling purposes. This approach will help detach the domain decomposition technique from the task-scheduling advisor, as described in the following. Further research will be conducted for multiple time steps as used in different length scales.

In order to investigate the feasibility, reliability, and application of the proposed method, several high performance computing benchmark studies should be conducted. The studies include (1) parallel performance speedup and efficiency to evaluate the behavior of high-performance computing, (2) detailed data communication (blocking, non-blocking, gather/scatter, one-site-communication effect using MPI2 features, parallel I/O, network impacts and network latency, load balance, etc.) among a group of processors and computational nodes to understand and reduce the communication loss, and (3) experiments of different HPC platforms and an analytical model of HPC scalability.

One-Dimensional Examples

To demonstrate the preliminary feasibility of the bridging domain multiscale method with high performance computing techniques, an experimental model has been developed. Similar to the previous example, the experiment is designed to observe the propagation of an imposed wave in a molecule chain passing from the molecular domain to the continuum domain. In this example, the bridging domain multiscale model contains 10,000 atoms and 10,000 finite elements. Each finite element contains 9 atoms. Therefore, the molecule chain has 100,000 atoms, and the length of the chain is around 20 micrometers. In this example, we mainly study the speedup of simulations due to high performance computing. The first computation is conducted on a local cluster (Microway 64-bit AMD Opeteron 32 processors). Figure 13(a) presents the speedup performance. The parallel performance increase exhibits a quasi-linear behavior. In order to test the parallel scalability, we also employed the algorithm on NSF's

TeraGrid (NCSA) system with 100 processors. The computational results are very promising, as shown in figure 13(b). Due to the large memory required for the finite element method, a superlinear behavior is observed. The preliminary study successfully demonstrates the feasibility and applicability of the proposed model. The achieved computations provide significant experience for future multi-dimensional studies.

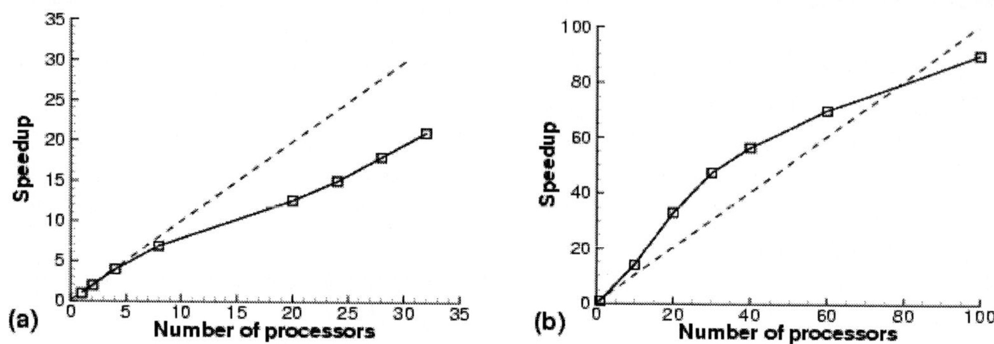

Figure 13. Speedup of 1D BDM method with high-performance computing; (a) on a 32-node UI-IA32-HPC cluster with low memory; and (b) on 64-node NCSA-HPC cluster with larger memory.

GRID COMPUTING TECHNIQUES IN MULTISCALE SIMULATIONS

Nano-Middleware

During the last several years, Computational Grids have been widely used to address computationally intensive problems in science, engineering, and commerce [24, 25]. Several disciplines have employed Grid computing to obtain solutions to computationally intensive problems by developing domain-specific middleware. This domain-specific middleware exploits characteristics of domain problems and aids the efficient use of Grids. In a similar manner, Grid application-specific middleware must be developed for multiscale methods to capture important method characteristics. This chapter develops a conceptual framework for multiscale methods that supports the location, allocation, and utilization of Grid resources to effectively and efficiently apply multiscale methods for nanotechnology applications. This middleware is referred to as *nano-middleware* in this chapter.

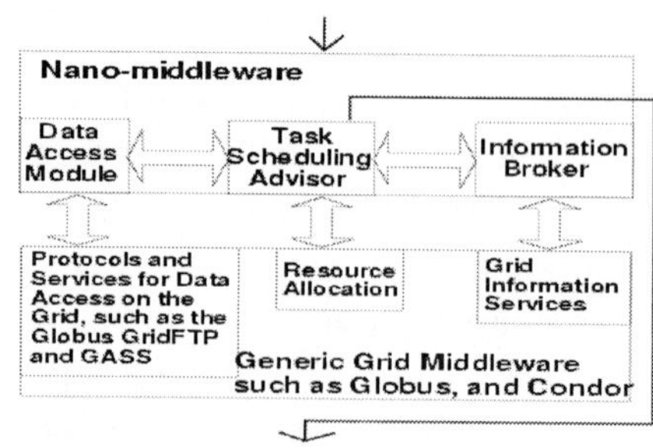

Figure 14. The proposed nano-middleware architecture.

The nano-middleware will be designed, as shown in figure 14, to enable Grid computing of the bridging domain multiscale method. Note that Globus is a software project, the purpose of which is to develop protocols and services for computational grids. Condor is a high throughput computing system. For a given problem, the nano-middleware will schedule decomposed domains to appropriate Grid resources (e.g., clusters) to achieve load balancing and efficient use of resources. The primary components of the nano-middleware are:

1) A task-scheduling advisor that takes the result of domain decomposition as input to produce scheduling plans and achieve high-performance simulation through load-balancing;
2) An information broker (IB) that leverages Grid information services to provide resource discovery [26, 27] functions to the task-scheduling advisor;
3) A data access module (DAM) that will manage the transfer, replication, and manipulation of data on the Grid.

Components 1 and 2 primarily deal with computing strategies. Data handling will be supported by existing generic middleware. The task-scheduling advisor is the key element of nano-middleware for its impact on performance gains. It should be noted that the concept of proposed nano-middleware can be applied to other multiscale methods.

Task Scheduling Advisor

Task scheduling is used to schedule subdomains to an appropriate set of Grid resources to achieve optimal performance—tasks are allocated in a way that balances computation across the selection of available resources. In the nano-middleware task-scheduling advisor, subdomains from the computational domain decomposition process are converted to tasks that are then placed in Grid resource queues in a specific order. In practice, queues will be managed by local resource schedulers, such as Portable Batch Systems (PBS) and Condor. The task-scheduling advisor is designed to achieve optimal performance by balancing tasks across available resources [28]. The advisor generates a scheduling plan that determines the correspondence between tasks and the available Grid resources to which they are submitted.

There are two general approaches to task scheduling: static and dynamic. When a static scheduling strategy is employed, the scheduling plan does not change until all tasks are completed [29]. In contrast, dynamic scheduling permits a plan to be altered while the set of tasks is being executed [30]. Using dynamic task scheduling for the BDM method is difficult to accomplish for two specific reasons [31]. First, the computation required to implement dynamic scheduling is much greater than for static scheduling; dynamic scheduling introduces additional overhead penalties created by network latency and the execution of the code that monitors the task status. Also, tasks are swapped between resources according to a dynamic performance evaluation. Second, fine granularity in individual subdomains, produced based on the BDM method, is desirable in order to achieve high levels of parallelism. Results for these subdomains can be inexpensive to compute even if scheduled to a Grid resource having a small capacity. Therefore, the time overhead that results from implementing dynamic scheduling on a task level may exceed the time required to compute results for an individual subdomain.

Consequently, static scheduling strategies are developed to assign tasks based on computational intensity information for each subdomain, as well as the variability in the computing capacity of each Grid resource. Two principles are used to guide the development of static scheduling algorithms: (1) Grid resources with greater computing capacity are used before those with less capacity; and (2) tasks that are more (less) computationally intensive are assigned to Grid resources with more (less) computing capacity [32].

CONCLUSION

Multiscale modeling and simulation has been at the forefront of nanotechnology research due to its ability to simulate larger systems than is possible with molecular dynamics. In this chapter, we first introduced the bridging domain coupling method, which can efficiently couple molecular dynamics and the finite element method. A more powerful multiscale method can be extended to bridge a number of length and time scales via the bridging domain coupling technique.

Recently developed multiscale methods, including the bridging domain multiscale method, still have limitations in length and time scales. This chapter proposed an alternative solution for the above problem: high performance computing techniques including Grid computing techniques. The speedup study demonstrated the advantage of implementing high performance computing techniques into the bridging domain multiscale method. Furthermore, the conceptual idea of Grid-based multiscale modeling and simulation will benefit from rapidly developing computer science technologies. Finally, the proposed research in this chapter can also be viewed as a framework for implementing high performance computing techniques in other potential multiscale methods.

REFERENCES

[1] S. Xiao and W. Hou, *Phys. Rev.* B 73, 115406 (2006)

[2] S. Xiao, D. R. Andersen, R. Han and W. Hou, *J. Comput. Theor. Nanosci.* 3, 142 (2006)

[3] T. Belytschko, S. Xiao, G. Schatz, R. Ruoff, *Phys. Rev.* B 65, 235430 (2002)

[4] J. S. Smith, D. Bedrov and G. D. Smith, *Comp. Sci. Tech.* 63, 1599 (2003)

[5] C. L. Rountree, R. K. Kalia, E. Lidorikis, A. Nakano, B. L. Van and P. Vashishta, *Ann. Rev. Mater. Res.* 32, 377 (2002)

[6] S. Xiao and W. Hou, *Phys. Rev.* B 75, 125414 (2007)

[7] E. B. Tadmor, M. Ortiz and R. Phillips, *Phil. Mag.* A 73, 1529 (1996)

[8] S. Q. Chen, W. N. E and C. W. Shu, *Mult. Model. Simul.* 3, 871 (2005)

[9] S. Xiao and W. Yang, *Int. J. Numer. Meth. Engrg.* 69, 2099 (2007)

[10] S. Xiao and W. Yang, *Comp. Mater. Sci.* 37, 374 (2006)

[11] F. Abraham, J. Broughton, N. Bernstein and E. Kaxiras, Europhy. Lett. 44, 783 (1998)

[12] J. Broughton, F. Abraham, N. Bernstein and E. Kaxiras, *Phys. Rev.* B 60, 2391 (1999)

[13] N. Choly, G. Lu, W. E and E. Kaxiras, *Phys. Rev.* B 71, 094101 (2005)

[14] G. J. Wagner and W. K. Liu, *J. Comp. Phys.* 190, 249 (2003)

[15] T. Belytschko and S. P. Xiao, *Int. J. Multi. Comp. Engrg.* 1, 115 (2003)
[16] S. P. Xiao and T. Belytschko, *Comp. Meth. Appl. Mech. Engrg.* 193, 1645 (2004)
[17] Y. G. Yanovsky, *Int. J. Multi. Comp. Engrg.* 3, 199 (2005)
[18] J. Ma, H. Lu, B. Wang, R. Hornung, A. Wissink, and R. Komanduri, *Comp. Model. Engrg. Sci.* 14, 101 (2006)
[19] I. Foster and C. Kesselman, The Grid: Blue Print for a New Computing Infrastructure. Morgan Kaufmann Publishers, Inc. San Francisco, CA, 1999
[20] I. Foster, C. Kesselman and S. Tuecke, *Int. J. Supercomp. Appl.* 15, (2001)
[21] T. J. R. Hughes, The Finite Element Method: Linear Static and Dynamic Analysis, Prentice-Hall, Dover, 1987.
[22] T. Belytschko, W. K. Liu and B. Moran, Nonlinear Finite Elements for Continua and Structures, Wiley, New York, 2000
[23] W. G. Hoover, *Phys. Rev.* A 31, 1695 (1985)
[24] I. Foster, *Phys. Today*. 55, 42 (2002)
[25] F. Berman, G. Fox and T. Hey, "The Grid, past, present, future," In *Grid Computing, Making the Global Infrastructure a Reality*, edited by R. Berman, G. Fox, and T. Hey John Wiley and Sons, West Sussex, England, 2003
[26] S. Wang, A. Padmanabhan, Y. Liu, R. Briggs, J. Ni, T. He, B. M. Knosp, Y. Onel, *Lec. Note. Comp. Sci.* 3032, 536 (2003)
[27] Padmanabhan, A., Wang, S., Ghosh, S., and Briggs, R. Proceedings of the Grid 2005 Workshop, Seattle, WA, November 13-14, 2005, *IEEE press,* pp. 312-317.
[28] J. Henrichs, *Proceedings of International Conference on Supercomputing.* 1998, 165 (1998)
[29] T. D. Braun, H. J. Siegel, N. Beck, L. L. Boloni, M. Maheswaran, A. I. Reuther, J. P. Robertson, M. D. Theys, B. Yao, D. Hensgen and R. F. Freund, J. Para. *Distrib. Comp.* 61, 810 (2001)
[30] O. Beaumont, A. Legrand and Y. Robert, *Para. Comp.* 29, 1121 (2003)
[31] H. J. Siegel and S. Ali, *Euromicro J. Sys. Arch.* 46, 627 (2000)
[32] S. Wang and M. P. Armstrong, *Para. Comp.* 29, 1481 (2003)
[33] S. Xiao and W. Hou, *Phys. Rev.* B 73, 115406 (2006)
[34] T. Kuzumaki, K. Miyazawa, H. Ichinose and K. Ito, *J. Mater. Res.* 13(9), 2445 (1998)
[35] S. Xiao and W. Hou, *Fullerenes, Nanotubes, and Carbon Nanostructures,* 14, 9 (2006)
[36] M. Yu, O. Lourie, M. J. Dyer, K. Moloni, T. F. Kelly and R. Ruoff, *Science*. 287, 637 (2000)
[37] C. A. Girifalco and R. A. Lad, *J. Chem. Phys*. 25(4), 693 (1956)

In: Cyberinfrastructure Technologies and Applications
Editor: Junwei Cao
ISBN: 978-1-60692-063-3

Chapter 15

CYBERINFRASTRUCTURE FOR AGRICULTURAL DATA AND KNOWLEDGE SHARING IN CHINA

Chunjiang Zhao[1], Yuxin Wan[2], Huarui Wu[3] and Wen Zhang[4]

[1] National Engineering and Research Center for Information Technology for Agriculture, Beijing 100097, P. R. China
[2] Department of Automation, Tsinghua University, Beijing 100084, P. R. China
[3] National Engineering and Research Center for Information Technology for Agriculture, Beijing 100097, P. R. China
School of Computer Science, Beijing University of Technology, Beijing 100022, P. R. China
[4] Department of Automation, Tsinghua University, Beijing 100084, P. R. China

ABSTRACT

During the last decade, billions of national investment has been spent in China on building agricultural information systems for data collection, integration, analysis and processing. Each province has built its own technology platform for information sharing and access. However, since data sources are separate and corresponding applications are developed by different organizations, cross-domain data and knowledge sharing becomes very difficult. A Cyberinfrastructure (CI) for agricultural data and knowledge sharing in China is proposed in this work and funded by the Ministry of Science and Technology of China under the national 863 high-tech R and D program. In this work, related work is summarized and our system structure is described in details. Heterogeneity of different operating systems and databases has to be addressed. System performance can be improved by avoiding large-scale data transferring by introducing an application server within each domain for local data processing.

[1] E-mail address: zhaocj@nercita.org.cn
[2] E-mail address: wanyx04@mails.tsinghua.edu.cn
[3] E-mail address: wuhr@nercita.org.cn
[4] E-mail address: wen-zhang05@mails.tsinghua.edu.cn

INTRODUCTION

During the past decade, billions of national investment has been spent in China on agricultural information system for data collection, integration analysis and processing. Up to now the information services has been widely used from crop production to stock breeding during the whole production management. Each province has established their own agricultural technology platform based on agriculture data resources. However, since these databases are separated in different locations and different information systems, it becomes very difficult to integrate all these highly distributed agricultural data and provide transparent access and massive processing supports on these data and knowledge. Major challenges are as follows:

- Data Heterogeneity. All these agricultural data are collected and maintained in different information systems. Data are stored in different databases, e.g. SQL Server, MySQL, and ORACLE, using different operating systems, e.g. Windows and Linux. Corresponding applications of these data are developed and maintained by different organizations and originally not motivated by data and knowledge sharing.
- Massive Data Transferring. Data are located separately and network situation may vary largely among these locations. Large-scale cross-domain data transferring becomes impossible and has to be avoided as much as possible.

There are already many technical mechanisms provided to solve similar problems, such as distributed database systems and data grid systems. In this chapter, a brief summary of related work is provided in Section 2.

A Cyberinfrastructure (CI) for agricultural data and knowledge sharing in China is proposed in this work and funded by the Ministry of Science and Technology of China under the national 863 high-tech R and D program. The project is a joint effort of National Engineering and Research Center for Information Technology for Agriculture and Tsinghua University. An overview introduction to project goals is included in Section 3.

A three-layered system structure for heterogeneous data integration is proposed in this work and related software techniques for implementing the system are introduced in details. A working procedure is given to explain how the system works. All details can be found in Section 4 and the chapter concludes in Section 5.

RELATED WORK

Distributed Database Systems (DDB)

A simple solution for data integration is using distributed database system and D-DBMS. DDB technology aims to provide mechanisms to manage and uniform access to those physically distributed but logically related information sources. There are two features of DDB. One is data independence, which includes logic independence, physical independence and distributed transparency. Second is DDB is a combined system that includes both centrality and autonomy. Centrality refers to centralized control which means a virtual center assort with each component DBMS (database management systems) and execute high-level applications. On the contrary, autonomy refers to control distribution and indicates the degree to which individual DBMS can operate independently. "There are three types of autonomy, tight integration, semi-autonomy, and full autonomy. In tightly integrated system, a single image of the entire database is available to users who want to share the information in multiple databases. Semiautonomous systems consist of DBMSs that can operate independently but have been designed to participate in federation to make their local data shareable. In fully autonomous systems, the individual components do not know the other DBMSs and do not communicate with them." [10] According to our scenarios, semiautonomous system is the most appropriate choice. The distributed data source system deal with their own business and a global control system is in charge of assorting with each data source and providing integrated application.

To all appearances, it seems DDB systems are a good choice for building agriculture data sharing systems. However, there are two main disadvantages of DDB. Obviously, a notable character of agriculture information system is the massive amount of data. However, DDB system is not appropriate for mass data integration. Firstly, DDB system need to make replication for safety but for agriculture data it is impossible. Second, DDB system has no mechanism for mass data transferring. The other shortage is that DDB is a close-coupling system, since once the whole system is constructed it's inconvenient to change, for example adding new data source nodes or removing existing nodes. However, agriculture information systems are dynamic systems since we cannot tell how many nodes there are at the beginning. The number of nodes will increase as the popularization of agriculture information systems. Also, it's almost impossible to build a DDB system from a huge number of data sources.

Although DDB is not suitable for our project, the idea of semiautonomy systems is very useful and we will talk about that lately in Section 4.

Grid Technologies and Applications

There are various efforts on large-scale data integration in the grid community for scientific applications in different domains [6]. For example, Grid Physics Network [11] and Open Science Grid [12] are grid testbeds in the US for data intensive computing on high-energy physics and gravitational wave astrophysics. There is no dedicated grid infrastructure in the world for agricultural data integration and processing, but some general-purpose tools provide similar functionalities, e.g., OGSA-DAI and SRB.

OGSA-DAI

OGSA-DAI [5] is a project conceived by UK Database Task Force and is working closely with the Open Grid Forum DAIS-WG. OGSA stands for "Open Grid Source Architecture" while DAI means "Data Access Integration". The main goal of the program is for scientific use "To contribute to a future in which scientists move away from technical issues such as handling data location …. instead focus on application specific data analysis and processing". This middleware bases on Globus and has provided control and access to relational and XML database management system currently. The framework, however, has been designed to allow other data sources such as file systems to be accessed through the same interfaces.

Figure 1 below shows the framework of OGSA-DAI and corresponding detailed descriptions can be found in the document [1].

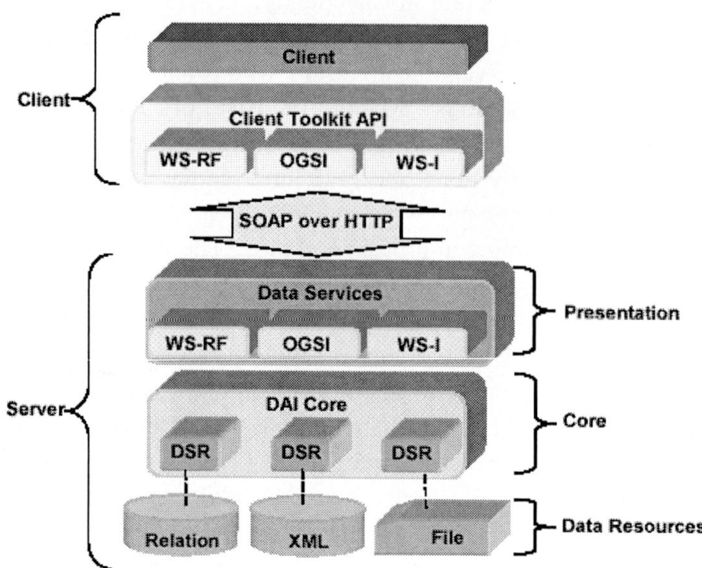

Figure 1. Schematic representation of OGSA-DAI.

- The presentation layer accepts message from clients and decide which Data Service Resource (DSR) in the Core Layer is the proper one to submit.
- The core layer consists of a set of DSR. Each DSR implements the core DAI functionality that includes overseeing the coordination of the activities for a specific Data Resource. A DSR may also expose additional capabilities such as data transport-related operations and can cache data for retrieval by third parties.
- The client toolkit (CTk) API has been refactored to abstract away the differences between the different messaging infrastructures.

Figure 2 below shows a typical OGSA-DAI interaction. The following interaction steps are cited from the document [2].

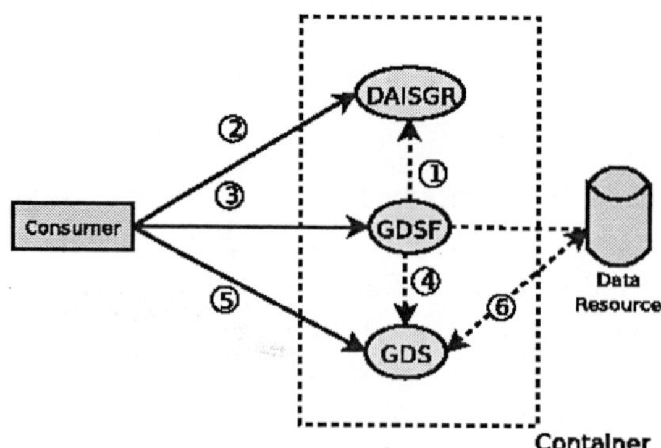

Figure 2. A typical OGSA-DAI interaction.

- Data Access and Integration Service Group Registry (DAISGR) is a service allowing other services to publish metadata about any data resources they represent and the capabilities they expose. A client can thus use a DAISGR to identify, by querying its registered metadata, a resource provider that best satisfies its needs.
- Grid Data Service Factory (GDSF) acts as a persistent access point to a data resource and contains additional related metadata that may not be available at a DAISGR. A GDSF creates GDSs to access and manipulate data resources.
- Grid Data Service (GDS) acts as a transient access point to a data resource. It is through a GDS that a client interacts with a data resource.

1. A GDSF may register its service handle with a DAISGR, along with sufficient metadata and capability information to allow service/resource discovery to take place.
2. Clients obtain information about available resources (represented by GDSFs) by querying a DAISGR. They can then ask for detailed information, e.g., the schema of the resource, at a particular GDSF of interest.
3. A GDSF, in effect, acts as a persistent Grid-enabled wrapper for a data resource but does not provide direct access to that data resource. Access to a data resource requires the creation of a GDS through the GDSF's Factory portType as specified in OGSI.
4. GDSs are transient services created at the request of clients who wish to access a data resource. Data resource access is done through a single document-based operation provided by the GDS.
5. A client submits a perform document to the GDS which contains the sequence of activities to be executed on that data resource or the resulting data.
6. The activities that can be executed by the GDS are defined when a GDSF is configured. The inner workings of a GDS are examined in more detail in the next section.

From above we can see that OGSA-DAI has constructed a uniform service format and provided a standard accessing criterion to distributed data sources. Especially, OGSA-DAI is a loose-coupling system. Data sources are not fixed to a special server but can be discovered instantly. OGSA-DAI project is based on Java and use JDBC technique for data integration. OGSA-DAI aim to providing not only database system access but also allow other data sources such as file systems to be accessed through the same interfaces.

Storage Resource Brokers (SRB)

Storage Resource Broker (SRB) [13] is a data grid middleware software system produced by the San Diego Supercomputer Center (SDSC) and commercialized by Nirvana that is operating in many national and international computational science research projects.

SRB provides a uniform interface to heterogeneous data storage resources over a network. As part of this, it implements a logical namespace (distinct from physical file names) and maintains metadata on data-objects (files), users, groups, resources, collections, and other items in an SRB Metadata Catalog (MCAT) stored in a relational database management system. System and user-defined metadata can be queried to locate files based on attributes as well as by name. SRB runs on various versions of Unix, Linux, and Microsoft Windows.

The SDSC SRB system is middleware in the sense that it is built on top of other major software packages (various storage systems, real-time data sources, a relational database management system, etc) and it has callable library functions that can be utilized by higher-level software. However, it is more complete than many middleware software systems as it implements a comprehensive distributed data management environment, including various end-user client applications. It has features to support the management and collaborative (and controlled) sharing, publication, replication, transfer, and preservation of distributed data collections.

Compared with the work described above, we adopts a different approach by introducing local applications servers so that cross-domain data encapsulation and transferring problems are avoid and system performance is approved.

Cyberinfrastructure (CI) for Agricultural Data and Knowledge Sharing in China

CI for agricultural data and knowledge sharing is a project in China funded by the Ministry of Science and Technology of China under the national 863 high-tech R and D program. The project is a joint effort of National Engineering and Research Center for Information Technology for Agriculture and Tsinghua University. Two other consortium members are China Agriculture University and Institute of Software of Chinese Academy of Sciences. Major challenges that have to be addressed by the project are as follows.

- Database integration, also known as data transparency. It's the foundation of our project, which means:
  - o Storage resource transparence - accessing data without knowing the type of storage. It is the bottom layer, aiming at combining different types of physical databases into a logical set of sources and offering standard interfaces for accessing.
  - o Storage location transparence - accessing data without knowing the location.
  - o Data identifier transparence - finding data without knowing the identifier. There are four types of data identifiers: unique name, descriptive name, collective name and physical name.
- User authentication. Each data source is a preserved data system and data accessing should be authenticated. Since our project goal is building a loosely coupled system, the authentication should not be too complicated.
- Easy access. Data sources should be easily added into our system.
- Massive data transferring. As we know, today's Internet provides a best effort service, although TCP/IP protocol defined some mechanisms for reliability and safety. However, some level of protection still has to be set at the application layer. We have to guarantee a reliable transferring through an unreliable channel.
- Data intensive computing. We cannot depend on a central processor dealing with all the query. Distributed computing is a necessary component in our program.

- Security guarantee. Since the program operates on databases directly, users should not have too much authority.

Among the challenges mentioned above, heterogeneous data integration and mass data processing are the top two questions [7]. In the section below, we will introduce our framework according to address these challenges.

HETEROGENEOUS DATA INTEGRATION

System Architecture

From down to top there are three layers of this system, as shown in Figure 3 below.

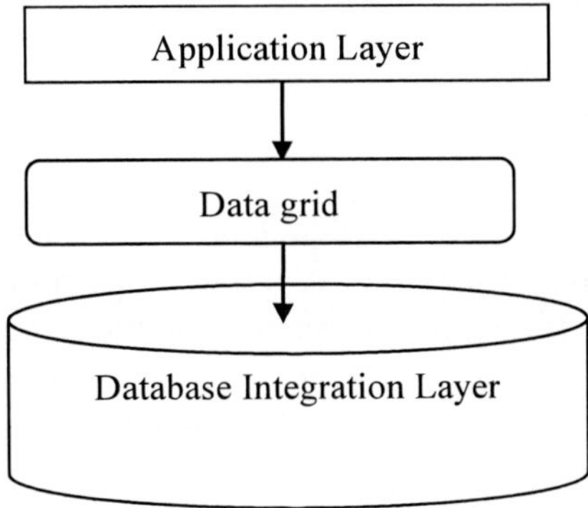

Figure 3. System architecture.

- The bottom layer is integrated data resources. This layer will integrate heterogeneous databases and provide standard accessing interfaces to different databases. Compared with OGSA-DAI, this layer accomplishes parts of the core layer function.
- On the second layer, we will establish a data grid system that enables mass data transferring and safe access of distributed database systems. Easy access and authentication mechanism will be provided in this layer. At the same time, this layer is the central processing unit of our system, according to user query collected from the application layer. It will operate on data sources and get corresponding information.
- On the top level, we will build a web server system and implement some web applications such as data searching. As the presentation layer in OGSA-DAI project, this layer accepts messages from clients and submits jobs to the right resource. In addition, some of the security control is done in this layer.

In this section, we provide an overview of our system, and in the following sections, we will present the implementation of our framework and technical details we used.

System Structure

According to the framework we describe above, the actual system structure is shown in Figure 4.

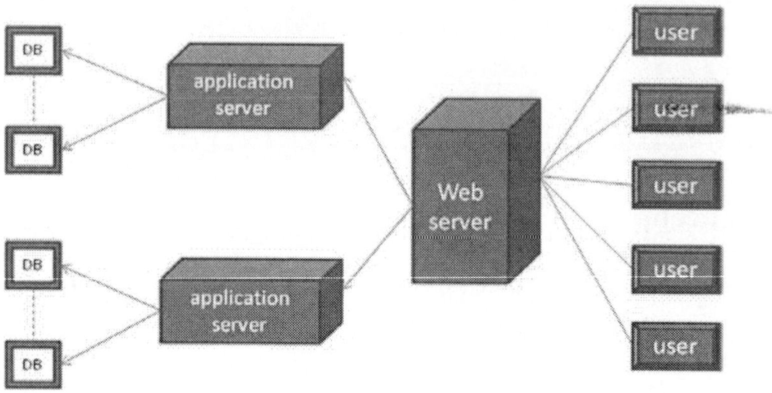

Figure 4. System structure.

Application servers and DBs, accomplish database integration and part of data processing. It is the whole database integration layer and a part of data grid layer. As we discussed above, one problem we need to solve is how to transfer and deal with massive data. If we merely rely on the Web server, it will be impossible in practice because of transferring and processing costs. The transferring cost depends on the Internet bandwidth while computation cost relies on processing ability of the Web server. Neither can we require the bandwidth nor hardware configuration so we have to adopt the concept of distributed computing. Each application server is the virtual center of nearby databases. It will gather the original data, make filtration and computing and then transfer them to the web server. We use ODBC techniques as the OGSA-DAI project for database integration. On each application server, we dispose a program that will get requests from the web server and make corresponding operations on databases.

The web server responds to clients and makes special requests to application servers, which is the center of the data grid. Actually, it is both the application layer and core of data grid layer. The communication between the web server and application servers are well protected which insures only legal clients can operate the database. On the other hand, as we cannot figure out how many application servers we need at the beginning so the web server provides a very simple and safe mode for application servers to plug in. It is a loosely coupled system. We use the grid middleware Globus to implement security of our system. At the same time, the web server need to get messages from users and make an interactive network interfaces. We use the Apache Web server and PHP techniques at the top level.

Technical Details

ODBC Techniques

This section introduces ODBC techniques that are adopted at the bottom layer of our system [3].

ODBC is an open specification for providing application developers with a predictable API with which to access data sources. Data sources include SQL Servers and any data source with an ODBC driver, such as Oracle, MySQL, etc.

There are four parts in an ODBC based system: user program, ODBC driver manager, ODBC driver, data source. The system structure is shown in Figure 5.

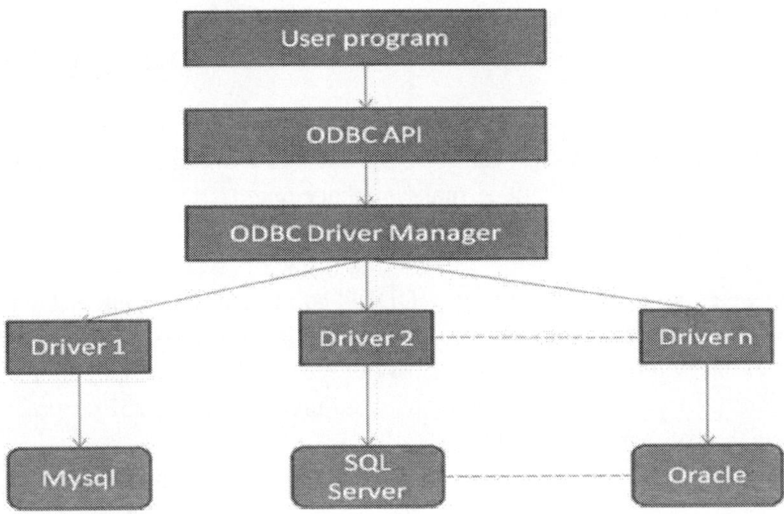

Figure 5. ODBC techniques.

User program is written based on standard ODBC functions and SQL functions. It contains:

- Connecting databases
- Sending SQL query to databases
- Allotting special memory and define data types for result returned from databases
- Getting result data from databases
- Making special processes with the data and return it to users, in fact it is exact the application server program.
- Disconnecting databases

ODBC driver manager is free software provided by both Microsoft (for windows systems) and unixODBC project (for unix systems). It manages communication between user programs and ODBC drivers. It contains loading ODBC drivers, selecting and connecting special drivers, manage data sources, checking using errors and recording using history of ODBC functions. Also you can set and make configuration and get data source or driver information from driver managers.

Driver program provides independence between user systems and database systems. Drivers contain the special code required to talk to the specific type of database you will work with. Drivers often come from the vendor of the database but may also be found in the unixODBC package. User programs do not connect with database directly, instead all the requests are submit to special ODBC driver program through the driver manager. Then the driver program will use corresponding functions connecting to data sources. If user program need to use different data sources, it has to connect different driver programs dynamically.

Data sources are the databases user finally access, it contains the location and type of the database. ODBC gives every data source a unique name called DSN (short for Data Source Name) and mapping all the necessary bottom layer software or library. During the connection, DSN is the agent of the real database system. Once ODBC and DSN are configured rightly, it will be transparent for users to connect databases.

There are three major advantages of choosing to code an application using the Unix ODBC API [4]:

- Portable data access codes. The ODBC API is available on all major platforms. Microsoft platforms include many enhancements to this specification; these enhancements are also supported by unixODBC.
- Dynamic data binding. This allows the user or system administrator to easily configure an application to use any ODBC compliant data source. Dynamic binding allows the end-user to pick a data source, i.e. an SQL Server, and use it for all data applications without having to worry about recompiling the application.
- All Unix ODBC development is distributed under GPL or LGPL, which means you, can use the software free of charge.

Globus Toolkit

Globus Toolkit is a fundamental enabling technology for the grid letting people share computing power, databases, and other tools securely online across corporate, institutional, and geographic boundaries without sacrificing local autonomy. The toolkit includes software services and libraries for resource monitoring, discovery, and management, plus security and file management. It is an open source toolkit, freely available in source code form for use by anyone, including both commercial and non-commercial purposes. And it is particularly useful to application developers and system integrators. Globus Toolkit aims to provide standard system components that can support a wide variety of highly customized applications and user interfaces without requiring a unique infrastructure to be developed for each application. And it does not provide a complete solution for Grid projects. The components in the Globus Toolkit have to be organized and fitted together according to a plan that fits the requirements of the project. In fact, the Globus Toolkit genuinely is a toolkit or a collection of tools [8].

The existing grid computing or network standards being capitalized on Globus Toolkit are IETF, W3C, OASIS and GGF etc. Some of the particularly useful standard mechanisms for our project used in Globus Toolkit are listed below.

- SSL/TLS. Secure Sockets Layer (SSL) / Transport Layer Security (TLS) are cryptographic protocols that provide secure communications on the Internet for such things as web browsing, e-mail, Internet faxing, instant messaging and other data transfers. This mechanism ensures the safe job submission from Web server to application server.

- X.509 Proxy Certificates [9]. Use of a proxy credential is a common technique used in security systems to allow entity A to grant to another entity B the right for B to be authorized with others as if it were A. X.509 Proxy Certificates forms a certificate profile for Proxy Certificates, based on the RFC 3280. Unlike unrestricted proxy, this profile defines a framework for carrying policies in Proxy Certificates that allows proxy to be limited through either restrictions or enumeration of rights. In addition, proxy Certificates will be with unique names, derived from the name of the end entity certificate name. This allows the Proxy Certificates to be used in conjunction with attribute assertion approaches such as Attribute Certificates and have their own rights independent of their issuer. X.509 Proxy Certificates is the core mechanism of our loose-coupling system. Only need to apply a Proxy Certificates from each newly found application server then web server will be able to operate all its date resources. In fact, server programs are disposed on every application server. When the web server gets their proxy certificate, it will be able to operate those server program as application server itself.
- GridFTP. GridFTP is a high-performance, secure, reliable data transfer protocol optimized for high-bandwidth wide-area networks. It is based upon the Internet FTP protocol, and it implements extensions for high-performance operation that were either already specified in the FTP specification but not commonly implemented or that were proposed as extensions by Globus Alliance. GridFTP guarantees the reliability and security of mass data transferring through the data grid system.

Globus Toolkit is adopted to solve the core authentication, easy plugging in and massive data process challenge of our project. It implements the whole data grid layer and part of data integration layer.

Apache and PHP

Apache HTTP server is open-source software and supports almost all operating systems. We use the Apache server and HTML to build an interactive interface from which to get messages from clients. PHP is a widely used general-purpose scripting language that is especially suited for Web development and can be embedded into HTML. Instead of lots of commands to output HTML, PHP pages contain HTML with embedded code that makes special operations. The code is executed on the server, and generated HTML pages are sent to the client. The client receives the results of running the script, without knowing what the underlying code is. For our project, we accept messages from clients by HTML and then use underlying PHP pages to execute our process program. The Apache server and PHP language are adopted to implement the top web server layer of our system.

Working Procedures

This system includes a web server deployed with Globus toolkit, an interactive Web page which user can submit requests, and a PHP page that includes the processing program. At

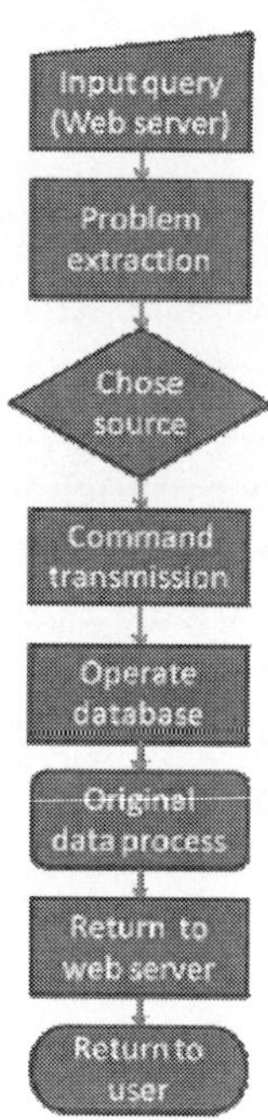

Figure 6. Working procedure.

least one application server is configured with Globus toolkit and Unix ODBC, and integrated with some heterogeneous relational database such as SQL server and Oracle. Those databases have been added into the ODBC configuration file. In this section, working procedures of our system are given below.

1. First step is that users input query requests. Clients visit our Web page and submit their requests.
2. Underlying PHP pages access html forms, deal with collected messages and make corresponding requirements. There should be at least one processing program deployed here. This program converts user messages into standard SQL sentences and decides which application server to submit the job.
3. Once the SQL query is compiled, PHP or the Web server program will transfer SQL query to the specific application server. As this process uses the Globus proxy certificate, there should be a recurrent shell program disposed on the web server applying for certificates once every 12 hours (The default remaining time of a proxy certificate is 12 hours).
4. When disposed program at the application server receives SQL query from the web server it will operate its databases. In fact, it is the web server itself that calls the program in the application server. Once the proxy certify is done, the web server is able to act as an application server and have the authority to operate its program. In this step, in order to achieve database transparency, there is a searching program in each application server that dynamically obtains all its managed database information, such as DSN, user name and password. When the application server program is connected to databases, those database info will automatically be added into the program.

5. The application server gets the original data from database and makes application specific data processing. This step is different according to different application scenarios.
6. The web server obtains the result data from each application server, combines results, and returns it to users.

CONCLUSION

This heterogeneous database integration system based on ODBC techniques and data grid middleware plays essential role in our CI project for agricultural data and knowledge sharing in China. The three layered system design come partly from OGSA-DAI and partly from distributed database systems. In fact, the bottom layer is a simple distributed database system while the second layer is composed with a loosely coupled data grid. The complete working procedures are similar to the OGSA-DAI project but some simplification is made. Compared to OGSA-DAI or other grid projects such as SRB, this system is far less complicated and much easier for implementation. However, because of limitations of ODBC techniques, this system can only be used for relational database integration. And also the overhead for authentication between the web server and application server is high sometimes and ongoing work includes performance optimization of the system implementation.

ACKNOWLEDGMENT

The work was funded by four grants from the Ministry of Science and Technology of China under the national 863 high-tech R and D program (contracts No. 2007AA10Z235, No. 2006AA10Z237, No. 2007AA01Z179 and No. 2008AA01Z118) and a grant from the Ministry of Agriculture of China (contract No. 2006-G63).

REFERENCES

[1] OGSA-DAI Status and Benchmarks. OGSA-DAI project group.
[2] The Design and Implementation of Grid Database Services in OGSA-DAI. OGSA-DAI project group.
[3] S. Wang, S. Sha, Conspectus to Database Systems, fourth edition. 2006.
[4] Unixodbc. http://www.unixodbc.org/
[5] OGSA-DAI project. http://www.ogsadai.org/
[6] Grid Data Management Systems and Services. VLDB Tutorial Berlin,2003.
[7] Feng He, Heterogeneous Database Integration in Grid Environment, Tsinghua University, 2007.
[8] Globus Alliance. http://www.globus.org/
[9] Internet X.509 Pubilc Key Infrastructure Proxy Certificate Profile.
[10] Distributed Database system: Where are we now? MT Ozsu, P Valduriez-Computer, 1991 IEEE

[11] Grid Physics Network. http://www.griphyn.org.
[12] Open Science Grid. http://www.opensciencegrid.org
[13] The Storage Resource Broker. http://www.sdsc.edu/srb/index.php.

INDEX

A

B

C

D

E

F

N

O

P

Q

R

S

T

U

V

W

X

Y